中外建筑科学发展理念与图说

Concept and Illustration of the Development of Chinese and Foreign Architectural Science

张祖刚 著

Writtern by Zhang Zugang

李 怡 译

Translated by Li Yi

中国建筑工业出版社

图书在版编目（CIP）数据

中外建筑科学发展理念与图说 = Concept and Illustration of the Development of Chinese and Foreign Architectural Science / 张祖刚著；李怡译.—北京：中国建筑工业出版社，2021.6
ISBN 978-7-112-25792-8

Ⅰ.①中… Ⅱ.①张…②李… Ⅲ.①建筑科学—学科发展—研究 Ⅳ.①TU

中国版本图书馆CIP数据核字（2020）第267561号

责任编辑：吴宇江　孙书妍
责任校对：赵　菲

中外建筑科学发展理念与图说
Concept and Illustration of the Development of Chinese and Foreign Architectural Science

张祖刚　著
Writtern　by Zhang Zugang
李　怡　译
Translated by Li Yi

*

中国建筑工业出版社出版、发行（北京海淀三里河路9号）
各地新华书店、建筑书店经销
天津图文方嘉印刷有限公司制版
天津图文方嘉印刷有限公司印刷

*

开本：787毫米×1092毫米　1/16　印张：34　字数：555千字
2022年5月第一版　　2022年5月第一次印刷
定价：300.00元
ISBN 978-7-112-25792-8
（37009）

版权所有　翻印必究
如有印装质量问题，可寄本社图书出版中心退换
（邮政编码　100037）

内容提要

本书内容包括上、中、下三篇。上篇主要介绍了中外建筑科学发展的六大理念,即创新发展理念、自然环境理念、历史文化理念、艺术美观理念、便民交通理念、公正社会理念;中篇是中国建筑科学文化精品图说,主要介绍西安、北京、上海、南京、苏州、杭州、扬州等全国各地的建筑文化,以说明上述理念;下篇是外国建筑科学文化精品图说,主要介绍开罗、雅典、罗马、威尼斯、巴黎、马德里、伦敦、莫斯科、华盛顿等世界各地的建筑文化,以说明这六大理念。

本书可供广大建筑师、城乡规划师、风景园林师、高等建筑院校建筑学、城乡规划学、风景园林学专生师生,以及广大建筑文化艺术爱好者等学习参考。

Synopsis

   The content of this book includes three parts. Part One mainly introduces six specific ideas about the development of Chinese and foreign architectural science, namely, innovative development, natural environment, history and culture, artistic beauty, convenient transportation and just society. Part Two is an illustration of Chinese architectural science and culture, which mainly introduces the architectural culture of Xi'an, Beijing, Shanghai, Nanjing, Suzhou, Hangzhou, Yangzhou and other parts of the country to elaborate the above ideas. Part Three is an illustration of foreign architectural science and culture, which mainly introduces the architectural culture of Cairo, Athens, Rome, Venice, Paris, Madrid, London, Moscow, Washington and other places around the world to elaborate the above ideas.

   This book can be used for reference by architects, urban planners, landscape architects, teachers and students majoring in architecture of advanced architecture schools, architecture, urban planning and landscape architecture, as well as fans of architectural culture and art.

前言

建筑科学文化是一门大学科，它包括城乡规划学、建筑学、风景园林学等学科，这是中国大科学家钱学森先生于1996年提出的，这一新观念需要大力宣传。钱学森先生还提出城市、建筑、风景园林的规划设计是一个层次，但其更高层次的指导思想是老子的自然宇宙观，即为人类建立生态文明生活环境的哲学观念。

中国传统建筑科学文化思想，源于老子的自然宇宙观。人类与自然万物的关系应是和谐的关系，这个观念是解决当今世界"生态环境危机"的重要理念。我们要继承和发展这一自然宇宙观，使其具有现代性，并在新的历史条件下探讨人类与自然、社会与自然的有机融合关系，以及生态文明中自然、经济、社会、文化的生态平衡内容。

本书提出的核心哲学观念和具体理念是作者通过亲身实践感悟到的，其中的中篇——中国建筑科学文化精品图说，是笔者1956年3月在国家城市建设总局城市设计院从事城市规划设计工作以来至今半个多世纪，到过除西藏外的百余座城市里选出的上百个精品实例，读者可以从中看到中国建筑科学文化1000多年来的发展变迁；而下篇——外国建筑科学文化精品图说，是笔者自1979年6月出访瑞士以来至今到过30多个国家的百余座城市中挑选出来的过百个精品实例。这些精品实例是"世界文化遗产"和各国的重点文物保护项目，它涵盖了近百年来的近代建筑。这些建筑都是中外著名建筑师创作的精品，代表着各个时代的创新发展。书中实例照片除7张署名外，其他均为本人的摄影作品，而书中的城市及其规划图则均为地方提供。通过对这些中外精品实例的分析，我们概括出建筑科学文化发展的6个具体理念，即创新发展、自然环境、历史文化、艺术美观、便民交通、公正社会。此6个理念也是解决当前中外城市弊端的思路和方法。

我们常说"古为今用，洋为中用"。本书就是期盼中外文化相互借鉴，以达到促进中外建筑科学文化的发展，这是本人编写此书的目的。

本书在编写和出版方面得到了香港霍丽娜女士的大力鼓励和鼎力资助，并承蒙李怡女士为此双语版著作作了翻译，以及中国建筑工业出版社首席策划、编审吴宇江先生和孙书妍编辑等对本书出版的支持与帮助，在此向他们表示衷心的感谢。

张祖刚
2020年10月

Preface

Architectural science and culture is an integrated subject, which includes urban and town planning, architecture and landscapes. This new concept was put forward by the great Chinese scientist Mr. Qian Xuesen in 1996, which needs to be publicized vigorously. He also mentioned that it is one level to plan and design cities, buildings and landscapes, while the higher level is to develop the philosophical concept for guiding the planning and design, that is, to study the relationship between architectural science and mankind and society. We summarize it as a natural understanding of universe that conforms to the objective development rules of all things and follows the law of nature.

In line with the law of nature, the guiding ideology of architectural science and culture originates from Lao Zi's natural world outlook, which points out the relationship between the development of human society and the objective development rules of all things. Mankind and all things in nature should be in a harmonious coexistence. This concept is important to solve the "ecological environment crisis" in today's world. We should inherit and develop this natural world view and make it modern. It means, under the new historical conditions, we should explore ways and practices for organic integration of human beings, nature and society.

The core philosophical concept and specific ideas in this book derive from personal experience. As for the domestic part of this book: In March 1956, I began to work on urban planning and design at the Urban Design Institute of the State Administration of Urban Construction. For more than half a century, I have been to over 100 cities except Tibet, and I selected hundreds of excellent examples from them, so that readers can understand the development of Chinese architectural science and culture in the past 1000 years. As for the foreign part of this book: I have been to over a hundred of cities in more than 30 countries so far after my visit to Switzerland in June 1979. I selected hundreds of excellent examples from them, which belong to the "World Cultural Heritage" or are the key cultural relics under protection in various countries. These modern buildings in the past hundreds of years are masterpieces created by famous Chinese and foreign architects, representing the innovation and development of each era. The example photos in the book are all my own works except for 7 photos signed by others and the planning maps of the cities in the book are provided by local departments. Based on the above-mentioned core philosophical concept of establishing ecological civilization and living environment for human beings, as well as the analysis on the examples both home and abroad. we have summarized six specific ideas about the development of architectural science and culture, namely, "innovative development", "natural environment", "history and culture", "artistic beauty", "convenient transportation" and "just society". They are also the ideas and methods to solve the defects existing in Chinese and foreign cities.

We often say that "to make the past serve the present and the foreign serve China". This book aims to combine and learn from the respective situations and features of China and foreign countries in order to promote the development of architectural science and culture both home and abroad. This is also the essence of writing this book.

In order to deepen the understanding of architectural science and culture, and enhance the mutual understanding and academic exchanges between China and foreign countries, this title received funding from Ms. Huo Lina of Hong Kong, and help from Ms. Li Yi and Mr. Wu Yujiang, senior editor of China Architecture and Building Press. Hereby, I would like to express my heartfelt thanks to them.

Zhang Zugang
October, 2020

目录

前言

上篇：中外建筑科学发展理念

2	序 言
6	一、创新发展理念
6	1.中国创新发展情况
7	1.1 "一五"计划期间我参加规划设计的四川省德阳市和属于初等发达地区的陕西省西安市的产业经济发展情况
7	1.2 深圳、珠海、汕头、厦门四个经济特区的产业经济发展情况
8	1.3 北京、天津、上海、香港、澳门和台湾六个发达地区的产业经济发展情况
10	2.外国创新发展情况
12	二、自然环境理念
12	1.自然环境是人人喜爱、向往的地方，但世界各地许多城市破坏了自然环境，需要加强环境治理，改善生活环境
13	2.创造自然环境，保证绿地面积，人与自然和谐共生，充分利用自然采光通风，体现节能、环保、舒适
16	3.适度控制城市空间容量，根据城市土地、水等资源的负荷量适度发展城市的规模，这是城市宜居、防灾、卫生的又一关键问题
18	三、历史文化理念
18	1.城市历史文化是城市发展的根，一定要有整体保护、合理发展的理念
21	2.要尊重历史建筑，发展新的现代建筑要处理好与历史建筑的关系，要继旧创新，发展建筑文化
24	四、艺术美观理念
24	1.创造城市标志，统领空间环境
27	2.控制尺度体量，协调城市肌里
28	3.掌握平衡、变化的艺术规律，增强城市建筑整体美
28	3.1 城市建筑有对比的空间变化，可达到印象深刻的整体完美
30	3.2 建筑对景、借景，丰富了建筑与城市的景观艺术
30	3.3 建筑的对称布局，可获得庄严、平衡、统一的效果
32	3.4 建筑形象与比例的变化，可以丰富意境，增强审美感知
32	3.5 建筑与城市的空间序列变化，可增强整体的和谐美
34	五、便民交通理念
34	1.在城市市区范围内限制小汽车行驶的数量
35	2.建立城市立体交通系统，大力发展人行道路系统

37	3. 尽量发展地铁交通，同时扩大地下建筑的范围	134	颐和园乐寿堂
		136	颐和园谐趣园
38	4. 恢复与发展自行车道系统，控制快速车行驶	138	实例11　北京天坛
		140	实例12　长城——八达岭
39	六、公正社会理念	142	实例13　北京四合院和鲁迅故居
39	1. 在城市居住方面要尽量搞好大众的住房	144	实例14　北京西四北六条幼儿园四合院
41	2. 在公共建筑服务设施方面，首先要考虑为大众服务	146	实例15　北京中国美术馆
		148	实例16　清华大学建筑设计研究院办公楼
42	3. 社会公正的理念是指导建筑科学发展的重要理念之一	150	实例17　北京国家体育场
44	结　语	152	三、河　北
		152	承　德
		152	实例18　避暑山庄

中篇：中国建筑科学文化精品图说

		156	四、山　东
		156	曲　阜
		156	实例19　阙里宾舍
99	一、陕　西	158	五、江　苏
99	1. 西　安	158	1. 南　京
100	实例1　西安秦始皇陵兵马俑	158	实例20　南京城墙
102	实例2　西安建章宫苑	160	实例21　玄武湖（公园）
103	实例3　西安慈恩寺大雁塔	162	2. 苏　州
104	实例4　西安城墙	162	实例22　"水乡"
106	2. 临　潼	164	实例23　拙政园
106	实例5　临潼骊山华清池	168	实例24　留园
		170	实例25　虎丘
108	二、北　京	172	3. 无　锡
108	实例6　北京总体规划与建设	172	实例26　寄畅园
111	实例7　天安门广场	174	实例27　蠡园
114	实例8　故宫（紫禁城）	176	4. 扬　州
118	故宫太和殿	176	实例28　扬州瘦西湖
120	故宫乾隆花园		
122	实例9　北京西苑（今北海公园部分）	178	六、上　海
126	北海团城	178	实例29　上海豫园
128	北海静心斋	181	实例30　上海里弄
130	实例10　北京颐和园		

182	实例31	上海体育中心
183	实例32	上海陆家嘴金贸区、上海东方明珠电视塔等
184	实例33	上海博物馆
185	实例34	上海大剧院
186	实例35	上海世博会中国馆
188	**七、浙 江**	
188	1．杭 州	
188	实例36	杭州西湖风景名胜区
192	实例37	杭州小瀛洲三潭印月
194	实例38	西泠印社
196	实例39	郭庄
198	实例40	杭州黄龙饭店
200	2．绍 兴	
200	实例41	绍兴水乡
202	实例42	绍兴兰亭
204	实例43	绍兴青藤书屋
206	实例44	绍兴鲁迅故居
207	实例45	绍兴三味书屋
208	**八、安 徽**	
208	合 肥	
208	实例46	合肥环城绿地及全市绿地系统
211	实例47	黄山风景名胜区
214	**九、福 建**	
214	厦 门	
214	实例48	厦门筼筜中心区
217	实例49	厦门康乐新村等居住区
218	实例50	厦门高崎国际机场候机楼
220	实例51	鼓浪屿
221	实例52	南普陀寺
222	**十、广 东**	
222	1．广 州	
222	实例53	广州矿泉客舍
224	实例54	广州白天鹅宾馆
227	2．珠 海	
227	实例55	珠海规划与建设
228	实例56	珠海渔女
228	实例57	珠海滨海情侣南路
230	实例58	珠海石景山景区
232	**十一、海 南**	
232	三 亚	
232	实例59	三亚天涯海角
234	**十二、湖 北**	
234	荆 州	
234	实例60	荆州古城
236	**十三、湖 南**	
236	长 沙	
236	实例61	长沙岳麓书院
238	**十四、广 西**	
238	1．桂 林	
238	实例62	桂林山水 漓江风光
240	实例63	桂林芦笛岩
241	2．三 江	
241	实例64	马鞍寨
244	3．龙 胜	
244	实例65	龙胜金竹寨
246	**十五、四 川**	
246	1．成 都	
246	实例66	杜甫草堂
248	2．都江堰	
248	实例67	伏龙观
250	3．乐 山	
250	实例68	乐山大佛

252	**十六、云　南**		282	**十九、香　港**
252	1．昆　明		282	实例89　香港城市交通
252	实例69　昆明滇池、西山与昆明规划建设		284	实例90　汇丰银行大厦
253	实例70　昆明地区民居建筑		287	实例91　中国银行大厦
254	实例71　筇竹寺五百罗汉塑像		290	**二十、澳　门**
255	实例72　昆明大观楼		290	实例92　大三巴牌坊
256	实例73　西山龙门		291	实例93　澳门博物馆
258	2．路　南		292	实例94　市政厅广场
258	实例74　石林		294	实例95　澳门岛沿岸城市轮廓
260	3．大　理		296	**二十一、台　湾**
260	实例75　南北城门楼		298	1．台　北
261	实例76　大理崇圣寺三塔		298	实例96　台北国父纪念馆
262	实例77　大理白族民居建筑		299	实例97　台北故宫博物院
264	实例78　白族村头小广场		300	实例98　圆山大饭店
266	4．丽　江		302	实例99　台北剑潭青年活动中心
266	实例79　丽江古城		303	2．台　中
268	实例80　丽江纳西族民居		303	实例100　东海大学校园规划和路思义教堂
270	5．西双版纳		305	3．鹿　港
270	实例81　傣族村寨民居		305	实例101　龙山寺
272	**十七、新　疆**			
272	1．乌鲁木齐			
272	实例82　新疆人民会堂			
273	实例83　新疆迎宾馆			
274	2．高　昌			
274	实例84　高昌故城		**下篇：外国建筑科学文化精品图说**	
275	3．交　河			
275	实例85　交河故城		308	**一、埃　及**
276	4．喀　什		308	实例102　开罗吉萨金字塔群
276	实例86　喀什民居		312	实例103　卢克索卡纳克阿蒙太阳神庙
278	实例87　艾提尕尔清真寺		315	实例104　埃及亚历山大城
280	**十八、甘　肃**			
280	敦　煌		316	**二、伊拉克**
280	实例88　敦煌莫高窟		316	实例105　巴比伦城

319	三、希　腊	375	2．杜关市
319	实例106　雅典卫城	375	实例130　弗雷斯诺国家当代艺术学校
		376	3．里尔市
322	四、意大利	376	实例131　里尔美术馆
322	1．罗　马		
322	实例107　罗马中心广场	378	六、西班牙
324	实例108　大斗兽场	378	1．马德里
326	实例109　罗马图拉真纪功柱	379	实例132　马德里埃斯库里阿尔宫
327	实例110　君士坦丁凯旋门	380	实例133　交通部门大楼
328	实例111　罗马圣彼得大教堂	381	实例134　马德里西班牙广场
331	实例112　罗马特雷维喷泉	382	2．巴塞罗那
332	实例113　罗马西班牙广场	382	实例135　巴塞罗那绿地系统和旧城
334	实例114　埃马努埃尔二世纪念碑	384	实例136　巴塞罗那蒙特胡依克山
335	2．佛罗伦萨	386	实例137　巴塞罗那国际展览会德国馆
335	实例115　佛罗伦萨市中心西尼奥列广场	387	实例138　巴塞罗那高迪宅邸建筑、圣家族教堂
338	3．威尼斯		
338	实例116　威尼斯圣马可广场	390	3．格拉纳达
342	实例117　威尼斯水乡风貌	390	实例139　阿尔罕布拉宫苑
344	4．庞　贝	395	实例140　吉纳拉里弗宫苑
344	实例118　古城公共建筑和住宅区		
348	5．巴格内亚	398	七、英　国
348	实例119　兰特别墅园	398	1．伦　敦
		398	实例141　伦敦国会大厦
350	五、法　国	400	2．布赖顿
350	1．巴　黎	400	实例142　生活环境
350	实例120　巴黎圣母院		
353	实例121　巴黎沃克斯•勒•维康特园	402	八、爱尔兰
355	实例122　巴黎凡尔赛宫苑	402	都柏林
364	实例123　巴黎卢浮宫	402	实例143　"三一学院"
366	实例124　巴黎杜伊勒花园	404	实例144　都柏林圣•斯蒂芬公园
368	实例125　巴黎协和广场		
369	实例126　巴黎明星广场凯旋门	406	九、德　国
370	实例127　巴黎蓬皮杜文化中心和"中国建筑、生活、环境展览"	406	实例145　法兰克福
372	实例128　巴黎新歌剧院	408	十、比利时
374	实例129　法国国家图书馆	408	1．安特卫普

408	实例146 安特卫普市旧城中心区	448	实例165 芝加哥伊利诺伊州政府大楼
412	实例147 鲁宾斯博物馆	450	实例166 芝加哥卡森·皮里·斯科特百货公司
414	2．布鲁日		
414	实例148 水乡	452	实例167 芝加哥弗兰克·劳埃德·赖特住宅和工作室
416	3．布鲁塞尔		
416	实例149 布鲁塞尔市中心大广场	454	4．波士顿
418	实例150 布鲁塞尔原子球	454	实例168 波士顿中心绿地公园区
		456	实例169 波士顿查理河两岸
419	**十一、俄罗斯**	458	实例170 波士顿市府办公楼
419	莫斯科	459	实例171 波士顿公共图书馆
419	实例151 莫斯科红场	460	5．费　城
422	实例152 莫斯科马涅什广场购物中心	460	实例172 费城中心区下沉式广场
		461	6．纽黑文
423	**十二、瑞　士**	461	实例173 纽黑文耶鲁大学
423	1．伯尔尼	464	7．洛杉矶
423	实例153 旧城保护	464	实例174 洛杉矶市区
424	2．日内瓦	466	实例175 洛杉矶亨廷顿文化园
424	实例154 湖滨保护园林化	468	实例176 洛杉矶迪士尼游乐园
426	3．阿尔卑斯山	471	实例177 洛杉矶加登格罗夫社区教堂（水晶教堂）
426	实例155 雪景		
428	4．北部和南方	474	8．西雅图
428	实例156 住宅	474	实例178 西雅图市中心区沿海滨建筑群
		477	实例179 西雅图公共市场中心
430	**十三、美　国**	478	实例180 西雅图北部展览馆建筑
430	1．华盛顿		
430	实例157 美国国会大厦	480	**十四、加拿大**
436	实例158 华盛顿纪念碑	480	1．蒙特利尔
437	实例159 华盛顿白宫	480	实例181 蒙特利尔市中心区
438	实例160 华盛顿国家艺术馆东馆	482	实例182 蒙特利尔Desjardins建筑
440	实例161 华盛顿史密松非洲、近东及亚洲文化中心	484	实例183 蒙特利尔加拿大建筑中心
		486	2．多伦多
442	2．纽　约	486	实例184 加拿大多伦多电视塔和滨湖建筑群
442	实例162 洛克菲勒中心下沉式广场等		
443	3．芝加哥	488	实例185 多伦多市中心区靠近湖滨休闲岛区
444	实例163 芝加哥滨湖建筑立体轮廓		
446	实例164 芝加哥密执安湖畔格兰特公园	489	实例186 多伦多伊顿中心

490	实例187 尼亚加拉大瀑布	504	实例194 东京都新都政厅
		506	实例195 银座
492	**十五、巴基斯坦**	508	实例196 上野公园
492	1. 拉合尔	510	实例197 浅草寺
492	实例188 真珠大陆旅馆	512	2. 名古屋
494	实例189 拉合尔城堡	512	实例198 名古屋市中心大街
496	实例190 拉合尔巴德夏希清真寺	514	3. 奈　良
498	实例191 拉合尔夏利玛园	514	实例199 奈良东大寺
500	2. 伊斯兰堡	516	4. 京　都
500	实例192 伊斯兰堡市区	516	实例200 京都平安神宫
502	实例193 伊斯兰堡费萨尔清真寺	519	实例201 金阁寺庭园
		520	实例202 银阁寺庭园
504	**十六、日　本**	521	实例203 龙安寺石庭
504	1. 东　京		

Catalogue

Part One: Concept of the Development of Chinese and Foreign Architectural Science

45	Foreword
50	Chapter 1 Innovative Development
50	1. Innovative development in China
51	(1) Industrial economy development of Deyang in Sichuan Province, which I participated in its planning and design during the "First Five-Year Plan" period, and Xi'an, which is now an elementary developed area.
52	(2) The industrial economy development of four special economic zones: Shenzhen, Zhuhai, Shantou and Xiamen.
53	(3) The industrial economy development of six developed areas: Beijing, Tianjin, Shanghai, Hong Kong, Macao and Taiwan.
55	2. Innovative development abroad
57	Chapter 2 Natural Environment
57	1. The natural environment is a place that everyone loves and yearns for, but many cities around the world have destroyed the natural environment, so it is necessary to strengthen environmental governance and improve the living environment
59	2. To create the natural environment, we should also ensure the green area and the coexistence between man and nature. By using the light energy and wind energy to save energy, protect the environment and live a comfortable life
63	3. Moderate control of urban space capacity and moderate development of urban scale according to the load of urban land, water and other resources are also very important for livability, disaster prevention, sanitation and improving natural ecological environment in cities
65	Chapter 3 History and Culture
65	1. The history and culture are the roots of urban development, that we must have the idea of overall protection and rational use and development
69	2. Respect historical buildings, deal with the relationship with historical buildings when develops new ones; inherit the old and create the new and develop architectural culture
73	Chapter 4 Artistic Beauty
73	1. Create city symbols and lead the space environment
76	2. Control scale and volume and coordinate urban texture

78	3. Grasp the artistic law of balance and change, and enhance the overall perfection of urban architecture
78	(1) Urban buildings have comparative spatial changes, which can achieve impressive overall perfection.
80	(2) The methods of opposite scenery and borrowed scenery have enriched the landscape art of buildings and cities.
81	(3) The symmetrical layout of the building can obtain the perfect effect of solemnity, balance and unity.
83	(4) The change of architectural shape and proportion can enhance coordination or perception of artistic conception.
84	(5) The spatial sequence change of architecture and city can enhance the harmony and perfection of the whole.
85	Chapter 5 Convenient Transportation
86	1. Limit the number of cars driving in urban areas
87	2. Establish a three-dimensional urban road traffic system and develop a large-scale sidewalk system
88	3. Try to develop subway traffic and expand the scope of underground buildings around the subway
90	4. Restore and develop bicycle lane system and control fast-moving vehicles
91	Chapter 6 Just Society
92	1. In terms of urban living, we should try our best to improve the housing of residents
94	2. In terms of public building service facilities, we should first consider serving the public
95	3. Social justice is one of the important concepts guiding architectural science development
97	Epilogue

Part Two: Illustration of Chinese Architectural Science and Culture

100	1. Shanxi	108	2. Beijing
100	1.1 Xi'an	108	Example 6 Master plan and construction of Beijing
100	Example 1 Emperor Qin's Terra Cotta Warriors	111	Example 7 Tiananmen Square
102	Example 2 Jianzhang Palace	114	Example 8 Palace Museum (Forbidden City)
103	Example 3 Big Wild Goose Pagoda in Ci'en Temple	118	Hall of Supreme Harmony
105	Example 4 City Wall	120	Emperor Qianlong's Garden
106	1.2 Lintong	122	Example 9 Xiyuan (now part of Beihai Park)
106	Example 5 Huaqing Pool in Li Mountain of Lintong	126	Round City

128	Heart-East Study	172	5.3 Wuxi
130	Example 10 Summer Palace	172	Example 26 Jichang Garden
134	Hall of Happiness and Longevity	174	Example 27 Li Garden
136	Garden of Harmonious Interest	176	5.4 Yangzhou
138	Example 11 Temple of Heaven	176	Example 28 Slender West Lake
140	Example 12 Great Wall – Badaling		
143	Example 13 Quadrangle Dwellings and Former Residence of Lu Xun	178	6. Shanghai
		179	Example 29 Yu Garden
144	Example 14 Quadrangle Dwellings of Beiliutiao Kindergarten in Xisi	181	Example 30 Lanes and Alleys
		182	Example 31 Sports Center
146	Example 15 National Art Museum of China	183	Example 32 Lujiazui Finance and Trade Zone and the Oriental Pearl TV Tower
149	Example 16 Office Building of the Architectural Design and Research Institute of Tsinghua University	184	Example 33 Shanghai Museum
		185	Example 34 Grand Theatre
		187	Example 35 China Pavilion of Shanghai World Expo
150	Example 17 National Stadium		
		189	7. Zhejiang
152	3. Hebei	189	7.1 Hangzhou
152	Chengde	189	Example 36 Scenic Area of West Lake
152	Example 18 Chengde Mountain Resort	192	Example 37 Three Pools Mirroring the Moon in Yingzhou
		194	Example 38 Xiling Society of Seal Arts
157	4. Shandong	196	Example 39 Guo's Villa
157	Qufu	198	Example 40 Huanglong Hotel
157	Example 19 Queli Hotel in Qufu	200	7.2 Shaoxing
		200	Example 41 Watertown
158	5. Jiangsu	202	Example 42 Orchid Pavilion
158	5.1 Nanjing	204	Example 43 Green Vine Study
159	Example 20 City Wall	206	Example 44 Former Residence of Lu Xun
160	Example 21 Xuanwu Lake (Park)	207	Example 45 Sanwei Study
162	5.2 Suzhou		
163	Example 22 Water Lane	209	8. Anhui
165	Example 23 Humble Administrator's Garden	209	Hefei
		209	Example 46 Green Space System
168	Example 24 Lingering Garden	211	Example 47 Yellow Mountain
170	Example 25 Tiger Hill		

214	**9. Fujian**	238	Example 62 Guilin Landscape and Li River
214	Xiamen		
215	Example 48 Yundang Center Area	240	Example 63 Reed Flute Cave
217	Example 49 Kangle New Village and Other Residential Areas	241	14.2 Sanjiang
		241	Example 64 Ma'an Village
218	Example 50 Terminal of Ximen Gaoqi International Airport	244	14.3 Longsheng
		244	Example 65 Jinzhu Village in Longsheng
220	Example 51 Gulangyu Island		
221	Example 52 Southern Putuo Temple		
		246	**15. Sichuan**
222	**10. Guangdong**	246	15.1 Chengdu
222	10.1 Guangzhou	247	Example 66 Du Fu's Thatched Cottage
223	Example 53 Guangzhou Mineral Water Guest House	248	15.2 Dujiangyan
		248	Example 67 Fulong Temple
225	Example 54 White Swan Hotel	250	15.3 Leshan
227	10.2 Zhuhai	250	Example 68 Leshan Giant Buddha
227	Example 55 Plan and Construction of Zhuhai		
		252	**16. Yunnan**
228	Example 56 Statue of Fisher Girl	252	16.1 Kunming
228	Example 57 Lovers Road	252	Example 69 Dian Lake, West Hill and Construction of Kunming
230	Example 58 Stone Scene Hill Scenic Spot		
		253	Example 70 Residential Area
		254	Example 71 Five Hundred Arhat Statues in Qiongzhu Temple
232	**11. Hainan**		
232	Sanya	255	Example 72 Grand View Tower
232	Example 59 Sanya Ends Area	256	Example 73 Dragon Gate of West Hill
		258	16.2 Lunan
234	**12. Hubei**	258	Example 74 Stone Forest
234	Jingzhou	260	16.3 Dali
234	Example 60 Ancient City of Jingzhou	260	Example 75 North-South Gate Tower
		261	Example 76 Three Pagodas in Chongsheng Temple
236	**13. Hunan**		
236	Changsha	262	Example 77 Residential Building of Bai Nationality
236	Example 61 Yuelu Academy in Changsha		
		265	Example 78 Little Square in the Village of Bai Nationality
238	**14. Guangxi**		
238	14.1 Guilin		

266	16.4 Lijiang	283	19. Hong Kong
266	Example 79 Ancient City of Lijiang	283	Example 89 City transportation
269	Example 80 Residential Area of Naxi Nationality	284	Example 90 HSBC Plaza
		287	Example 91 Bank of China Building
270	16.5 Xishuangbanna		
270	Example 81 Villages of Dai Nationality	290	20. Macao
		290	Example 92 Ruins of St. Paul
		291	Example 93 Museum of Macao
272	17. Xinjiang	292	Example 94 City Hall Plaza
272	17.1 Urumqi	294	Example 95 Outline of the Coastal City
272	Example 82 People's Hall		
273	Example 83 Guest House	297	21. Taiwan
274	17.2 Gaochang	298	21.1 Taibei
274	Example 84 Ancient City of Gaochang	298	Example 96 Sun Yat-sen Memorial Hall
275	17.3 Jiaohe	299	Example 97 Taibei Palace Museum
275	Example 85 Ancient City of Jiaohe	301	Example 98 Grand Hotel
276	17.4 Kashgar	303	Example 99 Taibei Jiantan Youth Activity Center
276	Example 86 Residential Buildings		
278	Example 87 Id Kah Mosque	303	21.2 Taizhong
		303	Example 100 Taizhong Donghai University Campus and Luce Memorial Chapel
280	18. Gansu		
280	Dunhuang	305	21.3 Lugang
280	Example 88 Mogao Grottoes	305	Example 101 Longshan Temple

Part Three: Illustration of Foreign Architectural Science and Culture

309	1. Egypt	320	3. Greece
309	Example 102 Cairo Giza Pyramid Group	320	Example 106 Athens Acropolis
312	Example 103 Luxor Amon Sun Temple		
315	Example 104 Alexandria	322	4. Italy
		322	4.1 Rome
317	2. Iraq	322	Example 107 Raman Forum
317	Example 105 The City of Babylon	324	Example 108 Colosseum

326	Example 109 Column of Trajan
327	Example 110 Arch of Constantine
328	Example 111 St. Peter's Cathedral
331	Example 112 Trevi Fountain
332	Example 113 Piazza Spagna
334	Example 114 Monument to Victor Emanuele II
335	4.2 Florence
335	Example 115 Signoria Square
338	4.3 Venice
338	Example 116 Piazza San Marco
342	Example 117 Watertown Landscape
344	4.4 Pompeii
344	Example 118 Public Buildings and Residence
348	4.5 Bagnaia
348	Example 119 Villa Lante
350	**5. France**
350	5.1 Paris
350	Example 120 Notre Dame de Paris
353	Example 121 Vaux Le Vicomte Garden
356	Example 122 Château de Versailles
364	Example 123 Louvre Palace
366	Example 124 Tuileries Garden
368	Example 125 Concorde Square
369	Example 126 Arch of Triumph
371	Example 127 National Art and Cultural Center of Georges Pompidou and the Exhibition of Chinese Architecture, Life and Environment
372	Example 128 Opéra de la Bastille
374	Example 129 National Library
375	5.2 Tourcoing
375	Example 130 Le Fresnoy National Contemporary Art Center
376	5.3 Lille
376	Example 131 Palais des Beaux-Arts
378	**6. Spain**
378	6.1 Madrid
379	Example 132 El Escorial Palace
380	Example 133 Transport Department Building
381	Example 134 Plaza de Espana
382	6.2 Barcelona
382	Example 135 Green Space System and Old City
384	Example 136 Montjuïc Hill
386	Example 137 German Pavilion of the International Expo
387	Example 138 Residence Building of Antoni Gaudi and Sagrada Família
391	6.3 Granada
391	Example 139 Alhambra Palace
395	Example 140 Generalife Garden
399	**7. United Kingdom**
399	7.1 London
399	Example 141 Parliament Building
400	7.2 Brighton
400	Example 142 Living Environment
402	**8. Ireland**
402	Dublin
402	Example 143 Trinity College
404	Example 144 St.Stephen's Park
406	**9. Germany**
406	Example 145 Frankfurt
409	**10. Belgium**

409	10.1 Antwerp	442	13.2 New York
409	Example 146 Central Area of the Old City	442	Example 162 Sunken Plaza in Rockefeller Center
412	Example 147 Rubens Museum	443	13.3 Chicago
414	10.2 Bruges	444	Example 163 Three-dimensional Outline of Lakeside Building
414	Example 148 Water Town	446	Example 164 Grant Park
416	10.3 Brussels	448	Example 165 Illinois Government Building
416	Example 149 City Center Square	450	Example 166 Carson Pirie Scott Department Store
418	Example 150 Atomiun	452	Example 167 Residence and Studio of Frank Lloyd Wright
419	11. Russia	454	13.4 Boston
419	Moscow	454	Example 168 Central Green Park
419	Example 151 Red Square	456	Example 169 Banks of Charles River
422	Example 152 Manezhnaya Square Shopping Center	458	Example 170 Boston Government Building
423	12. Switzerland	459	Example 171 Boston Public Library
423	12.1 Berne	460	13.5 Philadelphia
423	Example 153 Protection of the Old City	460	Example 172 Downtown Sunken Plaza
424	12.2 Geneva	461	13.6 New Haven
424	Example 154 Landscaping of Lakeside Protection	461	Example 173 Yale University
426	12.3 Alpine Mountain	464	13.7 Los Angeles
426	Example 155 Snowscape	464	Example 174 Downtown Area
428	12.4 North Part and South Part	466	Example 175 Huntington Cultural Park
428	Example 156 Residence	469	Example 176 Disneyland
431	13. United States	471	Example 177 Garden Grove Community Church
431	13.1 Washington	474	13.8 Seattle
431	Example 157 Capitol	474	Example 178 Seaside Buildings in Downtown Area
436	Example 158 Washington Monument	477	Example 179 Public Market
437	Example 159 White House	478	Example 180 Exhibition Hall at North Seattle
438	Example 160 East Building of National Gallery of Art		
440	Example 161 Cultural Center of Africa, Near East and Asian of Smithsonian		

480	**14. Canada**	500	Example 192 Downtown Area
480	14.1 Montreal	502	Example 193 Faisal Mosque
480	Example 181 Downtown Area		
482	Example 182 Desjardins	504	**16. Japan**
484	Example 183 Canadian Centre for Architecture	504	16.1 Tokyo
		504	Example 194 New City Hall
486	14.2 Toronto	506	Example 195 Ginza
486	Example 184 TV Tower and Lakeside Buildings	508	Example 196 Ueno Park
		510	Example 197 Asakusa Temple
488	Example 185 Leisure Island near the Lakeside	512	16.2 Nagoya
		512	Example 198 Central Street
489	Example 186 Eaton Center	514	16.3 Nara
490	Example 187 Niagara Falls	514	Example 199 Todaiji Temple
		517	16.4 Kyoto
492	**15. Pakistan**	517	Example 200 Heian Shrine
492	15.1 Lahore	519	Example 201 Courtyard of Kinkakuji Temple
492	Example 188 Pearl Continental Hotel		
494	Example 189 Lahore Fort	520	Example 202 Courtyard of Ginkakuji Temple
496	Example 190 Badshashi Mosque		
498	Example 191 Shalamar Bagh	521	Example 203 Stone Yard of Ryoanji Temple
500	15.2 Islamabad		

上篇：
中外建筑科学发展理念

Part One:
Concept of the Development of Chinese and Foreign Architectural Science

序　言

　　建筑科学是一门大学科，它包括城乡规划学、建筑学、风景园林学三个学科。而这三个学科又是三位一体的，其哲学指导思想是老子的自然宇宙观，即为人类建立生态文明的生活环境理念。

　　对于城乡规划、建筑学、风景园林三位一体的理念需要深化认识。这是因为城乡中的建筑与园林同城乡是一个相互联系的有机整体，你中有我，我中有你，三者缺一不可。城乡与园林的关系，主要是解决城乡生态环境问题，包括城乡外围的自然山水和城镇中的绿地系统以及城乡住区各类建筑内的绿地；建筑与园林的关系，可以说是建筑离不开花木，"建筑在园林中，园林在建筑中"，二者相辅相成，这既是提高生活环境质量所需要的，也是建筑与园林结合所追求的目的；城乡与建筑的关系，主要是讲建筑是城镇、乡村的主要内容，建筑也是城乡居民生活、生产所需要的实体。

　　关于城乡、建筑、风景园林三者综合为一体的关系，一些前辈著名建筑师是有认识的，他们具有这三个方面的基本知识，但都未能提出这是一门大学科的概念。20世纪50年代，大家公认的建筑界带头人——"南杨北梁"就是典型的实例。"南杨"，即杨廷宝先生，1924年获美国宾夕法尼亚大学建筑学硕士学位，曾任南京工学院副院长、建筑系主任，除其建筑学知识和作品出众外，对城乡规划与建设有深入的研究，在江苏省连云港、苏州、镇江、南通及福建省厦门等地都以报告和讲话的形式对这些城市规划建设提出了意见。他还重视风景园林学，对江苏省无锡市风景园林建设、福建省武夷山风景区规划与建设、上海古漪园整修等提出了宝贵意见。"北梁"，即梁思成先生，1927年获美国宾夕法尼亚大学建筑学硕士学位，曾任清华大学建筑系主任，他对建筑学、建筑历史有深厚的造诣。梁思成先生曾邀请苏联专家来清华大学建筑系讲授城市规划原理课，并将我们"建五班"同学（1951年至1956年初）最后一年的毕业设计分作城市、公共建筑、工业建筑三个专题组，后有10多位城市组毕业生被分配到国家城市设计院工作；梁思成先生和陈占祥先生合作提出了整体保护北京旧城的北京市总体规划。1958年建工部刘秀峰部长在青岛主持召开全国城市规划建设会议，梁思成先生在会上做了关于青岛市的历史与发展规划的报告，之后出版了此专题的著作。20世纪50年代初，梁思成先生还安排农业大学园林专业学生到清华大学建筑系来学习交流。

　　这一方面的实例还有不少。程世抚先生于1929年至1933年就读于美国哈佛大学、美国康奈尔大学景观建筑和城市规划专业。20世纪50年代，程世抚先生负责完成了上海市园林总体规划、苏州市旧城人民路街区的改建规划设计，后又参加了国内多个省会城市的总体规划。两院院士、清华大学吴良镛教授，通晓城乡规划、建筑学、风景园林，他对山东曲阜、福建厦门、广东深圳、海南三亚等市都提出了很好的规划设计。他从城市规划的角度改变原曲阜孔子研究院建筑的位置，设计建造出园林化的孔子研究院。他还从北京旧城原有面貌考虑，设计建成了北京菊儿胡同改建的四合院式住宅区等。广州市规划局总建筑师莫伯治院士，不仅精通建筑学与风景园林，还了解城市规划，从20世纪50年代至21世纪初，他先后设计了广州北园酒家、广州白天鹅宾馆等符合城市整体布局、建筑结合园林的工程项目52个，而且获奖数量位居全国榜首。同时，他还对珠海、合肥、敦煌、杭州等城市的规划提出了建议，这些建议皆刊登在《建筑学报》上。知识全面的前辈著名建筑师，其建筑创作适应城市规划与风景园林要求的，还有广州的佘畯南、上海的冯纪忠、成都的徐尚志先生，再晚一辈的著名建筑师有齐康、张锦秋、何镜堂等院士。

国外方面，世界著名建筑大师勒·柯布西耶1931年设计的"光辉的城市"（未实现）、1951年规划并建成的印度昌迪加尔城，都是城乡规划、建筑学、风景园林三位一体的实例。近一段时期，世界著名建筑大师诺曼·福斯特及其合伙人，于21世纪在阿布扎比（Abu Dhabi）规划设计了有9万人规模的新型Masdar生态城，并在德国改造了拥有50万人口的杜伊斯堡，创建了可持续发展的社区理念。他们还设计了德国法兰克福银行大厦、伦敦瑞士再保险公司总部大楼（Swiss Tower）以及北京首都国际机场3号航站楼等很多地标性建筑，将风景园林融合于室内外，为城市增色。这体现出他们重视城市、建筑与风景园林三者的有机结合，并创造出优美的生态环境。美国SWA事务所规划设计的香港屯门新城、菲律宾林韦斯特城市中心和美国亚利桑那州金牧场凯悦饭店等作品，都反映出了城市、建筑、园林成为一个整体的特点。

我们回溯中国历史，对于城市、建筑、园林三位一体的做法，已有2000多年的历史，从汉唐长安、唐洛阳、北宋开封、南宋杭州、元大都、明清北京以及南京古都的城市规划与建设来看，都反映了三者合为一体的思想观念，但是对此观念的实践都没能提出确切的科学名称。

1996年6月，我国大科学家钱学森先生提出了"建筑科学"的新观念，其理念即"城市、建筑、园林三位一体"。钱老将建筑科学列为11个大学科门类之一，这11个大学科门类是：自然科学、社会科学、数学科学、系统科学、军事科学、人类科学、思维科学、行为科学、地理科学、建筑科学、文艺理论。这个新概念是科学理论的新发展，符合其本身的特点，符合历史的发展变化，符合大自然的法则，特别符合城乡、建筑、园林三者合为一体的综合发展的需要，也提高了人们对城乡、建筑、园林的认识，它会对世界各国城乡建设事业的发展起着指导的作用。我们相信，在当前向知识社会发展的新阶段，与其相适应的是由学科综合向大学科发展，因而这一观念理论会逐步得到世界的承认。

关于钱学森先生提出的建筑科学文化最高层次的哲学指导思想，就是"建筑科学"大学科所包括的城乡、建筑、园林的规划设计与建设发展理念，即要以自然宇宙观作为指导思想。这一思想观念来源于中国老子。早在2500年前的春秋战国，老子就提出"故道大，天大，地大，人亦大。域中有四大，而人居其一焉。人法地，地法天，天法道，道法自然"。其核心思想是"道法自然"，认为宇宙万物（包括天、地、人）的演变都要服从其发展的客观规律，即"道"。否则，将阻碍发展或被淘汰。

从历史传统上看，这种自然宇宙观反映在我国城乡及其建筑、园林中，主要是堪舆学，掌握这门学问的被称为堪舆家。堪为天道，舆为地道，就是运用天文、地理、自然界及其发展变化的客观规律去进行城乡及其建筑、园林的规划设计与建设。对于我们的这种城乡、建筑、园林同自然共融的自然宇宙观，英国学者李约瑟在其编著的《世界大百科全书》中谈及"中国的建筑精神"里写道"再没有其他地方表现得像中国人那样热心于体现他们的伟大思想'人不能离开自然'的原则，这个人不是社会上可以分割出来的人，皇宫、庙宇等重大建筑物自然不在话下，城乡中不论是集中的，还是散布在田庄中的住宅也都经常地出现一种对'宇宙图案'的感觉，以及作为方向、节令、风向和星宿的象征主义。这是中国人和自然结合的象征主义及对'宇宙图案'的感觉"。这种崇尚自然的思想，还影响到20世纪世界著名建筑大师弗兰克·劳埃德·赖特（Frank Lioyd Wright）。弗兰克·劳埃德·赖特熟读过中国老子的《道德经》（德文版转译英文版），比较深刻地领悟了老子提出的"道"的自然宇宙观，认为它不仅是自然万物的表象，亦包括其内在的精神。弗兰克·劳埃德·赖特创作的建筑特点是建筑与自然的有机结合，并融为一体。弗兰克·劳埃德·赖特提出的"有机建筑论"，自我解释为"有机建筑"就是自然的建筑，它产生于事物内部的自然本质，建筑的本质就是自然。此外，还有世界著名的西班牙建筑师里卡多·博菲利（Ricardo E.Bofill），他讲自己有50年的设计实践生涯，直到设计巴塞罗纳港口酒店时（2010年建成）才感悟到"要改变对建筑学的态度，改变对于自然、生态、环境的态度，建筑就应保持对自然的尊重"。

关于自然宇宙观指导园林设计与建设问题，我国著名园林界前辈孙筱祥先生在笔者所著《世界园林发展概论》一书序言中，对18世纪前东西方两种截然不同的园林美学观念做了比较，他分析说："欧洲整形几何式园林，其美学主题为'人'是宇宙主宰，'人'是宇宙的目的，大自然必须按照人头脑中的秩序、规律、条理和几何模式来进行改造。而中国自然山水派园林的美学主题为大自然是宇宙的主宰，大自然是宇宙的目的，人不过是大自然芸芸众生中的一员。" 1986年，孙筱祥先生应邀在美国哈佛大学召开的"国际大地规划教育学术会议"上做学术报告，他在报告中指出："中国是世界上第一个以大自然为原型进行园林设计的国家。不仅如此，中国对大自然的深情挚爱、对大自然的领悟、对自然美的敏感，是极其广泛地渗透到哲学、艺术、文学、绘画的所有文化领域中……"。该报告受到大会主持人司坦尼兹教授的称赞，并被选为大地规划学科中首位最杰出国际教育典范。

我国大科学家钱学森先生多次盛赞反映自然宇宙观的中国园林，他说："我国园林号称'花园之母'，名园遍及全国各地，为世人所称颂……""我国的园林学是祖国文化遗产里的一颗明珠……""中国园林艺术是祖国的珍宝，有几千年的辉煌历史"。他还多次提出，要把自然的中国园林艺术融合到建筑和城市建设中，创造自然的中国山水城市。

老子的自然宇宙观，指明了人类社会发展与宇宙万物发展的客观规律，即人类与自然万物应是和谐的关系。它是解决当今世界"生态环境危机"问题的重要指导思想。我们要继承和发展这一观念，并使其具有现代性，在新的历史条件下探讨人类与自然、社会与自然有机融合关系的方式和做法，它扩展了生态文明中自然、经济、社会、文化的生态平衡内容，这些内容将在后面六章理念中加以说明。

建立生态文明生活环境既是科学也是艺术问题，本书所提的文化是"大文化"。从大文化的概念来看，它包括科学技术和文化艺术。一个文化圈或一个地域的文化，可以说是地域的特殊生活方式或生活道理，它包括这里的一切人造制品、知识、信仰、价值和规范等，它综合反映了社会、经济、科学技术、观念、习俗以及自然生态的特点。由此可以看出，建筑科学属于大文化的范畴，它既有文化艺术，又有科学技术。什么是科学技术？就是以逻辑思维、逻辑语言对大自然事物的探索、研究和认识，这是科学家、工程技术专家的事情；什么是艺术？就是以形象思维、形象感受对大自然事物的描绘、表现和传播，这是文学家、艺术家的事情。而包括城乡规划师、建筑师、风景园林师在内的建筑科学家是兼具科学技术与艺术并融为一体的专家，这是由建筑科学的性质、本质属性所决定的。因而，我们从建筑科学大文化的观点来分析研究中外建筑科学文化，形成它是为人类建立生态文明生活环境的综合科学和艺术的认识，以利于促进中外建筑科学文化的发展。

  通过改革开放40多年来的建设实践，中国城乡、建筑、园林事业取得很大的发展，积累了一些好的经验。但在一些地方还存在着缺陷：城市规模过大，容积率高，片面追求形式，不符合实际；水、气、垃圾、汽车污染环境；拆毁历史建筑和破坏自然环境，缺少整体保护规划；不重视发展城镇绿地，浪费资源，片面追求眼前利益；城市面貌没有特色，肌理无序；有些大城市盲目发展私人小汽车，道路变为停车场，交通混乱，影响居民生活与健康等。这些缺陷在外国亦存在。针对这些缺陷，结合我自1956年考察中国和1979年考察国外众多的城市后，选出有参考价值的中外著名的优秀实例，并吸取国外关于"现代主义"（Modernism）、"后现代主义"（Post-Modernism）、"新城市主义"（New Urbanism）、"批判地域主义"（Critical Regionalism）等理论中合理的观点，在为人类建立生态文明生活环境的哲学观念统领下，提出"创新发展""自然环境""历史文化""艺术美观""便民交通""公正社会"六大理念。这六大具体理念，可以说是过去、现在和未来关于城乡、建筑、园林综合体发展的正确轨迹和方向。如果中外城乡偏离这一发展方向，必将走弯路；沿着这一方向发展，就能走上生态平衡、科学发展、可持续发展的道路。下面具体阐释这六大理念。在具体分析各理念时，主要是举城市实例，极少数为乡村内容。

一、创新发展理念

发展产业经济，带动城市建设，重视创新发展，迈向更高层次。

　　从国内外这几十座城市的发展历程来看，都是因发展产业经济带动了城市、建筑、园林事业的发展，所以说发展产业经济是硬道理，经济是基础。产业经济包括第一、第二、第三产业。封建社会时期，以发展第一产业农业经济为主；到了资本主义社会时期，以发展现代工业经济为主；现阶段向知识信息社会转变。凡发达的地区，其第三产业经济都占到最大的比例，这是产业经济发展变化的总趋势。每个城市都要积极发展具有自身特色的产业经济。

1. 中国创新发展情况

　　21世纪初，中国学者在世界上提出了"第二次现代化"理论。中国科学院何传启研究员认为，第一次现代化是工业时代，第一位是经济发展；第二次现代化是发展知识时代，第一位是生活质量，知识和信息扩大了精神生活空间，精神和文化生活方式将高度多样化，从以工业化为主发展到以知识化为主，知识经济超过物质经济，知识社会初步形成。他还提出，不同地区有不同的特点，我国可分为北京、天津、上海、香港、澳门和台湾6个发达地区，辽宁等15个初等发达地区和甘肃等13个欠发达地区。发达地区应全力推进第二次现代化，发展知识经济，建设知识社会，加快"工业转移"和"城市扩散"，逐步实现第二次现代化；初等发达地区应协调发展两次现代化；欠发达地区则应该继续推进第一次现代化，同时关注第二次现代化。到21世纪中叶我国可基本实现第二次现代化，赶上当时世界中等发达国家的水平。

　　现在按中华人民共和国成立后的时序，说明一些有代表性的城市发展产业经济的概况。从20世纪50年代初期国民经济发展的"一五"计划开始，结合苏联援助156个重点工业项目，重点规划建设了北京、上海、南京、杭州、苏州、青岛、沈阳、昆明、成都、兰州和其他省会城市，以及东北、西南、西北地区的一些城市，这些城市都重点发展第二产业经济和部分第三产业经济，共同带动了这些城市的文化建设。20世纪70年代后，国际上出现了石油和美元危机，世界经济不景气，一些发达国家投资到中国香港和台湾等地区，促进了香港金融、商贸等产业经济的发展，也促进了中国台湾台北、新竹等市区高新技术产业经济的发展，中国香港、中国台湾、韩国与新加坡，被称为亚洲四小龙。20世纪80年代后，中国沿海的厦门、珠海、汕头、深圳成立经济特区，迅速发展了高新技术、电子、加工产业经济等，其产业经济的腾飞，促进了这四座城市建设和文化建设的快速发展。20世纪90年代后，上海开发了浦东新区，加快了上海金融、商贸产业经济的发展；北京建起了亦庄等新的经济技术开发区，发展了东三环的CBD和西城区的金融街；又由于北京筹备2008年的奥运会建设，上海筹备2010年世博会的建设，使上海、北京两座城市的产业经济迅速发展，同时建筑科学文化建设得到了进一步的发展。

以上是中国城市及其产业发展的概况，下面再具体介绍3个方面的情况。

1.1 "一五"计划期间我参加规划设计的四川省德阳市和属于初等发达地区的陕西省西安市的产业经济发展情况

四川省德阳市的规划设计始于1956年3月，这时我刚从清华大学营建系毕业，被分配到国家城市设计院工作，我有幸和同学齐立根同志等代表国家城市建设部门参加该市的总体规划设计。开始时，会同全国一机部、四机部、地质部等共同勘察、选择重型机械和电子工业等的厂址，并由苏联库维尔金专家指导这个城市的总体规划。此时的德阳人口只有1万人，其聚居的小城是在由南北向铁路、河流构成的长条形铁东区，铁路西面为开阔的平原区，主导风向为北风。因电子工业污染很小，所以将其布置在铁路东边北部的上风地段。此两项电子工业项目都是以代号相称，其具体内容保密。另外的重型机械厂，占地大，有污染，要求土地耐力强，故把它选在铁路西边南部的下风地段，以铁西区为主，以铁东区为辅，组成了德阳市统一的生活居住区。此重型机械厂到目前为止仍在生产全国最大的重型机械。德阳市以发展重型机械、电子、特种钢材等为主。60多年来，其第二、第三产业经济得到扩大发展，城市建设也得到大力发展。

属于初等发达水平的陕西西安市，是我国"一五"计划重点规划建设的八大城市之一，现已建成为我国新兴工业基地之一。自改革开放以来，西安市又调整了第二产业，扩展了电子、高新技术与航天科技等产业，同时作为重要的历史文化名城，又大力发展文化等第三产业，制定了老城"唐皇城"的文化复兴规划和具有西安市文化特点的区域规划，包括临潼国际旅游、蓝田美玉文化、长安生态居住、户县农民画、周玉老子文化、高陵现代农业等。西安市的观点和做法是正确的，重视城市产业经济的发展，协调发展第二产业经济和文化、服务等第三产业，以产业经济的发展促进西安历史文化名城的保护与发展。

西安市和德阳市，现应进一步协调发展两次现代化，以早日达到发达地区的水平，着手推进第二次现代化。

1.2 深圳、珠海、汕头、厦门四个经济特区的产业经济发展情况

深圳、珠海、汕头、厦门，位于我国东南沿海地带，20世纪六七十年代由于东南沿海地带与我国台湾地区近邻，以致经济上不能得以大力发展。到了改革开放的20世纪80年代，国际形势趋于和平，为了引进外资、发展对外经济合作，1980年国家批准成立深圳等4个经济特区，并给予经济特区一些新的政策，以加快中国国民经济的发展。实践证明，这一做法是正确、成功的。在此时期，我本人因在建设部所属的中国建筑学会工作，同这4个城市有多

次交往，深感其建设与发展的惊人成就。

深圳，其地名始于明代永乐八年（1410年），人口很少，一直是个小镇，1979年才设市，1980年被定为中国第一个经济特区，居民大都由外地迁来，其发展速度之快前所未有，创造了城市化、工业现代化的新记录，现已成为全国第四大城市。其工业以出口加工为主，发展有机械、电子、纺织、轻工、建材、医药、服装、食品等，近一段时期又大力发展信息、生物工程、金融、商贸、旅游、服务等第三产业经济，还为香港特别行政区提供农副产品。

汕头，是我国沿海对外开放的重要港口城市，可通往世界各地，其商贸产业经济发达。自成立经济特区后，汕头发展了高新技术产业开发区、珠池工业区、台商投资区、海洋工业集团工业区等。其北部临近历史文化悠久的潮州市，它是潮汕文化的发源地，包括潮汕音乐、潮汕歌舞、潮汕美食等深厚的传统文化产业，同时，带动汕头旅游产业的发展。

珠海，原是以发展农业、渔业等第一产业为主的小镇，1980年被定为首批经济特区，属其管辖的海岛有164个，范围甚广，水产业发达，并便于对外交往。其水产业和以出口加工工业为主的产业发展迅速，每两年还举办一次国际航空航天博览会，扩大了对国外的影响力。1984年6月，中国建筑学会应邀组织全国著名专家，前来对珠海经济特区总体规划进行研究，提出咨询建议，该市领导重视并采纳了专家的意见，现已将珠海发展成为经济发达、富有海滨花园风貌、适宜居住和旅游度假的城市。

厦门，亦是我国首批经济特区之一，是中国东南沿海对外贸易的港口。这里原是一个海岛，1955年鹰厦铁路海堤建成，后又建成海沧大桥，把厦门岛同大陆连接为一体。1958年中国国家城市设计院应邀组织工种齐全的城市工作组，前往厦门对其城市发展提出规划方案。由于当时厦门同金门岛关系紧张，该城市不适宜大力发展，建议本岛仅进行小规模建设和改善基础设施工作，重视保护好闻名遐迩的鼓浪屿，同时考察了厦门岛对面的杏林地区，这里有能承载万吨级以上船舶的深水港，可作为厦门未来对外经济贸易的重要港口城市。厦门大力发展以轻工业为主的出口加工工业和高新技术、信息、商贸、旅游产业，已成为开放的港口城市。其经济发展迅速，促进了城市建设，建成了环境优美的筼筜本岛市中心新区和园林化的大众新居住区，还获得了联合国人居环境奖，并重点保护发展了有"海上花园""音乐岛"之称的鼓浪屿。鼓浪屿于2017年7月被列为"世界文化遗产"。

这4个经济特区城市全都重视经济的"创新发展"和"自然环境"的保护，因而皆被列为"国家环境保护模范""国家园林城市""国家卫生城市"和"中国优秀旅游城市"。这四个经济特区的生产总值增长迅速，深圳市已超过何传启研究员提出的天津等发达城市地区的水平，由此可以看出，深圳、汕头、珠海、厦门现已发展到或接近发达地区的标准，应进一步发展知识经济，推进第二次现代化。

1.3 北京、天津、上海、香港、澳门和台湾六个发达地区的产业经济发展情况

北京，是中华人民共和国的首部，是全国政治、文化和国际交流中心。北京的经济是多元化发展，第二、第三产业协同建设。工业从钢铁、机械、电子、化工、轻工、纺织、印刷等

传统行业，逐步转向微电子、新生物医药、汽车、石化新材料、光机电一体化等现代高新技术产业。原北京市区内西部、东部的工业区已向外转移。近些年来，电子信息、信息网络、旅游文化、服务等产业迅速发展，第三产业所占比重最大。全市生产总值为3万多亿元。

关于文化方面，北京的传统文化丰富多样。已有多项列为世界文化遗产（故宫、长城、颐和园、天坛、大运河通惠河段、周口店北京人遗址等），是全世界拥有世界文化遗产最多的城市，其民俗文化还有胡同、四合院、京剧、风味美食和工艺美术品（景泰蓝、玉器、雕刻）等，这些传统文化是宝贵财富，它必然促进旅游产业和与其相关的交通、旅馆、餐饮、特色商品等的发展。北京也正在打造时尚购物之城、国际美食之都、全球旅游目的地、数字消费和新型消费标杆城市，力争早日建成国际消费中心城市。

天津，是中央直辖市，是中国重要工商业和开放港口城市。特别是改革开放后的迅速发展，现已成为中国重要的工业基地之一，纺织、汽车、电子、海洋、石油化工、海盐等是全国的重点产业。天津东部靠海地区，已建起天津滨海经济新区，包括高新技术开发产业区、先进制造业产业区、滨海化工区、中心商务区、海港物流区、海港休闲旅游区以及天津港保税区等。其中天津港保税区对国际贸易、现代物流、出口加工业的发展起着重要作用。天津港是中国华北地区最大的以进出口贸易为主的国际港口。天津亦是国家历史文化名城，有许多名胜古迹，其传统民俗文化有闻名全国的天津相声、杨柳青年画、泥人张彩塑以及餐饮文化等，这些都促进了天津旅游业的发展。

上海，是中国最大的综合性工业基地之一，工业门类多、规模大，有年产千万吨以上的冶金企业和造船、机械、汽车、石油化工等重工业，亦有电子、航天航空、计算机、通讯、医药、生物工程等高新技术产业，还有棉纺织、呢绒、丝绸、塑料和家电、服装、皮鞋、化妆品、钟表、玩具等消费品制造业，都具有一定的规模。20世纪90年代开发了上海浦东新区，建立起浦东对外经济开发区、陆家嘴上海金融中心。2013年8月，又设立上海自由贸易试验区，极大地促进了上海经济产业的发展。同时，上海港口、航空、铁路、公路的交通运输业也随之迅速发展，上海港口已成为世界第三大港，有60多个万吨级船舶位；上海航空港业已列为三大国际航空港之一；铁路也联系着全国东西与南北。上海的交通运输业和金融业、信息网络、保险业、房地产业等第三产业经济占了最大的比重。其生产总值已向4万亿元迈进。以上海为中心的长江三角洲城市群已成为世界公认的六大世界级城市群之一（另5个城市群为美国东北部大西洋沿岸城市群、北美五大湖城市群、日本太平洋沿岸城市群、英国以伦敦为核心的城市群、欧洲西北部城市群），其年经济总量为15万多亿元，在中国经济中的比重达到15%。这种"城市群化"是城市发展的必然趋势，应重视这一发展规律。

香港，中国已于1997年7月1日恢复对香港行使主权，设立香港特别行政区。香港的经济主要是出口加工、世界贸易与物流、金融业、旅游业、房地产业和服务业。出口加工业包括纺织、服装、电器、电子与钟表、玩具等。香港交通运输业十分发达，因它是自由贸易港，已发展成为亚州第二大航空港和国际与亚洲地区主要的航空中心，海运港口已发展成为世界第二大集装箱港，有通往世界100多个国家的20条远运航线，铁路有京九、广九联系全国各地，港珠澳大桥也已建成通车，陆路交通则更为畅通。

澳门，中国已于1999年12月20日恢复对澳门行使主权，设立澳门特别行政区。澳门是自由港，过去的经济以博彩业闻名于世。从20世纪60年代起，澳门经济开始转变，现已形成旅游博彩业、对外加工业、对外贸易业、金融业和房地产建筑业的多元经济结构体系。其对外加工业以中小型企业为主，包括针织、服装、电子、玩具等，这些产品主要销往欧美地区。

台湾地区，1895年被日本占领，1945年抗日战争胜利后归还中国，其省会为台北市。台湾地区工业以出口加工业为主，20世纪70年代前主要发展轻工业、纺织与食品加工工业；20世纪70年代后又重点发展了钢铁、机械、石油化工、飞机、船舶和汽车、电子、电器、电脑，以及体育用品、服装、鞋、玩具、手工艺品等。对于这些工业产品，注重不断采用高新技术更新换代，使其出口产品符合各个时期的国际标准。台湾的交通运输业发达，已形成环岛和贯通南北从基隆到高雄的电气化铁路网，公路形成高速公路、环岛公路网；海运形成以北部基隆港、南部高雄港和花莲港为中心的天然良港网；民航有北部的桃园机场、南部的高雄机场和其他10多处机场，对外通往中国香港特别行政区、东南亚地区、日本和美国等地。

以上这些城市都是根据自身条件特点，重视发展创新产业经济，从而带动了建筑科学文化的发展。

2. 外国创新发展情况

再看看我曾访问过的一些国外城市发展产业经济的情况。如美国的西雅图市，1916年在这里建立了波音飞机公司总部和下属4个子公司，其8万职工数占了全市职工总数的一半。1949年辟为自由港，成为美国西北部的最大港口城市。随着城市环境的改善，又发展了船舶和医疗设备制造业等，现已成为美国重要的飞机和船舶制造中心。其产业经济的发展，又带动了该市的建筑科学文化建设。美国东部城市波士顿，它是新英格兰地区最大的港口城市，文化历史较长，20世纪发展了机械工业、电子工业和金融业等，还建立起国家航空与航天研究中心，这些产业经济的迅速发展，带动了这座文化老城的文化建设。

又如欧洲地区，我曾多次访问过的法国，它是发达的资本主义国家，国内生产总值居世界第五位，工业主要有居于世界第三核发电量的核能业，有居于世界第三位出口量的汽车业，还有机械、造船、化学、石油化工、纺织业以及发展迅速的航天航空、海洋开发业等。其首都巴黎，是法国制造业中心，全国1/5的工业生产在大巴黎地区，巴黎的交通运输业发达，铁路、公路通往欧洲各地，航空通往各大洲，极为方便。2001年5月，我乘高速铁路从法国北部的巴黎至法国最南端的地中海城市马赛进行访问，当日就往返。巴黎的旅游业十分发达，其历史文化名胜古迹甚多，业已成为全球著名的旅游城市。

瑞士，是我1979年出国访问的第一个国家，它同样是发达的资本主义国家，其工业除冶金、机械、化工外，钟表、精密仪器、纺织、食品业等也很突出，并有"钟表王国"之称。首都伯尔尼大力发展钟表产业，以表都闻名于世。巴塞尔市亦以发展钟表和医药业为主，瑞士钟表业经过电子表带来的冲击后，现仍发展技术更为先进的机械全自动金表，这一高档产品享誉并畅销全球。伯尔尼、苏黎世、日内瓦和洛桑等城市的旅游业同样十分发达。

1995年，我访问意大利。意大利亦是发达的资本主义国家，外贸是其主要支柱，出口以汽车、钢铁、化工、机械设备、纺织和轻工业为主。意大利旅游业非常发达，历史文化悠久，有完整保留众多名胜古迹的首都罗马和文艺复兴发源地佛罗伦萨市以及水城威尼斯市等。意大利经济产业中旅游业占很大比重，如佛罗伦萨市，它的工业、手工业主要围绕着艺术品发展，包括皮革、珠宝、纺织、陶器、银器等。

1996年，我访问比利时。比利时属于发达的资本主义国家，工业有钢铁、机械、化工和有色金属业等。首都布鲁塞尔是北约秘书处与欧盟国际机构所在地，又有"欧洲首都"之称，是欧洲著名的旅游城市。比利时的安特卫普市是欧洲第二大港口城市，也是比利时钻石加工和贸易中心，设有250家钻石加工厂和7所钻石学校，其钻石加工量占全世界总量的一半。

1985年，我访问埃及。埃及是世界文明古国之一，也是非洲工业较发达的国家之一。其传统产业以纺织、食品加工等轻工业为主，近年来石油、钢铁、机械、化工、电力业有较快的发展，特别是石油工业发展迅速，已成为非洲第四大产油国。埃及首都开罗市，全国1/3的工业设在这里，市里保存有世界七大奇迹之一的吉萨金字塔群，它促进了该市旅游业的发展。位于埃及北部临地中海的亚历山大城，则是埃及最大的港口城市，其进出口货物吞吐量占埃及的90%，同时，它也是世界著名的棉花市场和避暑胜地。

1992年，我访问日本。日本是世界第三大经济体，工业高度发达，主要包括钢铁、机械、汽车、电机、电子、石油化工、造船、建筑、核能业等，其中核能工业较突出，已建有50多座核能发电站。其造船业也居世界前列。另外，造纸、印刷、照相机、食品的和陶瓷等轻工业产品也在世界上占有较多的市场，虽然许多原料靠进口，但进出口贸易发达。日本的东京、奈良、京都等城市的旅游业十分发达，如京都市，它是日本的古都（公元794—1869年为日本首都）。京都市还是日本的宗教与文化中心，早在1950年就被定为国际文化观光城市，其市中心的皇城、宫城是仿中国唐长安城、洛阳城格局建造的。京都市内保存的历史文物古迹最多，被列为日本国家级文物的建筑、园林、绘画、雕刻共21项，在本书实例部分介绍了其中的平安神宫、金阁寺、银阁寺、龙安寺石庭。京都市的产业经济以文化旅游业为主。1974年京都市同中国西安市结为友好城市。

纵观这些中外城市产业经济创新发展的进程，20世纪80年代美国率先从钢铁、汽车、纺织、建筑为主的传统工业产业转向信息、集成电路、软件、新兴材料、生物工程等高新技术产业的开发利用。日本和现代科技基础好的英国、法国、德国等欧洲国家奋起直追，我国和亚洲、非洲、拉丁美洲等发展中国家也开始重视这一高新技术发展新趋势，努力创新发展。这一时期中外发达的城市都在加快发展知识经济，广泛采用互联网、大数据、人工智能等高新技术和加强基础设施软硬件的建设，并取得了良好效果，从而引领生产、生活方式的全面创新，实现知识经济社会的到来。

上述中外城市的产业经济，都在不断地创新发展。创新是生产力，创新是推动人类社会向前发展的动力之源。如果我们的发展不重视创新、不采用高新技术，那么我们的发展必然会落后于时代。创新发展是21世纪的永恒主题，以期达到经济生态平衡。

二、自然环境理念

加强环境治理，人与自然共生，适度空间容量，宜居防灾卫生。

1. 自然环境是人人喜爱、向往的地方，但世界各地许多城市破坏了自然环境，需要加强环境治理，改善生活环境

对于城市、建筑与园林的发展，要逐步做到减少直至消除对水体、空气的污染，还要对垃圾做到无害化处理。一个多世纪以来，世界上的城市都在做这项工作，因为工厂、城市随时都在排放有害气体、污水和垃圾，机动车也在排放有污染的尾气，这四个方面（污水、有害气体、垃圾和交通）的环境治理，即使采用新技术，也需要大量的资金和较长时间。凡先进的适宜生活居住的城市，都重视并解决了这个环境问题。

我国许多城市正在逐步加强环境治理工作，如陕西西安市不仅保留了明代城墙，还保留了环城护城河水系，形成了环城绿化带，极大地改善了这一地区的环境。在西安市的总体规划中，将恢复与发展历史上的"长安八水"，即东面的灞河、浐河，北面的渭河、泾河，西面的沣河、涝河，南面的潏河、滈河。同时，还将投资几十亿治理环境。西安市规划局的领导同志曾对我说，治理大环境要有甘肃省配合，同时治理渭河上游水。浙江金华市是我国重视城市环境治理、发展生态城市较早的地方。截至2006年9月30日，金华市市区建沼气净化池5045个，总池容12.69万m^3，约可净化60万人排出的生活污水，它可以作为前处理系统，对金华市区已建的8万m^3的污水处理厂起到减轻负荷与压力的作用。另外，截至2005年底，金华市区已建生态公厕170座、生态建筑475幢，建筑面积183.9万m^2，屋顶绿化面积14.64万m^2。[①]

近一段时期，北京开始重视发展自然绿地空间，从2000—2003年底共造林绿化15.8万hm^2。从2000年开始实施京津风沙源生态治理工程，到2010年国家投资已达500亿元。在城区内再造600个3000m^2规模以上的小型绿地，使居民出行500m内都能享受到自然绿地空间。北京在城区河湖水环境方面的治理取得了突出的成果，城区内重点河道有18条，总长180km。2007年已完成120km的治理，其排污口已顺利截流，由新建的5座污水处理厂处理。中心城的污水处理率由39.4%提高到70%，河湖水质由过去的Ⅳ、Ⅴ类提升到Ⅱ、Ⅲ类。根据规划，北

---

① 以上情况是原金华市领导余义耕先生提供的。

京还在中心城区建5座污水处理厂，卫星城区建15座污水处理厂，到2008年城市河湖水环境基本变清，臭水不再过市，改善了首都居民的生活环境。

南京市近年来重点整治了秦淮河的环境，于2010年全部完成，涉及市政、水利、环保、交通、园林、旅游、社区、商业等多项内容，呈现出生态和谐的环境，因而在2008年获联合国人居环境奖特别荣誉奖。浙江绍兴市亦重视全面整顿环境，对老城区18条河道进行综合整治，建设地下排污管线和污水处理厂，关闭和外迁有污染的工厂120多家，并将城外活水引入老城区河道，改善水质，恢复原有水乡风貌，同时降低老城街道人口密度，保护文化景点与文化遗存，协调建筑风格，亦获得2008年联合国人居环境奖特别荣誉奖。

我国"十四五"计划要继续抓污染物减排、环境治理、源头防控，还要大力推动生态保护和修复，扩大生态空间、容量，促进产业结构调整，为建设生态文明环境打好基础。

发达国家一些城市的环境治理要比我国早半个多世纪至一个世纪。1993年，我在美国西雅图市听到规划委员会介绍，西雅图在半个多世纪以前城市环境污染严重，交通拥堵，居民生活不安定。为了改变这种不良现象，他们进行了反思，首先治理了位于市中心区东北面的华盛顿湖，对全地区的污水进行监控排放，使湖水变清，并治理全市水系，严格处理下水道系统，同时改善交通，发展步行和大众运输的道路系统，发展免费的城市公共交通，恢复了西雅图原有的自然面貌。西雅图市环境的改善带动了经济的发展，而经济的发展又促进了环境的进一步改善，走上了科学发展的道路，成为人们公认的"宜居城市"。

1979年，我访问瑞士时，看到这个国家的城市与乡村环境是清洁卫生的。穿过首都伯尔尼的莱茵河支流阿勒河，只见河水清澈见底。沿日内瓦湖畔经洛桑到日内瓦市，湖水清澈碧蓝。苏黎世市所依的苏黎世湖亦经过了长期的治理，湖水逐步变清。这样的环境是瑞士各地加强环境治理、严格管理的结果。在垃圾处理方面，瑞士各城市早已采取分类处理的办法，既废物利用，又清洁环境。政府鼓励建筑使用太阳能，给予优惠政策，既节能又环保。瑞士是欧洲发展太阳能建筑最多的国家之一。

1996年，我来到比利时西北部的布鲁日城。布鲁日城素有"北方威尼斯"之称，真是名不虚传。旧城河道成网，河水晶莹碧清，整座城市整洁有序。我们了解到，城市管理部门对河道水质、垃圾处理、烟气排放都有一套符合环保要求的标准和管理办法，以保障居民生活和旅游的环境质量。

2. 创造自然环境，保证绿地面积，人与自然和谐共生，充分利用自然采光通风，体现节能、环保、舒适

保证绿地面积是建立适宜人们生活、工作和有利于防灾的生态环境基础。自然风景区、公园和街道、住区、公共建筑的园林绿地等，是城市的重要组成部分，是保证人与自然共生、创造良好生活环境的基本因素。因此，我们要重视城市外围大的生态环境、城市本身

的生态环境（绿地系统）以及建筑内外和道路的生态环境与绿地建设。

关于大、中、小生态环境与绿地建设，合肥市就是一个突出的优秀城市实例。1949年时合肥是一座小城市，旧城面积5.3km^2，人口5万，现为安徽省省会。1992年合肥市成为中国首批3个国家"园林城市"之一，已建起围绕旧城周长8.3km的一环状绿带，在此绿带的四角结合古迹发展了4个30多公顷的公园。在距市中心9km的西部大蜀山上开辟了550hm^{2}的森林公园。在离市中心17km的东南郊巢湖计划发展成为具有特色的风景旅游区，并拟由巢湖至市区建起东南方向的引风林带，顺城市的主导风向，把新鲜空气引入城市中心区。环绕西部、东南部这两大片绿地的边缘，拟逐步建成二环与三环绿化带，将全市绿地连接起来，形成合肥市自然生态绿地系统。2006年至今又增加了大量的绿地。这一自然绿地空间系统非常可贵，它对改善城市生态环境起着重要的作用。中国改革开放以后，我看到广东省中山市的规划建设亦反映了这一理念。中山市1996年获得了国家"园林城市"称号，它有一个更为完整的自然绿地空间系统规划，现正在逐步实现中。2003年，中国城市规划设计研究院（简称中规院）在昆明主城核心区概念规划国际竞标中获胜，其他参与竞标的还有日本的黑川纪章、美国的SASAKI、澳大利亚的PTW三家。中规院方案的空间模式是，将组团以河流为主干的自然生态绿地空间加以分隔与联系，形成城市组团与自然共生，依山傍水、城绿交融的春城特色。昆明市、合肥市、中山市的绿地覆盖率都达到了50%以上。

国内许多少数民族聚居的村寨，给我留下了村寨完全融于大自然的深刻印象。如广西三江马鞍寨，是一个侗族山水村落，背靠青山，寨前溪水环抱，知名的程阳桥和平岩风雨桥跨水横卧，沟通着寨外交通。广西龙胜金竹寨，是一个壮族干栏民居村寨，建在山腰上，山上翠竹成荫，山下溪水流淌，整体与大自然环境融为一体。又如云南大理地区的喜洲、周村等白族聚居的村落，位于苍山洱海之间，气候温暖湿润，林木繁茂，其民居建筑群完全融合在大自然的优美环境之中。在云南最南部的西双版纳，属热带气候，它盛产竹木，且水资源丰富。这里的傣族村寨，每家每户占一小方格用地，有多棵高大的椰树为其遮阴。村寨一般建在溪旁或江河边，村口的大青树是傣族人民崇拜的图腾，整个村寨就在竹林、椰林和碧水环绕之中。这些少数民族的村寨，其绿地覆盖率远远超过了城市，真正达到了人与自然的共生。

在建筑创作中重视利用自然采光、自然通风和太阳能、地热等，这是节约不可再生资源、保护环境、创造舒适环境的一项重要内容。随着建筑技术的发展，人们过于依赖空调等电气设施，不仅增加能源和资源的耗费，而且不利于身体健康。我国著名建筑大师莫伯治先生对这个问题早有认识，其设计建造的建筑多以利用自然光和自然通风为主，并将自然花木、山水同建筑空间有机结合在一起。莫伯治大师是我国建筑走向自然的引导者，其早期的作品有广州北园酒家、泮溪酒家、白云山庄等。20世纪六七十年代建成的广州矿泉客舍则是一个典型实例，其主楼前为一水庭，楼后为小溪流过的小花园，主楼底层为支柱层，自然通风极为良好，所有客房也都是自然采光、自然通风。莫伯治先生关于建筑室内外空间设计要同树木花卉融为一体的理念，至今并未引起中国建筑师的重视。

20世纪80年代中期建成的香港汇丰银行，其底层通透，将建筑主要支撑结构与垂直交通系统放在了两侧，南北通透，突出了自然采光与通风，其设计师是英国著名建筑师诺曼·福斯特。诺曼·福斯特十分重视利用太阳光，还在中庭顶部装有镜片，把阳光折射到广场。21世纪初，北京清华大学建成的清华大学建筑设计研究院办公楼，一改那时的常规做法，于朝南面设遮阳板，朝西面做成实墙并留有空隙，顶部南北两边各设一排高天窗，组成了一套自然采光、自然通风与防晒的系统，取得了明显的节能、环保和舒适的效果。这种利用自然、重视生态、节能环保的设计理念应是21世纪建筑创作的正确方向。

生态环境建设突出的国外优秀城市实例，我所看到过的要属巴基斯坦的首都伊斯兰堡。伊斯兰堡于1961年始建，1970年基本建成，人口20多万。伊斯兰堡北依玛格拉山，东临拉瓦尔湖，南有夏克帕利山，山上、湖旁林木丛生，城市就在这自然绿色中呈东西向长条状、棋盘式布局发展。东部为行政办公区和公共事业区，西部为住宅社区，西南部为轻工业区和大专院校区。这些区内的建筑都没有高层，居住社区住宅为1～2层，建筑容积率在1以下，各区内的建筑都融于绿化之中。路网绿带将城市的绿地、花园、公园同周边山上、湖畔绿地连接为一个整体，建成区范围内的绿地覆盖率高达70%，完全是一座花园城市。

西班牙巴塞罗纳是一座滨海城市，背山面海，背面扁长的山体上连续的丛林构成天然绿色屏障，平行于山体的海滨有浓郁的林木连绵不断。沿海西部矗立着蒙特胡依克（Montjuïc）山丘，它的东面紧邻着旧城，经过几百年的建设，这里有著名的城堡、博物馆、展览馆、植物园以及1992年举办奥林匹克运动会的主体育场等，这些新老建筑隐没在苍翠绿海的环抱之中。1995年12月，我专访了这一片景区，同时还对北山脚下东北部的园林进行了考察，这里有高迪设计的极为自然的桂芦（Güell）公园、18世纪末建造的赖伯润特（Laberint）历史名园等。1996年7月，国际建筑师协会大会期间，我还参加了在巴塞罗纳北山脚下西北部的花园举办的学术活动。国际建协大会之后，我们还游览了滨海东南部的公园和小游园10多处，并徒步观赏新城的林荫大道。这些东西向、南北向以及对角线放射形的林荫绿带将上述4个方位的60多个绿色斑块连接成为一个整体，形成极具地域特色的绿地系统，它镶嵌在大自然的青山碧海之间，为当地居民创造了一个极好的生态环境。巴塞罗纳的绿地覆盖率已达到50%。

1989年，我访问了美国首都华盛顿。华盛顿市中心区位于波托马克河和阿纳科斯蒂亚河两河交汇处，两条河畔的大片绿地环绕着它。由国会大厦至林肯纪念堂3.5km长的中轴线及两侧绿地与分散不高的建筑组成华盛顿市中心区，这里突出的是大自然的绿地环境，是一个园林式的中心区，它体现出首都的政治文化氛围。华盛顿市中心区外围布置有多处大片的森林绿地，居住区也多为低密度低层住宅。全市无高层建筑，规整的路网绿带将这些森林绿地、公园、各区内绿地、花园连接在一起，使得全市的建筑融于绿地之中，其绿地覆盖率达到了60%。

此外，重视发展城市中心绿地的还有法国巴黎吐洛里花园、爱尔兰都柏林圣·斯蒂芬公园、美国波士顿中心绿地公园区和芝加哥格兰特公园、加拿大多伦多中心区近湖滨休闲岛区、巴基斯坦拉合尔真珠大陆旅馆等，它们都有相当大的绿地面积，创造了人与自然和谐共生的美好环境。

另外，举一个创造自然环境的高层建筑实例。2014年10月，意大利米兰市中心的伊索拉区建成了世界上第一对"森林塔楼"，即在塔楼各层平台上种植树木与花卉，以改善塔楼的生活环境。我国大科学家钱学森先生在20世纪90年代初倡导山水城市建设。他提出要在高层建筑中安排植被，创造人与自然共生的"空中花园高层建筑"，这种理念在米兰伊索·拉区的"森林塔楼"中得以实现。还有一个建筑内外与树木花卉融为一体、创造自然环境的实例，即苹果阿文图拉专卖店，它位于美国佛罗里达州的迈阿密，设计者是诺曼·福斯特及其合伙人。苹果阿文图拉专卖店的室内树木和桌子一直延伸到了双层通高的大厅内，并与室外供游人休息的花园遥相呼应。这种创造具有自然环境的"花园建筑"的做法值得中国建筑师重视。

在建筑创作中重视自然采光、自然通风等，一些世界著名建筑师对此已早有认识，并在其设计的建筑中也大都以利用自然采光通风为主。如世界著名建筑大师赖特设计的住宅，包括为自己设计的芝加哥住宅和工作室，都是"自然的建筑"，特别是自然采光、自然通风，且与室外自然环境组合为一个整体，就连其平面布局都体现着自然与灵活。对于窗的细节，赖特反对推拉窗，而采用平推窗，以利于自然通风。又如著名美籍日裔建筑师山崎实，他崇尚自然的光和水，他设计的西雅图北部展览馆建筑和其他一些低层建筑，都体现出光亮、洁白和反影，自然采光、自然通风、并伴有水池喷泉，其倒影清透，设计新颖醒目，建筑节能环保。

1979年，我访问瑞士北部的巴塞尔、中部的伯尔尼、南部的博瑞格，其住宅、体育馆、室内游泳池等都是利用自然采光和自然通风的典范，说明这个国家很早就重视节能环保和人体的舒适。1990年，我访问加拿大的蒙特利尔市，这座城市的建筑装饰简洁而实用，如蒙特利尔Desjardins建筑，它位于市中心东部片区，此建筑的地下层将东部南北两个地铁站之间的建筑群连接起来，形成地下空间网，特别是这个建筑的地下层和地上层都布置有花木、水池喷泉，创造出自然舒适的环境。更巧妙的是，在地下层内引进了自然光线，而地上层大厅更是开敞明亮，它充分利用自然光和自然通风，并取得了节能、环保、舒适的效果。

3. 适度控制城市空间容量，根据城市土地、水等资源的负荷量适度发展城市的规模，这是城市宜居、防灾、卫生的又一关键问题

目前，我国许多城市的建筑容积率过高、城市人口过密，已成为城市生态环境质量不够理想的一个重要原因。这不仅给城市带来混乱和矛盾，也降低了城市与建筑本身的使用与经济价值，而一旦遇到像2003年的"非典"传染病和2020年的新型冠状病毒肺炎传播时，城市人口密度过大的地区将十分被动，这些地区的环境卫生亦受到负面影响。

节约用地无疑是正确的，但要从城市可持续发展和防灾、卫生基本要求来考虑，应该找出城市空间容量的合理限度。在城市空间容量合理的限度范围内，又因其度的不同而分出不同的等级。

如合肥市于20世纪90年代建成的琥珀山庄住宅小区，用地11.4hm², 总建筑面积11.7万m²，其建筑容积率为1左右；珠海拱北花园新区，用地6.92hm²，总建筑面积11万m²；昆明春苑小区，用地15.34hm²，总建筑面积18.66m²，绿地占42.4%；上海三林苑小区，建筑面积毛密度为1.32万m²/hm²，居住建筑密度为1.2万m²/hm²，绿地率37%；厦门康乐新村等居住区，其建筑密度为1.4万m²/hm²左右；这些新建居住小区的建筑容积率大都在1.5以下，其空间容量比较适度，有利于防灾和卫生。

关于建筑容积率，从中国城镇的平均水平来看，我们认为一般住宅区应在1.5左右，高层次的住宅应为1左右或更低。中心商业区建筑容积率为2～4，只有个别地段或大城市建筑容积率在4以上。当前，中国的建筑容积率普遍偏高，其环境质量上不了档次。厦门、珠海、常州、无锡、合肥、扬州等城市的多数建筑容积率适宜，如厦门于20世纪90年代建起的筼筜中心区等，创造了较好的生活环境。

关于人口密度，就一个城市总体来计算，生活居住用地今后要发展到200人/hm²，每人占地50m²左右；再发展应为100人/hm²，达到每人占地100m²。这里所提的是生活居住用地，不包括工业、仓库、铁路、机场、军事用地等。这些分析重点说明我们要有"宜居、防灾、卫生、适度空间容量"的理念，不能盲目扩大城镇规模和提高建筑容积率。

关于适度控制城市空间容量问题，这里举几个我访问过的国外实例来说明。一些城市因度的不同而分出不同的等级，如美国西雅图市，分为4个等级：中心区（Urban Center Villages）居住密度为每英亩15～50居住单元；核心区（Hub Urban Villages）为每英亩15～20居住单元；住宅区（Residential Urban Villages）为每英亩10～15居住单元；邻里区（Neighborhood Villages）为每英亩8～10居住单元。为了保证城市的环境质量，西雅图市采用都市村落的理念。当然，西雅图市的标准较高，我们应根据我国地少人多的实际情况制定出适度的城市空间容量，但西雅图的理念与做法有参考价值。

国外一些历史文化名城中心区，由于人口密度比较高，为了改善环境采取了减法，以降低建筑容积率。法国巴黎在其旧城核心部分适度调整了空间容量，将住宅建筑容积率由4～5降低到2.7以下；把经营多种活动的10、11区的建筑容积率降为2；在旧商业区内，不准提高建筑容积率，有些建筑容积率还要降低到2以下，以保留19世纪形成的8、14、17区的城市结构。

又如美国西海岸的洛杉矶市，其城市布局是分散式的，仅中心城市区集中修建了一些高层建筑；其余地方建筑容积率不高，住宅多为1～2层木构房，空间容量低、生活环境舒适，有利于防灾与卫生。这一实例的建筑容积率标准高，但其分散布局的模式有参考价值。再看看意大利的佛罗伦萨市，它是历史文化古城，也是欧洲文艺复兴的发源地。为控制城市的空间容量，仅容许佛罗伦萨大教堂高107m、大教堂旁的钟楼高84m，以及西尼奥列广场旧宫塔楼高95m，其余的皆为低层建筑，也没有任何新建的高层建筑，其城市规模与空间容量适度，城市环境舒适。

本章所阐述的：加强环境治理，创造自然环境，保证绿地面积，利用自然采光通风，适度控制城市空间容量等，其目的是为了达到自然生态平衡。

三、历史文化理念

整体保护名城，合理使用发展，建筑继旧创新，发展建筑文化。

1. 城市历史文化是城市发展的根，一定要有整体保护、合理发展的理念

  我国保护历史文化古城做得比较好的城市，如所述的安徽省合肥市，其旧城5.3km²基本保留完好，特别是围城的水系充分保存，形成富有特色的环状绿化带，并以旧城为中心发展成为园林化的大城市。又如陕西省西安市，其明代的古城9km²，古城南北、东西中轴线以及建筑总体上基本完整保留，这在中国历史文化名城中是极其少有的一个优秀实例。此外，还有少数较小的历史文化古城被整体保护下来的，如山西省平遥古城等。中国历史文化名城总的情况是，20世纪上半叶因战乱毁掉了许多，下半叶因建设又拆掉不少。20世纪90年代以后，逐步树立起保护历史文化名城的理念，很少再拆毁历史文化名城的街区与建筑，采取了保留老居民与老字号的做法，取得了较好的效果。下面举一些我所见到的实例。

  中国历史文化名城发展到中等规模以后，除其古城已缺少历史文化之外，或为确保古城作为历史文化的中心，大多数的历史名城都应考虑多中心的布局，即将行政、经济、商贸、科教等中心迁出，转移到新城区。西安市的总体规划就作了分散、多中心的安排，现将西安市行政中心转移到古城外的北部地区，带动了新区发展。多中心的转移还可提高古城的自然生态环境质量，缓解古城的矛盾和压力，确保西安历史文化街区面貌的恢复。云南丽江市的总体规划建设也是个优秀实例。为了完整地保护丽江古城历史文化风貌，丽江市政府行政管理机构迁至南面的八河新区，经济、商贸等中心安排在西面的几个新区，确保了丽江古城得以整体保护，而不受新建设的破坏。

  我国一些历史文化名城的中心区，修建了高大体量的新建筑，破坏了整体尺度的和谐，应积极实施"减法"整治。

  北京带头采取这一做法，值得宣传。北京已将中轴线上的地安门百货商场减去了两层，把南池子东侧的市房管局办公楼顶拆掉了3层，以保证这两个地带建筑尺度的和谐。对于景山公园与北海公园之间不和谐的较大建筑，也将拆除整治。西安市钟楼广场的西南角、东北角矗立着两个大体量的建筑，也应实施"减法"，以突出钟楼、鼓楼的形象，保持西安古城核心区的整体面貌。南京古城已制定出保护和更新的规划，突出保护和发展古城文化，总的原则是好的，但南京新街口中心区却修建了许多高层建筑，并拟再建更高的超高层建筑，这个做法欠妥，其核心区应实施"减法"，以使旧城建筑群整体尺度和谐，并减轻旧城中心区人口、交通、基础设施等的压力。

  采取"微循环法"改善历史文化名城的街区也是个好方法。福建泉州市中心区中山路的

改善，没有大拆大改，没有插建高大体量的新建筑，没有拓宽成大马路，而是重点改善了基础设施，改善建筑及其环境，基本维持原有建筑和街道的尺度，这就是"微循环法"，它可确保旧城原有的历史面貌、生活方式与文化习俗，又可改善、提高旧城的生活环境，符合我们建设小康社会的要求，值得各地政府效仿。北京2005年提出了故宫缓冲区保护规划，对缓冲区内的旧街区采用"微循环"和"有机更新"的方法，严格保护区内的胡同、四合院，原则上不成片拆除，主要街巷原则上也不再继续加宽，对区内基础设施积极改善，而不是全部废除或重建新的。采取同样做法的，还有上海旧城里弄住宅保护、云南昆明旧城"一颗印"住宅保护、新疆喀什旧城民居保护。

  对于被保护的旧城街区或建筑项目，要充分利用并有所发展，使其更具活力。有的增加绿地及生活内容使其成为文化休闲区，如西安大雁塔，在塔的北面开辟出一个现代绿地广场。大雁塔作为广场中轴线的终点，旨在突出慈恩寺及唐文化，并在绿地广场的中轴线上布满水池及喷泉，为西安人民和中外游人提供了一个极好的文化休闲场所。又如西安临潼骊山华清池，这里有古迹华清宫，是唐玄宗同杨贵妃沐浴之处。此处风景优美，1959年中华人民共和国成立10周年前夕，在原"女汤"西北面修建起5300m^2的"九龙池"绿化园林区，并建有飞霜殿、晨旭亭、晚霞亭以及传统庭院式的宾馆。1990年又根据新发掘遗址，建起"唐代御汤遗址博物馆"等，整个规模比原来扩大了11倍，使其成为文化风景游览区。还有的增加文化研究内容，成为文化研究区，如敦煌莫高窟。在南北长1600m的敦煌莫高窟窟区对面，新建了一座大型博物馆，并扩大了敦煌文物研究所，使这里发展为研究敦煌文化的基地。在湖北荆州古城，于城西开元观旁扩大并发展了荆州博物馆和三国绿地公园，于城东修复宾阳城楼，并建起历史碑苑，使其成为荆州古城文化研究区。也有的增加社会活动内容，成为文化与社会公益活动区，如福建厦门南普陀寺。在南普陀寺寺院范围内，发展佛教教育和社会公益事业，并办起佛教教育学院。佛教教育学院分为男女二部，男部设在寺内。同时，南普陀寺还建立慈善事业基金会和义诊医疗机构，为社会群众服务，受到市民的欢迎。又如台湾地区鹿港龙山寺，它是福建泉州"安海龙山寺"分灵割香而来的，现是台湾省内最大最佳的龙山寺。龙山寺除祭诸神灵外，这里还是一个社会救济机构，深受广大民众的欢迎。

  充分利用并发展历史文化建筑的实例，还有西安秦始皇兵马俑、南京玄武湖公园、上海豫园、杭州西湖风景名胜区（包括三潭印月、西泠印社、郭庄）、绍兴青藤书屋、厦门鼓浪屿、长沙岳麓书院、成都杜甫草堂、新疆喀什艾提尔清真寺、澳门市政厅广场等，它们都增加了绿地或博物馆，显得更富有生气和活力。

  欧洲古老的城市较多，其整体保护的实例也多。我到过的城市，整体保护最好的要数意大利首都罗马城。罗马城已有2000多年的历史，从公元前建起的古罗马中心广场遗迹一直到20世纪初为意大利开国国王建成的埃马努埃尔二世纪念碑，都完整地保留着。罗马这座历史文化名城已变成一座巨型的历史博物馆城，没有什么损坏。首都新城在其南7km外另建起来。这种分开新旧城的做法，使历史文化名城得到了整体的保护。

  另一个保护历史文化老城的好实例是法国巴黎。巴黎是法国的首都，其做法是采取"成片保护、分级处理"的整体保护原则,这个原则是在1977年巴黎议会通过的巴黎旧市区改建规划中提出的。巴黎旧市区,是指1845年修筑的城墙以内的地区,包括东、西两个森林公园,面积为105km²。这个整体保护原则,是将旧市区分为成片的两圈,椭圆状的内圈是保护传统的核心部分,它属于一级保护区。巴黎旧市区整治有三条具体原则:一是保护老的住宅区,并加强居住功能,提高其舒适程度;二是保存各种各样的功能,改进公共活动空间,包括文化、娱乐、商业等设施的改进;三是保持19世纪建筑面貌的统一。为什么把内圈作为保持传统的核心部分呢?这是因为巴黎城的产生与发展都在这个区,著名的城市中轴线贯穿这个区,城市的传统文化、商业区、街道以及闻名的古建筑和传统居住街坊等也都在这个区。外圈属于二级保护区,在这个外圈中又分为成片的3类。各类处理的具体原则是:一类,为确定的有保存价值的街坊,保持原建成区的特点;二类,维持原有居住的功能,对已有建筑重新控制,并适当改建建成区;三类,发展居住功能,对已有建筑重新控制,可以改变原有的特点。对于建筑密度、建筑高度都有严格的控制数据,建筑容积率控制数在前一章已有部分说明,其内圈核心部分"屋顶"的高度控制在25m,外圈最高建筑控制在37m以下,外圈内需要保留的街坊、广场和某些风景地区同样控制在25m。巴黎旧城规划这一"成片保护、分级处理"的原则,确实保持了巴黎旧城历史文化的特点。

在西班牙的巴塞罗那,最早的老城在滨海的西面,19世纪城市扩展时就被完整地保存着。新修的开阔城市干道沿老城东侧和西南侧连通,老城的南北向中心道路Rambles路与新城区南北向干道通过加泰罗尼亚广场衔接。在19世纪、20世纪发展的新城区里,其近代建筑,如高迪设计的教堂和米拉住宅等都被完整地保留着,突出在重要的街道上。类似巴塞罗纳整体保护历史文化旧城的实例,还有西班牙首都马德里、瑞士首都伯尔尼与日内瓦、比利时的安特卫普、意大利的佛罗伦萨和水城威尼斯等,在欧洲各国普遍存在。

对这些整体保护的旧城街区或建筑项目,要充分利用并有所发展。有的增加生活内容,如意大利威尼斯圣马可广场,在市政大厦底层和广场北边的小街内,开设各类商店、餐馆等,在广场东面设一些室外咖啡座,方便了生活,使游客感到它是一个"漂亮的生活客厅"。日本东京都的浅草寺,此寺一年内要进行许多节日活动,在这个被保护的历史建筑周围,逐步建起商场、游戏机室等设施,并扩大了花市,形成了保留老东京历史风俗的繁华商业区,现已成为外国游客访问东京必去的观光处。其他如西班牙马德里的市长广场、美国西雅图城市中心滨海区、德国法兰克福罗马人广场等,都是在历史建筑保护群内及其周围增加生活设施,使这些地方增添生活情趣和活力。还有的增加园林绿化内容,如日本东京都的上野公园,在原有的博物馆建筑旁建起绿地和动物园,将这里组合成一个占地52.5hm²的大公园,使之成为人们参观博物馆或观赏动物并能休闲的绿化场所。美国洛杉矶亨廷顿文化园,这里原是亨廷顿先生的住处,有一个藏有珍贵书籍的图书馆和一个艺术品收藏馆,后又发展成占地130英亩的植物园,共建了12个花园,现已成为洛杉矶市的文化和教育中心,可称其为"亨廷顿文化园"。人们到此园内可阅览珍贵的原稿,欣赏历史艺术珍品,还可在不同类型的花园之中游览或休息。其他如西班牙马德里埃斯库里阿尔宫、马德里西班牙广场、巴塞罗纳蒙特胡依克山、巴塞罗纳国际展览会德国馆,比利时安特卫普的鲁宾斯博物馆,巴基斯坦拉合尔的城堡、巴德夏希清真寺和夏利玛园，日本奈良的东大寺与鹿苑以及京都的平安神宫、金阁寺、银阁寺和龙安寺等，都是在原有历史建筑中增加绿地花园，使其更富有吸引力和生气。

2. 要尊重历史建筑，发展新的现代建筑要处理好与历史建筑的关系，要继旧创新，发展建筑文化

西安市在新旧建筑和合共生方面，做了大量的实践探讨，其中中国建筑西北设计研究院总建筑师张锦秋院士的规划设计成果最为出色。比如，1998年建成的西安市钟鼓楼广场，其规划设计力求突出两座14世纪的古建筑形象，沿着"晨钟暮鼓"这一主题向古今双向延伸，在钟楼、鼓楼之间安排绿化广场、下沉式广场、下沉式商业街、地下商城、商业楼，既将钟鼓楼呼应在一起，很好地保护了古迹，又解决了旧城发展的生活需求，使这一中心地段兼具观光、休息、购物、餐饮等多项功能，成为名副其实的市民广场，为古城西安提供了一个颇具地域文化特色的"城市客厅"。1962年建成的中国美术馆，由于它位于北京故宫博物院、景山、北海公园的东面，同这些皇家宫苑相对，在这重要的历史建筑范围内如何和合共生发展新建筑呢？著名建筑大师戴念慈院士，他采用了新结构、新材料，只在中心部位做成重檐黄色琉璃瓦大屋顶和入口歇山式屋顶，而在东西两侧的展开部分，即底层与最上一层做成空廊式，下大上小，其比例尺度近似于西边的传统建筑，整体造型舒展、虚实对比有序、色彩明快亮丽，又有松竹花木配合，这样的处理同故宫皇家建筑协调一致，使新旧建筑和合共生在一起。1998年建在澳门大三巴牌坊东侧山冈的澳门博物馆，它依大炮台而建（大炮台建于公元1616年，是为击退来犯的荷兰人），设计师充分利用地形，将部分建筑置于炮台山内，在博物馆入口处有一段400年前砌筑的挡土墙，此墙被保留在厅的侧面。澳门博物馆为3层，第三层展厅的出口处就是大炮台顶，在平台顶上可观赏澳门全景。澳门博物馆、大炮台以及大三巴牌坊和合共生在一起，成为一个和谐的整体。

在新建筑创作时，要考虑继承传统优秀建筑的精神，以此来发展地域建筑文化。1968—1972年，由中国台湾著名建筑师王大闳先生设计建成的台北孙中山纪念馆，采用变异的中国传统黄色琉璃瓦大屋顶形式，这是一个继承创新并探求"中国现代建筑风格"的重要实例。台北孙中山纪念馆已选入《20世纪世界建筑精品集锦》（东亚卷）中。同台北孙中山纪念馆创作思想相类似的还有台北"故宫博物院"建筑。20世纪80年代，新疆建筑设计研究院率先重视地域建筑文化的继承与发展，如乌鲁木齐新疆人民会堂。该会堂吸取传统的方圆组合形式，以主楼为方，副楼为圆，主楼四角各建一圆形塔，此角塔为管道间，塔顶以不锈钢做成小穹顶，外墙柱间饰以连续拱形构件，檐部采用深金黄色琉璃瓦，这些内容都体现出新疆伊斯兰传统建筑的风格，但同时又是新的结构、新的面貌，使我们感受到浓郁的新疆地域建筑文化特色。又如1985年建成的乌鲁木齐新疆迎宾馆，将功能、结构、地域传统风格结合在一

起，多处使用传统尖拱，并把水塔构筑物转化为表现维族文化美的形象之物。20世纪80年代后期，在杭州西湖风景区宝石山的北面建成了黄龙饭店，该建筑设计师是程泰宁大师。程泰宁先生首先采取内向庭园式的布局方式，此举在大格局上有了中国园林的传统特点，其中心庭园花木、桥石与池亭的组合完全是江南式样，具有地域园林特点；入口处的大堂面对的是类似中国横幅长卷山水画的真实园林空间，体现着中国传统园林主要厅堂正对主景的布局特点；在黄龙饭店庭园可观赏到远处宝石山上的保俶塔，这又体现出中国传统园林借景、对景的布置手法；黄龙饭店南面是大堂、北面是餐厅、东部是多功能厅、西部是健身房，当你游走在这南北东西不同的空间环境时，可感受到空间序列的变化；各单体建筑的墙面为浅色，再配上深绿色的瓦顶，就类似于江南民居的粉墙黛瓦色调了；但这组建筑群采用的是新结构、新材料、新设施，其各个建筑空间根据新的需求设计，显得宽敞、新颖、简洁，所以这是一个发展了地域建筑文化的创新实例。由佘畯南、莫伯治院士设计建造的广州白天鹅宾馆也是一个继旧创新、富有地域文化特点的新建筑。该建筑首先追求自然意境，其建筑与自然景观结合，突出了作为"羊城八景"之一的"鹅潭夜月"意境。另外，设计建造出高百米的白色腰鼓形主楼，其形态有如白天鹅在戏水，而且还创造了一个具有岭南园林风格的中庭。中庭四周为开敞的廊道，与休息厅、餐厅等相连，环廊遍植垂萝，庭园由山石池水、亭桥花木组成，具有诗情画意。中庭上部与天相通，顶为藻井式的玻璃顶棚，阳光洒进此庭园极富生气，特别是"故乡水"的点题，引发了海外游子思念祖国故乡之情。环绕此中庭廊道还可到三层的中餐厅，在其入口处就可以看到一个能望到天空环廊水池的岭南庭园，此庭园旁的大小餐厅皆为中式装修和中式桌椅，古朴典雅；主楼客房的装饰挂屏和部分家具等亦为中式，而且是中西结合；此建筑采用新技术，空间组合也有创新，建筑内外与园林融为一体，发展了岭南地域的建筑文化。

中国台湾台北剑潭青年活动中心、台中东海大学校园及其路思义教堂、上海体育中心（包括1975年建成的体育馆与1997年建成的体育场）、1996年建成的厦门高崎国际机场T3航站楼及2014年12月投入使用的T4航站楼、2010年建成的上海世博会中国馆以及2019年建成的北京大兴国际机场航站楼等都是采用新技术、低能耗的优秀建筑。

国外在重要的历史建筑范围内是如何和合共生地发展新建筑呢？爱尔兰都柏林"三一学院"给我留下了深刻的印象。该学院位于都柏林市中心区，创建于17世纪，是爱尔兰最著名的高等学府。学院由两个庭院组成，中间立一高耸钟楼，将议会广场与图书馆广场分开。其新建筑的发展是在图书馆广场的侧面开辟一个学友广场，学友广场的一面为老图书馆，紧挨老馆建一新图书馆，并在其对面建一新艺术馆，这些新馆的建筑高度、体量、色彩、材料等都尊重老馆，同老馆协调一致，只是采用新技术将墙面门窗分割装饰得比较简洁。学友广场中心铺以草坪并立一新雕塑，形成了一个完整而又和谐的大院落，学生们在此席地而坐，有的读书，有的交谈，生机勃勃。美国的耶鲁大学，创立于公元1701年，它是美国最古老的高等学府之一，校园建有希腊、罗马古典复兴式的庭院建筑群。20世纪下半叶耶鲁大学增建了一些新建筑，如1977年建成的由世界著名建筑师路易斯·康设计的耶鲁英国艺术中心，完全是新材料、新结构，但设计师路易斯·康将建筑外墙线条的分割同周围环境协调一致，使新旧建筑融为一体。还有扩建的图书馆新馆，采用白色大理石墙面，控制建筑高度，使其同周围砖石建筑相协调。整个校园的最高建筑是建于1917年的哥特式哈克尼斯钟楼，楼高67m，它统领着耶鲁大学的建筑群空间，新旧建筑和谐一体。还有法国里尔美术馆，原老馆建于1895年，是座古典复兴式建筑，到了20世纪70年代才扩建此馆。首先恢复原有老馆，并在其对面建一采用新技术的新馆。新馆的建筑上附一薄片，薄片外表面是透明玻璃，中间的小镜片设计成方格网，镜面反映出老美术馆的风貌，新与旧和合生长在一起。其他如法国著名建筑师屈米设计的杜关市弗雷斯诺国立当代艺术学校，世界著名华裔建筑师贝聿铭设计的华盛顿国家艺术馆东馆，美国著名建筑师菲利浦·约翰逊设计的波士顿公共图书馆新馆等，都是尊重原有历史建筑，同其周围环境协调，并组合成新旧和合生长的建筑。

关于继承传统精神、发展地域建筑文化的国外优秀建筑实例，如巴基斯坦首都伊斯兰堡的费萨尔清真寺。费萨尔清真寺的建造时间是1970—1986年，设计师是土耳其建筑师V.达洛开依，他是通过国际设计竞赛中标的。这组清真寺建筑包括礼拜堂、沐浴区院落和一所国际伊斯兰大学，它突破了传统常规的拱券、穹顶、伊斯兰图案的模式，应用新材料、新结构以体现传统清真寺的精神内涵。其礼拜堂屋顶以钢筋混凝土空间构架形成巨大的帐篷形式，大堂四角还立有4座高耸的拜克楼，礼拜堂前布置一个大院落，"麦加朝圣墙"和"圣龛"位于堂和院的主轴线上，此建筑创新发展了清真寺的建筑文化。又如美国洛杉矶南郊的加登格罗夫社区教堂，建于1980年，设计人是美国著名建筑师菲利浦·约翰逊。这个设计创新体现在教堂的平面布局和空间环境，而且继承了教堂要与天呼应的精神和老教堂自然通风、采光的做法。为了使每个座位接近圣坛，将古典希腊十字形教堂平面改为四角星状，圣坛在长边的中心，其前为长方形座池，东西厅和南面设挑台座池，并采用白色钢网架新结构和反射玻璃新材料，仅使8%的光线透入，形成晶莹透明、无顶无墙并与天对应的宁静空间。建筑内部无空调设施，完全利用自然通风和自然采光。此外，如法国巴黎国家图书馆、美国波士顿市政府办公楼、美国华盛顿纪念碑、加拿大蒙特利尔建筑中心等，都是采用新技术并与原有传统建筑相结合，从而创造出适合所在地的新建筑，并发展了这个地域的建筑文化。

新建筑的创新性是十分突出的，早期的如建于1904年的美国芝加哥卡森·皮里·斯科特百货公司，它开创了现代框架结构的特点。又如1955年建成的具有开创性的美国洛杉矶迪士尼乐园，它是世界闻名的卡通式大游乐园，后被许多国家纷纷仿建。还有1958年为在比利时布鲁塞尔举办的万国博览会而设计建造的原子球建筑，它表达了人类和平利用原子能的美好前景。此外，1977年建成的法国巴黎蓬皮杜文化中心，其建筑外部像是一座工厂，被漆成不同颜色以代表不同功能，其内部为大柱网，可以灵活变动使用功能。还有1985年建成使用的美国芝加哥伊利诺伊州政府大楼，它在高科技基础上运用古典穹顶模样，围绕公众使用的高大中庭布置开放式办公室，以表达现代州政府大厦的公众性含义。

中外保护完整的代表性历史建筑与风景园林，中国的诸如北京故宫、天坛、西苑与颐和园、长城（八达岭），承德避暑山庄，南京城墙，苏州拙政园、留园与虎丘，无锡寄畅园和蠡园，扬州瘦西湖，绍兴兰亭、鲁迅故居与三味书屋，安徽黄山风景名胜区，三亚天涯海角，桂林漓江山水、芦笛岩，四川都江堰伏龙观、乐山大佛，云南昆明西山龙门、昆明筇竹寺五百罗汉塑像、路南石林，新疆高昌故城、交河故城等；国外的诸如埃及开罗吉萨金字塔群、卢克索卡纳克阿蒙太阳神庙，伊拉克巴比伦城，希腊雅典卫城，意大利罗马中心广场、大斗兽场、图拉真纪念柱、君士坦丁凯旋门、西班牙广场、埃马努尔二世纪念碑、特雷维喷泉、圣彼得大教堂和佛罗伦萨中心西尼奥列广场、庞贝古城公共建筑与住宅区、巴格内亚兰特别墅园，法国巴黎沃克斯·勒·维康特园、凡尔赛宫苑、卢浮宫、协和广场和明星广场凯旋门，西班牙马德里交通部门大楼、格拉纳达城阿尔罕布拉宫苑，英国伦敦国会大厦，比利时布鲁塞尔中心广场，俄罗斯莫斯科红场，瑞士阿尔卑斯山雪景和加拿大尼亚加拉大瀑布风景区等。

上面所述中外城市、建筑、风景园林皆具有强烈艺术美的特点。

本章重点强调，要整体保护名城，合理使用与发展，特别是新建筑创作要尊重历史建筑，发展地域建筑文化，以求继旧创新，实现文化生态平衡。

四、艺术美观理念

创造城市标志，统领空间环境，掌握艺术规律，协调城市建筑。

前面所讲历史文化的实例，都包含着艺术美观的内容。在这一章里，作者将阐述、强调建筑科学文化如何从三个方面去创造艺术美观。

1. 创造城市标志，统领空间环境

每个城市都有自己的特色，都想要创造出符合自己特色的标志性建筑，以便在其城市中心和周围地段形成优美的建筑立体轮廓线，这些标志性建筑将起着统领城市空间环境并使人感受到美的作用。

1995年建成的上海东方明珠电视塔，位于浦东黄浦江的弯道处，三面为江水所环绕，隔江面对着西岸外滩，正对着繁华的南京路，同时又是北京路、福州路、延安路、四平路等重要街道的视觉焦点。上海东方明珠电视塔塔高468m，塔身构架是3个圆筒体，以大小不同的11个球体作为连接体，构成积极向上、圆球圆柱的空间造型，因而这座高大雄伟壮丽的电视塔成为了上海市新的标志性建筑。坐落在香港中环花园道的中国银行大厦，高达368m，其节节高的突出造型形象控制着中环地区的空间环境，成为了香港中环的标志性建筑；位于香港湾仔港湾道的中环广场大厦，高375m，是一个三角形构造形体，它控制着香港半岛东部的空间环境，也是一个标志性建筑；香港本岛东、西两边的这两座标志性建筑都不够高，中间修建一座600m高的大厦，就像美国芝加哥中心区于Grand Park后拟建一座高667m的螺钉形住宅大厦一样，如此超高的建筑形体统领着整个地区的空间环境，就能突出城市中心区的立体轮廓线。

保存完整的西安古城，其中心钟楼是西安市的标志性建筑，该城的组织是古代府城十字道路交叉、由城墙围成田字形的格局，南北向主轴线和东西向轴线的道路通向南关、北关、东关、西关四个关城门，而钟楼就坐落在两条轴线交叉的中心处。钟楼建于明洪武十七年（公元1384年），下部为35.5m见方、高8.6m、四方开圆洞券的砖墩，四面圆洞券直对四方的关城，砖墩上建有多重檐三滴水攒尖顶木构钟楼，总高36m，是古城最高的建筑，且体量、造型、色彩也最为突出优美；另一标志性建筑——鼓楼，则位于它的西北面。明清时期，西安城内的建筑均为5～8m高的低层建筑，钟鼓楼同四周高12m的城墙及其关城门楼，共同构成了西安古城有节奏的立体轮廓，其中城市中心的钟楼统领着全城的空间面貌。现沿街和内部一些建筑虽然增加了层数，但钟楼仍能够控制住古城的空间环境（只是钟楼旁的两幢较高建筑需要采取减法）。

北京现存古城，建成于明永乐十八年（公元1420年），修建的城墙有里外3层。北京古城内宫城称紫禁城，中套皇城，外是京城，亦称内城。明嘉靖三十二年（公元1553年）北京古城又建外城，形成凸字形。北京古城内城有9个城门，外城有5个城门，城门处建有30多米高的门楼与箭楼，在内城中部还建起高60多米的西苑白塔、景山万春亭和高50.86m的妙应寺白塔，城内大多为高5～8m的一二层居住建筑等。其中西苑白塔、景山万春亭、妙应寺白塔，这三个最为突出的高点，连同高30多米的大殿与城门楼构成了北京古城极有韵律的立体轮廓。这是一幅如诗如画的画面，一般建筑与高点建筑比为1∶5～1∶8，以突出作为标志性的高点建筑，统领着古城的空间环境。从北京古城南北向中轴线来看，最南端的是高近40m的永定门城楼，往北前行又分别是高40m的正阳门楼与箭楼、毛主席纪念堂、高37.9m的中国人民英雄纪念碑和高近40m的天安门城楼，再通过中心高33m的故宫太和殿，面对的是景山最高的万春亭，再往北是中轴线最北端高40多米的鼓楼和距其100m远、高47.95m的钟楼。这些建筑都是标志性的，这一组中轴线上的建筑仍起着控制北京中心区空间环境的作用，人们只要看到其中一个标志性建筑就能知道自己所在的方位。这些标志性建筑都具有高度大、造型优美的特征。沿着北京古城南北向中轴线，出城延伸向北便可来到北京奥林匹克体育中心，在此延长的中轴线东边是国家体育场，该项目是通过国际设计竞赛选定的，建筑为椭圆形，外部造型似鸟巢，故俗称为"鸟巢"，其建筑体量巨大，壮观新颖；中轴延长线西边是北京国家游泳中心，呈方形，又称为"水立方"。这一圆一方、一刚一柔对比鲜明的大体量新建筑，成为统领北京中轴延长线北部体育中心地区空间环境的标志性建筑。

广东珠海渔女雕像，位于珠海市香洲和拱北区中间的滨海公园北面的海水中，有桥与情侣路相连，其造型简洁朴素、神态亲切、尺度宜人。优美的大自然山水背景使得这个制作精细、神情动人的雕像成为这一中心地区的视线焦点，成为控制珠海香洲连接拱北地区空间环境的标志性作品。此外，还有广西三江马鞍寨的鼓楼，它是侗族的标志性建筑，体量大，布置在全寨的中心位置。其平面呈方形，木结构，中心为一公共大厅，有4根10余米高的主承重柱支撑着楼顶鼓亭，外轮廓是7层呈密檐的方形锥体。鼓楼前设一大广场，这是全寨居民聚会和交往的地方，诸如议事、礼仪庆典、迎宾送客、休息娱乐、谈情说爱等。鼓楼及其广场，侗族人称其为"寨胆"，即寨子之魂的意思，它是全寨居民社会活动的中心地。

关于标志性建筑起着统领城市空间环境作用的国外实例，如加拿大多伦多市中心区的电视塔和湖滨建筑群。在沿安大略湖滨西边建一高553m的加拿大国家电视塔，在此高塔旁建有屋顶可开启的大型体育场馆，以这组有对比关系的建筑统领着湖滨的高层建筑群，形成了一个富有节奏韵律、空间优美的立体轮廓线，并成为多伦多城市的标志。又如美国芝加哥市中心区高层建筑群，在密执安湖滨长条形公园绿地后面，高层建筑排列有序，其南部是最高的深色西尔斯塔楼，中部有第二高的浅色石油大楼，北部是第三高的深色汉考克大厦，这3幢标志性的高层建筑之间又布置了其他较低建筑，组成了高低错落的建筑群。从湖岸由东向西望去，呈现在眼前的是一幅有节奏和韵律的建筑立体画面，它已成为芝加哥城的标志。芝加哥高层建筑的布局是有规划的，1993年6月我访问SOM建筑师事务所时看到了这个规划，几十年来该所一直负责这项工作，他们是从用地平衡、使用功能和改善环境等方面进行规划建设的，这种做法值得大家参考。再看看美国西雅图市中心区海滨建筑群，这里从北到南分为3段：北段是低层建筑群，但立有高185m的西雅图最高标志性建筑"太空针塔"；中段沿湖滨保存着原有的低层公共建筑，其后为新建的高层建筑群，包括金融区、商贸中心、通讯和市政厅等标志性建筑；南段又是低层区，有开拓者广场和历史保留区，还新建了一幢低平但体量巨大的供娱乐与运动使用的圆形穹顶建筑Kingdome体育场，其后为国际区。这北、中、南三部分有节奏、有韵律、有层次地构成了西雅图城市中心区立体轮廓线。其布局沿海边建筑低，往后退依山势排列高层，这样既保存了历史文化建筑，又做到了功能分区合理、城市用地平衡。

意大利佛罗伦萨西尼奥列旧宫，现为市政厅，在雉堞式檐口上矗立着总高95m的塔楼，极具中世纪城堡的特点。它是一个突出的标志性建筑，控制着佛罗伦萨市中心区，包括它前面及其左右的露天雕塑博物馆，以及左面阿尔诺河边的空间环境。

法国巴黎圣母院，也是一个突出的标志性建筑，它是世界闻名的天主教堂，建于公元1163—1345年，坐落在巴黎市中心塞纳河中西岱岛上。其内部大厅高32.5m，可容9000人进行宗教活动，为哥特式教堂结构，其中13吨重的大钟挂在正厅顶部，高达90m的尖塔与前面一对塔楼成为人们视线的焦点。巴黎圣母院之所以成为标志性建筑，不仅因为它的建筑体量高大、外观精美壮丽，还因为它已成为一个政治社会活动的中心。西班牙巴塞罗纳新区于20世纪初始建巴塞罗纳神圣家族教堂，该教堂由世界著名建筑师高迪设计，是座典型的标志性建筑。它具有8个高耸入云的圆形塔，体现着高迪创造的曲线、曲面"自然式建筑"空间，人们称其为神圣家族教堂博物馆。英国伦敦国会大厦，同样是典型的标志性建筑，矗立在泰晤河畔。国会大厦立面长280m，大厦西南角有高104m的维多利亚塔楼，东北角立有高98m的方塔钟楼，此楼即是世界闻名的威斯敏斯特钟，顶部装有21吨重的特大时钟。巴黎圣母院、巴塞罗纳神圣家族教堂和伦敦国会大厦这三处标志性建筑皆起着统领各自城市中心区空间环境的作用。

通过对中外建筑的分析，我们可以得出城市标志性建筑要具备位置、高度、体量、艺术造型、视线聚集等条件。

2. 控制尺度体量，协调城市肌里

关于这一问题，在第三章历史文化理念中已有论述。这里仅再强调建筑师要重视、掌握与控制建筑的尺度与体量，以使城市肌里具有整体的和谐美。

这方面做得比较好的实例，如山东曲阜阙里宾舍。设计师戴念慈院士十分注重建筑尺度和体量关系，他将宾舍建筑高度控制在二层，其中客房为合院式，布置水池与花木。同时在立面分割、门窗大小等方面都很考究其比例尺度。此宾舍建筑群同曲阜孔府、孔庙建筑协调一致，并保持了这一中心地区原有历史文化风貌和城市肌理。类似的做法，还有珠海宾馆，它位于珠海市石景山风景区旁。该景区满山是花岗岩的巨石蛋，千姿百态，并有一个白莲大水池。设计师莫伯治院士同样采取低层分散庭院式的布局，其尺度、体量完全与景区协调，庭园同景区融合为一体，并丰富了这里的景观。

又如建成于20世纪90年代的上海博物馆，它位于上海人民广场的中轴线南端，同北面上海市政府大厦遥遥相对。上海博物馆总建筑面积3.85万m^2，设计人是著名的建筑师邢同和。为适应这里开阔的广场空间，并要与周围高层建筑相呼应，设计师重视控制建筑的尺度与体量，采用横向伸展的做法，既舒展又平稳，避免压抑感。将建筑下部做成方形，呈两级台阶状，似为基座；顶部做成圆状，又似中华铜镜。设计造型寓意"天圆地方"。上海博物馆内部功能齐全，建造技术先进，是一座尺度与体量适宜、具有地域文化特点、城市肌里和谐的新建筑，该作品已被选入《20世纪世界建筑精品集锦》（东亚卷）中。位于上海市人民广场西侧的上海大剧院，总建筑面积6.8万m^2，设计单位是法国夏邦杰建筑师事务所和华东建筑设计研究院，建于1998年。整幢建筑重视尺度和体量，采用分散设计、部分伸入地下的做法。在基座上布置适度的观众厅和舞台，最上层屋顶向天空展开，象征上海市民对世界文化艺术的热情追求，并充分运用现代高科技与新材料来营造上海大剧院轻巧透明、典雅壮丽的形象。

国外历史文化名城尺度、体量做得比较好的建筑实例，如法国巴黎新歌剧院。巴黎新歌剧院坐落在象征法国革命的巴士底广场上，设计方案是通过国际竞赛选出的，其建筑形式注重体量的分割与对比，以此同原有街区、广场的尺度相适应，建立一种整体、协调的美。又如巴黎卢浮宫扩建工程，20世纪80年代因扩大规模需要，决定扩建卢浮宫，最后选用了著名华裔建筑师贝聿铭先生的金字塔式扩建方案。该方案扩充的建筑面积完全被放在地下，而地面上玻璃金字塔入口的建筑尺度、体量与原有建筑空间环境协调一致，同时还保持着原有城市肌理。类似的做法，还有美国华盛顿史密松非洲、近东及亚洲文化中心工程，该工程于1986年建成，建筑面积为62057.2m^2，建在1849年修建的维多利亚式史密松学会总部及其下属两个历史文化博物馆之间的一块1.7万m²的地段上，这一地段处在华盛顿国会大厦左侧前众多的博物馆之中，非常瞩目。设计者大胆采用地下方案，只将3个入口亭建在地上，把96%的面积放在地下。入口建筑亭总高12m，体量小，保持了地面原有建筑的历史文化风貌和城市肌理。该建筑受到美国各界的好评，并获得1988年托克优秀设计奖（Tucker Award Excellence）。再有1997年完成的俄罗斯莫斯科马涅什广场购物中心工程。马涅什广场位于红场的北面，它与克里姆林宫墙相邻，为了改善这里的生活环境，20世纪90年代对此广场进行了改造，并修建了这个地下4层的多功能购物中心。由于该购物中心建在地下，因此保持了原有的城市肌理。

再看位于美国华盛顿国会大厦右侧前新建的国家艺术馆东馆，它紧靠国家艺术馆老馆的东面，设计师是著名美籍华裔建筑师贝聿铭。设计师重视控制尺度，国家艺术馆东馆高度、体量同老馆协调，外观简洁而有变化，建筑以实墙体为主，开窗尺度、比例适宜，特别是新旧馆的连接，主要通过地面广场、喷泉、水斗、天窗和地下通道水景组合在一起。新馆内部则完全根据新的功能需要，采用新结构、新材料和新的三角形空间构图组织，该建筑1978年建成。卡特总统出席国家艺术馆东馆开幕剪彩时，称赞这座建筑与城市协调，是一个公共生活同艺术结合的象征性建筑。还有美国西雅图公共市场中心区，它临近普吉特（Puget）海峡水面，并保留着一层供应市民以海鲜食品为主的老市场。在老市场对面以及周围地区建筑群的建设，比较重视控制尺度、体量，采用与老市场和谐并同中心区东部行政、商贸等高层建筑相适宜的高度与体量，仅修建了3~6层高、体量适中的建筑，并取得了城市肌理整体和谐的效果。

3. 掌握平衡、变化的艺术规律，增强城市建筑整体美

艺术形式美是城市建筑的重要内容，也可以说是重要的组成部分，这是因为建筑的功能包括使用功能和精神功能，艺术的形式美对使用者的精神起着很大的影响。形式是包含于功能之中的，它是任何建筑不可缺少的内容。我们进行城市规划、建筑设计创作时，应将功能与形式、内容与形式结合为一体来考虑，重视建筑外部与内部空间同城市和谐一致的形式创作。下面讲的几个方面的关系，都考虑到了形式与内容、形式与功能的一致。

3.1 城市建筑有对比的空间变化，可达到印象深刻的整体完美

从北京颐和园东面入口进入以仁寿殿为中心的院落，这里过去是皇帝上朝处理政务的地方，是个庄严的封闭空间，建筑、院落陈列华贵多彩。从此院落向西行，经过渡空间便可看到对比强烈的万寿山、昆明湖山水景观，这里不仅空间大，而且整体为青绿色调的自然景色，这一小一大、庄严与自然、色彩浓厚与淡雅的对比，给人以整体美的感受；在进入山水

景区之后，多处都存在着这种对比的空间变化，如沿万寿山脚下东面的乐寿堂院落、西面的听鹂馆院落和此山坡中东边的景福阁、西边的画中游景点等。回到北京城内的西苑（北海公园），当你从南边正门进入看到的局部景色，和你过桥登上琼岛白塔高地后极目远望到四方开阔的城市空间环境大有不同。这种空间对比变化，在西苑（北海公园）内也多处存在，如你从北边后门进入静心斋这组空间封闭的园林景点后，再登上叠翠楼观景或走出静心斋后都可观赏到开阔的西苑全景或更远的景山景观。

当你去看杭州西湖风景区，经白堤到孤山北部丛林，走到上面西泠印社后部封闭的高台地空间后，再前行进入四照阁向南俯视，可望到开敞辽阔的西湖全景，这种突然的景观空间对比变化，让你心旷神怡；你乘车西行来到西湖北面的郭庄园林景点，先进入封闭的"一镜天开"主景区，再进入此景区主体建筑景苏阁，从其二楼上可欣赏到宽阔自然的西湖美景，如此类似的空间对比变化，使你获得同样完美的感受。如果到四川都江堰去游览，最好能去两水交叉宝瓶口处的伏龙观，此观为3层台地院落，你可从封闭空间一层台地老王殿院进入二层台地铁佛殿院，再登上三层最高台地玉皇楼院，在此楼上向北、向西眺望，可观览到宽广的岷江、横跨的安澜索桥、苍绿丛林中的二王庙和赵公山、大雪山等优美的山水全景，此景点辽阔自然、古朴幽静，对此种整体完美的空间对比变化，人们印象深刻，难以忘怀。

国外有此空间对比变化感受的亦有多处。比如，意大利威尼斯，我曾两次去圣马可广场观赏，需经长长的几米宽的商业窄巷才能通过圆形拱门进入长175m、东面宽90m、西边宽56m的梯形大广场。梯形大广场连通着南面的一个小广场，在连接两个广场的转角处矗立着高100m的钟塔，给你两个空间对比变化的感受，其中一个是窄巷转入圣马可大广场的对比；另一个是高百米塔楼与四周只有三四层高的教堂、总督府、新旧市政楼等的对比，这增强了我对这个世界闻名的圣可马广场整体美的印象。西班牙巴塞罗那的蒙特胡依克山，它位于巴塞罗那旧城西面，17世纪建有城堡，1929年为举办国际博览会在山丘北建起博览会艺术品陈列馆，在此馆旁与馆内封闭较小空间向北俯视远望，则是一幅开阔、富有层次、建筑与自然结合的动人画面。其中近景是长条形的博览会绿色广场；中景是有6条路交叉于此的西班牙广场，广场中心立有精美的三角亭式雕塑喷泉，广场东北角是19世纪建起的能容纳2.6万观众的大竞技斗牛场；远景是作为城市背景的绿色横向山林景观。游览过此名胜，其空间对比强烈的印象至今存留在我的脑海中。我在美国波士顿中心区偏西南的汉考克大厦旁的教堂参观过，其内部是一般常规的空间和精美的装饰，随后我来到高240m的汉考克大厦顶部的观览厅，向东眺望，这里以市中心大片绿地的公园、高层建筑为前中景，远处以低层建筑、水面作为背景的开阔画面；向北眺望，前景是绿带穿插其间的低层住宅区和查理河，中景为查理河北岸沿岸的马萨诸塞理工学院主楼等建筑，画面绿色开敞。由教堂的封闭空间环境，再到大厦顶部可观赏到宽阔的波士顿全市绿色空间环境，其对比无比强烈，让人永记心上。

3.2 建筑对景、借景，丰富了建筑与城市的景观艺术

本书实例部分介绍的中国长安城建章宫苑太液池，它是"一池三山"园林形式的起源，池中有象征东海三神山的蓬莱、方丈、瀛洲三山，此三山与岸上主体建筑是对景的关系。后来杭州西湖模仿太液池采用"一池三山"的做法，在湖中立有三潭印月、湖心亭和阮公墩三个小岛。北京颐和园总体布局也完全仿杭州西湖风景区的格局，在昆明湖中建有西堤、支堤，将水面划为一大二小，在这3个水域中各建一岛，象征东海三神山，与杭州西湖中的3个小岛同孤山景点和宝石山上的景点关系一样，此3个岛同其北面万寿山佛香阁景观都是对景关系。这种对景，从池湖中三岛可分别看到山上的主景，从山上主景又可望到池湖中具有三山的景色。又如北京西苑（北海公园）的琼岛白塔景点，同其北部五龙亭景点也是对景关系，彼此可相互观赏。在苏州拙政园中，主体建筑远香堂面对北部山丘上的雪香云蔚亭，别有洞天直对梧竹幽居亭，玉兰堂面对见山楼，香洲面对荷风四面亭等，它们皆是对景关系。苏州拙政园内的对景大都在50m左右，都可观赏到清楚的景观。扬州瘦西湖中部的五亭桥、白塔同吹台亭是对景，可从吹台亭中的两个圆洞门望到白塔和五亭桥明亮的景观，从五亭桥上又可观赏到以吹台亭为中景、四桥烟雨为远景的多层次画面。这些对景都起到了丰富景区的效果。

关于借景，中国实例也很多。如北京颐和园将玉泉山宝塔和远处的峰峦借入园内，其景色十分深远；又如无锡寄畅园，把锡山上的龙光塔借入园内。这两处借景都扩大了风景园林的空间环境，也丰富了风景园林的艺术景观。

在国外，有对景关系的建筑亦不少。如意大利罗马西班牙广场顶端的三一教堂和其下端老泉水池喷泉以及直对的窄长时装街道都互为对景。罗马西北面巴格内亚兰特别墅园内，最底一层的正方形大水池，其中心圆形岛上立有雕像喷泉与二层台地圆形喷泉互为对景；三层斜面连续水流景观与四层台地的中心海豚喷泉景观也互为对景。美国华盛顿市中心高169m、顶部为金字塔型的华盛顿纪念碑与东边国会大厦、西边林肯纪念堂、北边白宫等都是明显的对景关系。以上这些对景布局，都起到了丰富建筑与城市景观的艺术作用。

3.3 建筑的对称布局，可获得庄严、平衡、统一的效果

对称布局，在中外历史文化名城与建筑中应用得最多。中国古代唐长安城、元大都、明清北京城等都是对称的格局。明清北京内城，前朝后市，左祖右社，紫禁城居中，中轴线突出，连方格网形的道路通向南北东西的九大城门都是对称的布局。其东面有东直门、朝阳门，西面有西直门、阜成门，南面居中为正阳门，其两旁为崇文门、宣武门，北面为德胜门、安定门。现故宫（原紫禁城），就是以太和殿（最核心）、中和殿、保和殿三大殿为中心，左右对称的总体布局，它体现帝王至高无上的尊严。

新中国成立后，北京于1959年改建天安门广场，以天安门、正阳门为中心线，广场西侧建起人民大会堂，东侧对称地建起中国革命和历史博物馆，广场中心建立一座中国人民

英雄纪念碑。天安门广场面积为40万m^2，它是新中国的象征，气势磅礴，建筑色彩统一，新老建筑和谐，形成一个完整的建筑艺术活动空间。此外，北京四合院居住建筑也是严整对称布局，一般是正房3间布置在主轴线上，正房两侧各有一间较低的耳房，正房前边左右对称安排东西厢房，并与南墙、垂花门形成中心院落。中国的居住建筑，许多也是采用对称式院落或天井布局。还有中国的寺庙，多数亦是选用对称式格局。如台湾地区鹿港龙山寺，建筑总布局为轴线对称三进院落式，一进院落为前殿；二进院落主殿居中高耸突出，为重檐歇山式，供奉主神观音菩萨。主殿前两侧为对称的厢廊和两棵茂盛的大榕树，显得非常壮观；三进院落为后殿。这种有节奏的空间对称布局，给人留下了庄严、平衡、统一的印象。

国外对称式建筑属欧洲中世纪建起的宫苑最为突出，如法国巴黎沃勒维孔特园（Vaux Le Vicomte）。该园始建于1656年，由著名造园家勒诺特（Le Notre）设计。设计师采用了严格的中轴线对称布局，庭园南北长1200m、东西宽600m，居中高台上的主体建筑对面布置着对称的花坛、水池、装饰的喷泉和建筑小品，并在末端有横向运河相衬，因而这条明显的中轴线控制着人心，让人感到主人的威严，此设计达到主人富凯的要求。富凯是路易十三、路易十四的财政大臣，他当时专政，建此园以显示自己的权威。该园1661年建成后，路易十四应邀来观园赴宴，感到富凯有篡权可能，于是借题将其下狱问罪，判无期徒刑。之后，年仅23岁的路易十四将沃勒维孔特园的设计师勒诺特（Le Notre）、建筑设计师勒沃（Le Vau）和室内设计师勒布兰（Le Brun）召至宫中，让他们负责规划设计凡尔赛宫苑，并提出宫殿要超过西班牙马德里的埃斯科里亚尔宫，园林要超过沃勒维孔特园，要造出世界上未见过的花园。经过几十年的建设，设计建造满足了路易十四为了显示其君主权威的要求。这个占地800hm^2、建筑面积11hm^2、中心园林100多公顷的规模宏大的凡尔赛宫苑，其均衡对称的布局突出了纵向大轴线、大运河的宏伟气势，被人称为"勒诺特式宫苑"，这种模式影响欧洲各国整整一个世纪。在德国、英国、奥地利、西班牙、瑞典、俄国都建造了大轴线突出、体现君主威严的对称式宫苑。巴黎凡尔赛宫苑大轴线对称格局是：宫苑外3条放射线聚焦在中心位置的主体建筑凡尔赛宫上，宫后第一层高台对称布置着两个大水池与雕像，并安排左右对称的北园和南园，沿中轴线向下是低一层的拉托娜（Latona）雕像水池喷泉台地园，再往西沿中轴是长条草坪绿地皇家大道，大道两旁对称布满雕像与树丛，两边树丛后有对称的多个花园，接着皇家大道的是阿波罗（Appolo）雕像水池喷泉和1.56km长的十字形大运河。大运河东端水池旁设有对称的放射线，西北向通到大小特里亚农宫，西南向通至动物园，这个景色深远、严整气派、雄伟壮观的宫苑，体现出炫耀君主权威的意图。

这里再分析一下世界上最大的天主教堂——罗马圣彼得大教堂。它建于公元1506—1626年，历时120年，整体布局突出严格的对称式。中心穹顶直径41.9m，内部顶点高123.4m，外部顶尖高138m，它是罗马全城的最高点，由对称的4个墩子支撑这个穹顶。教堂前由一个对称的椭圆形加梯形的柱廊组成，柱高19m，很有气势。广场中耸立一座方尖碑，其两侧对称各设一水池喷泉，整体显得格外壮观。

3.4 建筑形象与比例的变化，可以丰富意境，增强审美感知

北京天坛祈年殿是一个有象征意义的建筑，其基座是高6m、直径90m的圆形汉白玉3层须弥座平台，直径30m的圆形殿立于基座上。天坛祈年殿高38m，三重檐攒尖顶，原上、中、下檐分别为青、黄、绿色，乾隆十六年（公元1751年）重建时改为三檐蓝色琉璃瓦，其圆形、蓝色更为突出天的象征，它是天坛中轴线北端最高、体量最大、造型风格最为完美的建筑。又如1996年建成的厦门高崎国际机场候机楼，采用钢筋混凝土屋架式顶棚结构，创造出闽南曲线大屋顶意境，其造型如鸟展翅飞翔。候机楼上部几层高侧窗，自然采光、自然通风，它所创造出的中庭式大空间，可随着需求变化，这是个建筑造型独特、节能高效的创新建筑。

关于建筑间距与高度的比例关系，在园林或其他建筑中以平视视角舒适为宜。如苏州拙政园主景远香堂距离对景雪香云蔚亭为50m，雪香云蔚亭及其背景林木高20m，从远香堂平视雪香云蔚亭，视角为18°～27°，此角度是最好的视角，其距离为高度的2.5倍。居住建筑之间的比例要低一些，应是1.7～2倍，以便冬季时阳光从南面以30°角照进北面住宅底层窗内。在公园或广场，其比例可为6～8倍，也能获得较好的视觉感受。掌握此比例后，可取得协调舒适、整体完美的视觉与使用效果。

在国外，埃及开罗吉萨金字塔，其三角金字向上寓意法老化身的灵魂永存、朝天向上的神圣之意。希腊雅典帕提农神庙，其顶部三角形山形墙代表尊天向上之意，它的外立面分为有节奏的3段，底部阶级宽度、柱子、楣梁山形墙，都与柱径有一定的比例关系，十分协调。柱头采用多力克（Doric）式，代表男性，它是一座具有尊天刚强之意、造型完美的古典建筑。

美国西雅图市为1962年举办"21世纪博览会"而修建了高185m的太空针塔（Space Needle），将高高的塔顶部做成飞碟形象，其创作寓意是象征21世纪的宇宙航天。当我观赏了这座轻巧、高耸、美观的太空针塔后便体会到了它的象征意义。日本京都于15世纪建造的龙安寺石庭，在白砂地面上布置了15块精选之石，依次按5、2、3、2、3比例摆放成5组，象征5个岛群，并按三角形的构图原则布置，达到均衡完美的效果。龙安寺石庭底部白砂摆成水纹条形象征着大海，这组抽象雕塑寓意着大海、群岛的大自然景观，其形象和比例的变化确实丰富了意境，增强了人们的审美感知。

3.5 建筑与城市的空间序列变化，可增强整体的和谐美

苏州拙政园的序列就很有参考价值，进拙政园园门，正对假山是一个以石山屏障为主的前导小空间，再跨小桥、绕假山进入一个有山水景观的半开敞小空间，然后进主体建筑远香堂便可观赏前面最为开敞的自然山水景观和空间。从远香堂向西行便可看到以小飞虹为主的曲折变化的小空间，再从柳荫路进入见山楼，登临而望的是一个开敞的大空间。出见山楼到北部花径，是一个具有水乡特点的小空间，后折入池中山来到雪香云蔚亭，可观赏到以远香

堂为中心的全园大空间,下山后过桥还可看到以枇杷、海棠为主题的小空间。苏州拙政园这一空间序列可简化为小、小、大、小、大、小、大、小和闭、半敞、敞、半敞、敞、闭、敞、闭,此同诗词中的平仄韵律是类似的,同文章的引言、描述、高潮、转折变化、结尾是相仿的,这构成了富有韵律的流动空间,也增强了整体的和谐美。

广西的三江马鞍寨,这里有知名的程阳桥,建筑也甚为壮丽。通过此桥长长的通道,进入后山前河水围绕的开阔的村寨大空间,此处靠近后山的高大鼓楼及其前面的广场,并控制着全寨的空间环境,从此中心大村寨东面沿河岸可看到寨边起伏的山水景观以及半开敞的建筑空间,最后通过较小的平岩风水桥到寨外的大自然山林地段,其空间序列是封闭、大开敞、半开敞、封闭、开敞,同样使人感到空间序列的和谐与整体完美。

建于公元前14世纪的埃及卢克索卡纳克太阳神庙,它是埃及最为壮观的神庙,也已有明显的空间序列。其入口从窄门道进去,前院是一横向开敞的大空间,侧面立有拉姆西斯二世雕像,往前行再进入有16列共134根高大密集的石柱组成的列柱大厅,其中间两列12根圆柱最为高大突出,这里形成高耸、但并不开敞的大厅空间,列柱大厅后为立有方尖碑较开敞的空间。此神庙空间序列的变化是:封闭、开敞、封闭、开敞和小、大、特大、小大,依然使人感到它的空间环境是有节奏的整体美。再看西班牙格拉纳达的阿尔罕布拉宫苑,它建于公元1238—1358年,建筑与庭园是西班牙式的伊斯兰园,人们称其为帕提奥(Patio)式。先进入中等规模的桃金娘庭院,此院正殿是皇帝举行仪式接见各国大使之处,殿前有水池,两侧种植桃金娘绿篱。由此院东行进入狮子院,该院是后妃住所,院中布置了十字形水渠,水渠交叉中心立有圆形盆状的水池喷泉,喷泉下为12个精细石狮雕像,四周是由124根细长柱拱券廊围成,这里是宫苑中最为精美的中等规模院落。出此院往北是一小规模的林达拉杰帕提奥式小庭园,再东行就到了不属于帕提奥式的开敞帕托园。阿尔罕布拉宫苑空间序列是中、中、小、大,半敞、半敞、闭、敞,增强了整体的和谐美。最后分析一个具有城市序列空间的法国巴黎市中心区,其起点是主体建筑卢浮宫,它面对东西向下沉式长条形杜伊勒里(Tuileries)花园。该花园是17世纪路易十四时期由著名造园家勒诺特改扩建的,这是一个半开敞的园林空间,沿其市中心轴线园路向西行,连接此园的是一个南北横向展开的、开敞的协和广场,广场中心耸立一埃及方尖碑,两侧为仿意大利罗马圣彼得大教堂前建造的一对喷泉水池,沿此广场中心轴线便走进东西向的长条形香榭丽舍商业街,此街西端便是著名的巴黎凯旋门及其圆形大广场,四周放射之路皆聚焦于此,形成向心的、外部放射的开放空间环境,这4处沿巴黎市中心轴线的连续空间序列分别是半开敞的园林空间、开敞长方形的广场大空间、不开敞的街道空间、开敞放射形的广场大空间,这种极富韵律感的序列空间,我曾多次去观赏,深深感受到它整体空间变化的和谐美。

五、便民交通理念

改变汽车主宰交通的观念,构建人行和自行车系统,联网公共交通。

为了方便城市居民出行,城市交通不能被汽车所主宰,要构建起人行和自行车道路以及公共交通系统,以此三者联网为主,改变以汽车主宰城市交通的观念。

在一个城市的中心区,特别是旧城中心区,都要逐步构建人行街道系统,使得这些为广大市民生活服务的街道空间环境不受汽车交通的干扰,让人身安全,也使街道人气旺盛,真正成为满足城市市民生活需要的繁华街道。这是城市繁荣和发展的重要措施,也是促进消费增长、经济发展的重要举措,它体现着城市现代化的水平以及城市兴旺和对城市居民关怀的程度。随着城市汽车的发展,特别是私人小汽车的快速增长,城市交通被小汽车所主宰,城市道路交通拥塞。现在发达国家的城市正在扭转这种混乱局面,他们控制城市中心区汽车总量,构建起人性化的步行道路系统,并与公共汽车网和地下铁路交通网连成一体,从而丰富了这一地区的活动内容,并提高了其空间艺术水平,同时也方便并吸引更多的居民到此活动,由此提升了城市生气。

1. 在城市市区范围内限制小汽车行驶的数量

在这方面,中外城市都采取了多种办法。有的城市中心区划出一定的范围,在此范围内不准小汽车驶入,只许人行和公共交通通行。如1979年我访问瑞士时,看到瑞士的一些城市就是这样做的,在城市中心区范围外设有停车场,而在城市中心区范围内还保留着原有无污染的有轨电车和公共汽车交通。有的将城市中心区的名胜古迹与建筑周围划为步行区,不准任何车辆进入,如意大利佛罗伦萨开辟步行区区域(Pedestrian Precinct)的做法。该市将中心区西尼奥列广场及其周围地区,北至圣·玛丽亚教堂、钟楼,南至阿尔诺河、旧桥,全划为步行区,在城内其他名胜处也划出一定的范围作为步行区,不准汽车驶入,这方便了旅游参观者。还有的将市中心区著名建筑前的广场道路改为步行街,如英国伦敦中心区著名的特拉法加(Trafalgar)广场。该广场于2003年完成了改造,它将广场前的车行道路封闭,并将下沉式广场的高差大台阶改造成国家美术馆入口的大台阶,把美术馆建筑同广场连为一体,并在广场两侧与前面组成一个人行道路系统,人行与车流分行。通过这样的改造,既方便了居民,也方便了游客,前来广场的人流比以前多了13倍,大大提升了这一地区的人气。此外,2007年伦敦市长肯·利文斯通在伦敦市政委员会上说,他计划把伦敦市中心区几条繁华的街道和公园,如摄政公园、牛津街、购物区等改造为步行街,只允许行人和自行车通行,并希望这一区域的这些街道建成像西班牙巴塞罗纳Rambles

步行街一样的林荫路，此项改建工程将分期完成。西班牙巴塞罗那旧城中心Rambles步行街，是欧洲闻名的步行街，它还是个鲜花市场，人称"花市大街"。这里已成为各国游客必到之处，其步行街出入口都与城市地下铁路网相连，方便出行。还有的以固定的时间改作步行街的，如日本东京都银座，它是东京最繁华的商业区，从1970年8月起，每逢星期日下午、节日午后，一律禁止车辆通行，商业街改为步行街，街心还设有咖啡座。步行街上游人更多，更活跃，也更安全。此银座大街北起京桥，南至新桥，处处都有地下铁路网与其连接，方便人们到此游览。

我国一些城市已开始重视这个问题。如北京20世纪90年代开辟了王府井步行街，其范围在王府井南口至八面槽一段，今后应扩大范围，并与北部地下铁路网连通。北京前门外大栅栏街，早已是步行街。现在前门外大街已全部改为步行街，它四周的许多街道和胡同也改作步行区，并形成步行区域。上海闻名的南京路商业街在20世纪90年代也变为步行街，公共交通和机动车从其南北两侧街道通过。天津市中心区劝业场旁的滨江道，开辟为步行街也已多时，深受广大市民欢迎。

2. 建立城市立体交通系统，大力发展人行道路系统

城市道路交通立体化，就是将城市中的人行道、自行车道、汽车道、地下铁道，还包括地上高架行车道等，同建筑物相连接，最后形成完整的道路交通立体化系统。中外城市大都已重视连接问题，但能构成系统完整的却是少数。最早重视道路连成系统的城市，要算美国的费城（Philadelphia）。费城市中心区规划是有计划地逐步发展立体的道路系统，它把原有历史性建筑、全市商业性服务建筑、文化娱乐性建筑同人行道连接在一起，于1963年形成连接的整体，1973年又与地下铁、公交汽车道连接发展成完整的系统。其设计师是美国著名规划师、建筑师埃德蒙·N.培根（Edmund N. Bacon）。该平面图中的中央黑块为独立大厦，灰格网是中心地带密度最高的活动区，黑线表示街道下面的人行系统，白线表示街道上的人行系统，这些人行系统连接着各部分的商店。立面展开图中间是独立大厦，前面为开阔的、不受车辆干扰的下沉式广场。广场左边联系服务大楼，广场右边连接市场街，市场街上有带着玻璃顶的3层建筑。这些建筑的底层连接在一起，形成了步行系统，建筑的第二层与通行汽车的街道连接。在中央下沉广场的下面，设置地下通道，这样人们去市中心区就十分便捷，而且在中心区活动时也不受汽车干扰，有安全感。城市中心区市街的设计，同样强调空间的不同处理。在步行系统中，街下面布置成花园式的活动空间，这个空间又与地下铁相连，步行的端头设公共汽车终点站和停车场，这样不仅交通方便，而且从街上还可以看到不同的空间层次和丰富的景观。

采取这种连接成立体交通系统的简单实例，如加拿大多伦多市中心新建的伊顿中心。它是一个大商场，室内步行街长274m，其周围布置3层楼高的300家店铺，此室内商业街通过天

桥和地道，也同周围建筑相连，构建起大范围的人行系统，并与地下铁路和地上公共汽车道路相接。又如规模大的立体化道路交通系统的实例，以加拿大蒙特利尔市中心区为例。该市于1960年从发展玛丽城大厦（Vile Marie Tower）地下空间开始，有机组织、引导发展这种系统，其中心区有3片连接的综合建筑群，即德雅尔丹片（Desjardins）、玛丽城片（Place Ville Marie）和皇家山商场（Les Cour Mont-Royal），这3片建筑群同中心地区地下铁路环状网相连通，并与中心区主干道凯瑟琳（Catherine）路、勒维克（Rene-Levesque）路垂直交叉。经过28年的建设后，这个城市的道路交通立体化系统基本完成，每天有近30万人使用。从3片6个地铁车站来统计，它们联系着170万m²的办公室空间、1400个时装用品商店、3个音乐厅、2个百货商店、3800个旅社房间。商店的街道入口大量地转为室内入口，从1961年的2.7%增加到如今的36%，鞋店从0增加到如今的48%，女服装店从1.8%增加到如今的67%，室内步行街、广场的发展促进了街面商业文化设施的改进，增加了电子游戏、快餐、酒吧等，使之与室内步行街连通的知名室外商业街凯瑟琳、迈松纳夫（Mai-Sonneuve）同步繁荣，发挥了道路交通立体化的作用。

北京已基本形成了这种立体化道路的交通系统，北京中心核心区天安门广场已有天安门东、天安门西两个地铁站与其连通。天安门广场前面的前门外商业文化区，已形成步行区网，并联系着前门、珠市口两个地铁站。北京中心核心区后面的鼓楼南大街商业文化区，有北海北、南锣鼓巷两个地铁站口同它连接。北京中心核心区东面的王府井商业文化区、王府井大街已改为步行街，街内东边的具有室内步行意味的东安市场已把其东面的街道连接起来，在此街南端与王府井地铁站相连，只是街的北端距离东单北大街灯市口地铁站口较远。北京中心核心区西面的西单北大街商业文化区，有西单和灵境胡同两个地铁站连接。以上这5片重要区域皆同地上公交网连接，并与地面长安街、前门外大街、鼓楼南大街、王府井大街、西单北大街等连通，基本组成了北京中心区的立体化交通。将来还需要增加一条从永定门至地安门的地铁线，并改善鼓楼南大街与西单北大街的人行道路，以及增强道路交通与建筑之间的连接，北京的立体化交通系统早晚得以实现，它将方便市民与中外旅游者的出行。

这种连接成立体化交通的模式，确实能解决以下几个主要问题：

- 避免人车交叉的矛盾，使城市中心区的居民有安全感。
- 满足多功能要求。商业、文化、娱乐，甚至办公、旅馆等连接在一起，使用十分方便。
- 提供良好的环境，有自然采光或良好的人工照明，并有绿化、喷泉、座椅等，便于在此活动、会友或休息。
- 改善交通，道路网、地铁网、人行道路网与重要建筑物和公共设施等纵横交织在一起，来去方便。
- 同人防工程相结合，地下街、地下广场、地下通道三者连成的系统就是最好的人防工程。比如莫斯科的地下铁路网，它在苏联卫国战争中发挥了巨大的作用。

3. 尽量发展地铁交通，同时扩大地下建筑的范围

中外大城市、特大城市的中心区普遍存在着交通紧张与拥堵的现象。经过近几十年的不断改进，这些城市的交通已有所改善。凡城市中心区交通得到缓解的大城市、特大城市，大都是发展了地下铁路，并组织得比较好，如上节所述的道路网连接系统。现概括起来，发展地下铁路可分为3种情况：

第一种情况是，地铁交通网同地上棋盘式道路网或环形放射式道路网相衔接，在城市中心区主要地点都有地铁站出入口，去往城市中心区的交通十分方便，具体城市有巴黎、伦敦、莫斯科、纽约、北京等。

第二种情况是，在城市中心区核心区部分组成地铁环状网，另有几条城市干道地铁线路通过中心环状网。如华盛顿市有从四方穿过中心区并组成中心核心区5条线路。芝加哥市中心区地铁线路同地上高架桥交通线结合组成中心核心区环状网，并通向北、西北、西、南几个方向。加拿大多伦多市中心区地铁环状网比较小，另有几条地铁线接上这个中心环状网。

第三种情况是，在城市中心区只有一两条地铁线路通过，主要是地面上组织好公共汽车和汽车交通线路，如美国旧金山市等。

这3种情况，第一种做法效果最好，交通畅通；第二种做法虽不如第一种方便，但从出发地通过地铁可直达市中心区环状网，基本上是方便的，只是有些地方要步行一段路程才能进入地铁站；第三种做法，其效果不如前两种。

采用第一种做法的巴黎，其地下铁路网最为完整，而且它同地上的建筑衔接也最方便。1982年我第一次到访巴黎时他们就告诉我，你到各处去办事或开会，想准时到达最好是乘地铁，如选择地上交通就没有把握了。巴黎地铁网的东西向主轴线上人流最为密集，他们加开了高速车，又缩短了时间，很受欢迎。北京地铁网，是从20世纪60年代开始修建的。第一条地铁线沿东西向主轴线（长安街两头）延长发展，接着又建了二环路地铁，再后建起向四面八方放射的地下铁路，东北向至顺义、东南向至亦庄、西北向到昌平西山、西南向到良乡与苏庄，构成了一个整体成网的地下铁路系统，方便了市民的出行。中国的情况是人多地少，大城市、特大城市人口密度大、建筑容积率高，所以北京、上海、天津、广州、西安、武汉等城市都应该向第一种做法发展。

关于扩大地铁周围有关地下建筑和地上建筑连接的问题，加拿大蒙特利尔中心区的建设是个好实例。从蒙特利尔市中心地下空间系统图中可以看出黑色的地铁周围发展了较多的地下建筑，并充分利用了地下空间，减少地面建设用地的压力，同时地下成片建筑亦与地上扩大的建筑群连接，更大范围地方便了居民的生活。日本名古屋中心大街，它是该市南北向主轴线，布置了宽100m的绿化带，在此中心大街下开辟了地下商业街，在各个与东西向道路的交叉路口都设有宽阔的地下商业文化广场建筑，此处亦为地铁站出入口，联系着地上的建筑群，市民与游客来去都非常方便。这种扩大发展地下商业文化服务范围、形成大面积的地下

空间建筑的做法，起到了节约用地、方便生活、避免风吹日晒与雨雪困扰等作用。中国应提倡和重视这种做法，搞城市地铁的工程师要同城市规划师、建筑师一起研究地铁和地下建筑的开发、建设与利用，以取得多方面的良好效果。

4. 恢复与发展自行车道系统，控制快速车行驶

20世纪50—80年代，我国城市的道路交通基本上是正常的，各地采用三块板的道路模式比较多，中间行驶机动车，其两侧设绿化隔离带，绿化带旁边为自行车道。在5～6km的距离范围内，许多人都选用自行车出行。改革开放之后，我国私人小汽车发展很快，在大城市、特大城市里道路的三块板逐渐被汽车所占用。北京市就是个典型的实例，原两块板的自行车道，一部分划为汽车转弯行车道，还有很大一部分作为小汽车的停车场，有的地段还被取消了自行车道。用这种压减自行车道路面积来临时满足超过容量的小汽车需要，这是北京道路交通需要重点改善的一面。超速的摩托车、电动自行车在自行车道上飞驰行驶，威胁着骑车人的安全，管理部门应对这些车辆加以控制。城市道路交通是为广大市民服务的，合理使用它是居民的基本权利。北欧许多国家和德国等地的城市，不断发展自行车道系统，其自行车出行量约占整个城市交通量的30%。如德国西部的费尔德贝格市就奖励推行骑自行车上下班、购物和休闲活动。德国著名建筑师托马斯·赫尔佐格告诉我，他有两辆汽车，但这两辆汽车只是到城郊或其他城市时才使用，而要到市中心区上班，他则选择骑自行车。托马斯·赫尔佐格认为这对城市有利，可减少交通拥挤，也不污染空气，又节能环保，还锻炼了自己的身体。在上海世博会，有自行车王国之称的丹麦在其丹麦馆内设置了自行车道，并推广这一健康理念。我国所有的城市都要重新认识这个理念，恢复和发展自行车道系统，控制超速车行驶。城市增加小汽车数量要同发展城市道路交通以及停车场相适应。发展一辆私人小汽车，城市要增加许多道路和停车场面积，而不能只看生产一辆汽车能增加多少国民经济产值。国家发展改革委和各地城市领导要有综合的观念，认真解决这个问题，不能不顾城市建设的需求，单方面追求眼前大量发展小汽车的利益。

除上述几个方面外，城市规划设计也要重视发展城市的混合社区。城市的混合社区不仅具有居住功能，还可安排一些无污染的第三产业，这样可以方便居民生活，减少道路交通压力和人们上下班的出行。同时，还要重视加强城市道路交通的管理工作。

我们要重视便民交通这一理念，因为它是衡量城市是否宜居、城市是否先进的一项重要内容，而且它同实现生态文明同样也有着密切关联。

六、公正社会理念

落实社会公正，关怀弱势群体，开放城市空间，服务城市居民。

社会公正在建筑科学文化中的反映，就是要关怀广大城市居民及其弱势群体，公正地缩小城市生活与工作环境里的贫富差距。政府管理部门要从城市生活居住用地中所包括的居住、公共建筑、道路交通、绿地等各个方面给予弱势群众以关照和优惠，而不应为了经济利益，扩大富人和弱势群体的差距。同时，还要不断扩大城市开放空间，使广大的城市居民能够平等地享受各项社会文化服务设施，而且政府管理部门也应该更加贴近并服务城市居民。

中国已于2020年末实现了全民脱贫，达到小康社会的标准，这对世界各国都有重要意义。有关中外城市绿地建设和道路交通等在前面也已介绍。这里仅就居住和公共建筑两方面内容和一些优秀实例作简要地介绍，可供大家学习和参考。

1. 在城市居住方面要尽量搞好大众的住房

福建省厦门市为满足低收入者与弱势群体的需求，建成了六七层楼高、环境舒适宜居的住宅区，房价给予优惠与关照。2004年，厦门市荣获联合国"人居环境奖"，以特别表彰该市为弱势群体解决住房问题。从2007年起，中国各地都在增加经济适用房的建设量，以满足收入较低弱势群体的需求。目前，负责住宅的政府管理部门提出，要通过发展租赁住房，促进解决城市困难群体、农民进城落户和新就业大学生等的住房问题。

再举一个解决住房问题的优秀实例。1990年前完成的北京小后仓危房改建工程，该工程位于北京西直门内，占地$1.5hm^2$，共有住户298户，共1100人，人口密度近800人/hm^2，原有住宅都是简易平房，居民大都是低收入者和弱势群体。工程设计师是北京市建筑设计研究院的知名建筑师黄汇，她对这里的298户都做了仔细的调查，并同住户交谈，倾听他们的要求，最后做到原住户搬回原地居住，住户的居住标准都得到了提高，诸如面积扩大、朝向调整、功能完善、绿地增加等，从而改善了此区的生活环境，同时保持了北京的原有面貌，更满足了大众的居住要求。黄汇设计师告诉我，此区建成后一位老太太还抓住她的双手，反复激动地说"以前我很穷，我要感谢政府，让我住上了能晒太阳的房子，这都是党给我的阳光、党给我的温暖，请把这话跟政府说一下"。这说明，关心弱势群众并帮助解决住房问题，这是非常得人心的。此项改建工程刊载在《建筑学报》1991年第7期上。我曾对设计师黄汇女士说，你专为大众做设计，并获得大众的好评，你可以称作"人民的设计大师"。

香港特区解决大众住房的情况是：政府投资修建公屋和居屋两类，公屋是多户居民共用厕所、厨房的低标准模式，以低廉租金租给弱势群体；而居屋相当于大陆的经济适用房，但其标准比内地要低很多，政府以优惠的房价卖给低收入者。

在社会公正方面，国外城市一些好的做法是：老城区保留老居民，如法国巴黎老城区、比利时布鲁日水乡老城区、西班牙巴塞罗纳老城区、美国西雅图老城区，都保留并改善原有居住建筑，其大部分老居民还在，而不是拆旧建新，把老居民迁至郊区，而使得老城区变成有钱人的居住地。这些改造老城区住房、保留老居民的做法，值得大家肯定。这不仅是经济问题，而且还是社会问题，它保存了城市的历史文脉。只有社会公正，才有社会的稳定。具体实例如英国伦敦市区内的一个大众住区改建工程，设计师是英国著名建筑师哈克尼。哈克尼关照这些低收入的住户，充分利用原有建筑材料改善空间环境，他自己组织施工改建，费用也不多，但取得了很好的效果，颇受大众称赞。哈克尼后来还当选了英国皇家建筑师学会（RIBA）主席和国际建筑师协会（UIA）主席。还有埃及著名建筑师哈桑·法赛设计、建造的生土建筑居住区，它符合埃及当地生活习惯和外貌风格，空间完整、功能齐全、生活舒适、造价低廉，切实为穷人解决了住房问题，此设计获得了国际建筑师协会第十五届大会的金奖。

这里再介绍一些国外修建大众住宅的情况，此资料是黄汇建筑师提供的。1987年，"国际住房年"的口号是"给无家可归者以住宅"，这个口号在一些国家是不切实际的，但仍引起了很大的反响。给大众以住宅，这已成为全世界瞩目的焦点。由于各国的政治、历史、经济、自然条件不同，这样解决大众住宅问题的途径与手段就不同了。一是许多国家对住宅建设的关注有明显的侧重。政府侧重于低收入住户，建大众住宅，而私营房地产商则侧重于高收入住户，建高档住宅。大众住宅的投资来自政府，建成后以极优惠的租金向低收入公民出租（或出售）。若住户家庭总收入提高到受资助资格定额以上时，就停止优惠待遇，或迁出，或付高租金，或以高价购买原住房，法国等国家就是这样做的。二是向世界银行或国内银行贷款。政府搞开发建设，土地有时由政府提供，有时由私人提供。菲律宾首都郊外的奎松城有1/3的人口收入很低。1987年，由住宅及城市发展协会协助，并向世界银行贷款。后来在1.8hm²土地上建设了一个居住点；1990年又在卡瑞格兰村规划了40hm²的居住区，这均由世界银行贷款。泰国距曼谷市中心20km处的汤松洪小区建设了3000多套住宅，其投资的50%借自世界银行，43%为本国贷款，7%从政府的公共福利基金中提取，贷款由住户在20年内还清。三是住户向国家申请低息贷款，由地方发展组织负责开发。如斯里兰卡的艾瑞玛亚小区，在1985年靠贷款建成建筑主体和道路以及配套设施，设备部分由住户自筹自借，住户应在15年内还清贷款，市政设施及赤贫家庭无力偿还的由政府资助。四是大企业投资建职工住宅区，向低收入职工优惠出售或出租，如巴基斯坦的一个钢铁工人区建在靠钢铁厂的地方，可以减轻工人上下班路途疲劳，有利于劳动生产率提高。五是私人集资、合作建房。有些国家把居民组织起来，合作建房，部分资金来自银行贷款，居民选出住房委员会负责建设计划的实施，并组织住户参与设计。荷兰在组织住户参与设计方面有较完整的工作体系。各国大众住宅的面积标准和设施也有不小的差别，发达国家住宅的内厨房、厕所等设施较为完善，而发展中国家的住房还有许多住户合用一个公共厕所。

发达国家建设的大众住宅，虽然其面积设施标准高，但欧美发达国家城市中还存在不同种族的贫民住区。从2020年新冠病毒流行可以看出，许多贫民住区，因环境条件差、缺医少药，其感染率和死亡率也最高。所以，西班牙著名医疗专家曼努埃尔·弗朗哥于2020年呼吁：要关注社会不平等。由于世界各国贫困居民感染新冠病毒人数和死亡人数均高于富人许多倍，因此，全社会今后在住房等方面要更加关注弱势群体。

2. 在公共建筑服务设施方面，首先要考虑为大众服务

在公共建筑与文化生活方面，如合肥、厦门、珠海、上海等城市都很重视修建满足大众生活需求的各种公共设施。在新开发社区与旧有社区里，除了满足居民日常生活服务的衣食商店外，还有幼儿园、托儿所、中小学校，以及为居民看病的医疗卫生所、文化体育等服务设施。上海市规划局编制的《上海市菜市场和公共厕所规划布局纲要》中提出，到2020年全市中心城区将新增200多个菜市场、800多座公共厕所，菜市场以500m为服务半径，公厕是以300m为服务半径。这种方便大众生活的服务思想，值得大家学习、效仿。

在国外，社会公正也体现在公共服务设施的妥善安排上。如美国西雅图市中心区中部海边依然保存着为一般市民服务的海鲜市场，建筑为原有的低层房屋，但整洁有序；美国东海岸现代化城市波士顿，在其中心区市政府大楼的后面仍保留着为广大市民服务的老商场，受到居民们的欢迎，因为它不是借新建市政府大楼之机，拆去老商场而建新的商业大厦，以损害大众利益去获得自身的钱财。这些好的做法都值得我们学习和肯定。为方便大众生活，应多开辟一些城市室外公共活动的小型广场空间，如美国纽约洛克菲勒中心下沉式广场、西雅图中心区中部高层建筑间的休闲广场、芝加哥市民中心广场、英国布莱顿住区生活广场等。在这些小型广场中布置有咖啡馆、花木雕塑、冰场、跳舞场等，它已成为市民休闲、游乐的公共场所。

开放城市空间，就是要政府所属机构对公众开放，这是城市发展的新趋势，以便城市空间更贴近城市居民。20世纪90年代，广州市规划局的建筑布局与城市管理是开放的，时任副局长的林兆璋建筑师对这种发展趋势还比较了解。20世纪90年代的厦门市中心大会堂建筑设计方案，在评选时选中了上海市建筑设计研究院的设计方案，其重要一点就是它的开放性，这座建筑早已建成并投入使用，其城市建设展览经常对大众开放，很受广大市民欢迎。

开放城市空间，还有一点值得各地参考，就是各地政府办公楼对公众开放。如美国波士顿市政府办公楼，平时市民可随便进入这座办公楼，市民都知道此楼突出部分为市长办公室，十分开放，这种做法拉近了市政府与市民的距离。又如美国芝加哥伊利诺伊州政府大

楼，其布局体现出公众性，平易近人。伊利诺伊州政府大楼中间为公众活动的中庭，围绕中庭的低层部分是商店，供大众使用，政府办公楼层在上面，均敞向中庭，为开放式办公室，每日来此参观游览的游人与市民络绎不绝，给人的感觉是：这里的州政府工作人员贴近城市居民。1989年我们到达美国费城，看到的费城市中心市政厅塔楼同样是市民可以随便进出，足见其开放程度。

3. 社会公正的理念是指导建筑科学发展的重要理念之一

《求是》杂志1988年第9期发表了我国大科学家钱学森先生的文章——《建立意识的社会形态的科学体系》。钱学森先生在文章中指出："夏衍同志曾讲到'两个70年'，从马克思、恩格斯1847年写《共产党宣言》到1917年十月革命胜利是第一个70年，从1917年十月革命到1987年我们党的十三大提出社会主义初级阶段理论，这是第二个70年。我们再加一个70年，就是到2057年，看我们能否完成社会主义初级阶段的各项任务。"这说明为达到理想国的共产主义社会还需要很长时间。

最后，我要进一步讲老子的"道法自然"哲学观。"道法自然"是老子哲学思想的核心。老子认为，世界万事万物的进步发展都要符合自然发展的客观规律。

老子在2500年前称社会领导人为圣人，他在其所著的《道德经》中指出，圣人对待百姓要"无为"，其"无为"之意是不能有私心、私欲，而"有为"则是为百姓谋福利，促进社会发展。老子提倡的"道法自然"哲学观是一体两面的，这是由"无"与"有"两大哲学元素构成的辩证统一观。其"无"与"有"象征一切事物包含的对立关系，两者又统一形成万事万物。国外很重视老子《道德经》的哲学思想，早在几个世纪前就有译本。19世纪、20世纪初的德、英译本曾流行于欧美，20世纪末美国多家出版社出版了《道德经》英译本，影响很广。世界著名建筑师弗兰克·劳埃德·赖特的"有机建筑论"，就是受老子哲学思想启发后提出的。1988年，时任美国总统里根在其国情咨文中曾引用老子《道德经》中"治大国若烹小鲜"之句。"烹小鲜"就是根据小鱼本身的特点，要采取保持它鲜美的做法；而"治大国"就要认识所处时代、社会背景，采取符合社会和人民进步发展的政策。其中心意思是说：无论做小事，还是管大事，都要遵循事物发展的客观规律。2011年，时任联合国秘书长潘基文在演说中亦引用老子《道德经》中"天之道，利而不害；圣人之道，为而不争"之句。这是老子《道德经》的最后一章，其意思是：圣人要顺应历史发展潮流，引领社会不断进步，而不要以自己的主观意愿去引导他人，从而背离客观规律。潘基文借此意说明要认真贯彻《联合国宪章》，以求各国的一致发展。

　　这些事实说明，世间万事万物都要遵循自然规律。人类的终极目标是建立公平公正的法治社会，社会公正的理念也是指导建筑科学发展的一个重要理念。

　　这一部分所谈的关怀弱势群体、开放城市空间、保留老城居民、安排大众公共设施、方便大众生活等社会公正问题，全都关系社会的安全与稳定。我们期盼通过社会各界的共同努力达到社会生态的平衡。

结 语

  建筑科学这门大学科发展的指导思想是老子的自然宇宙观，即为人类建立生态文明生活环境的哲学观念，这一观念也是贯穿本书的一根红线。

  通过前面6部分的分析，我们对中外建筑科学文化发展的感悟是：在自然宇宙观为人类建立生态文明生活环境的哲学观念统领下，要有"创新发展"的理念，重视发展具有自己特色的产业经济，特别是数字技术经济的创新发展，不断促进建筑科学文化迈向更高层次；要有"自然环境"的理念，加强水、气、垃圾、交通的环境治理，建设城市绿地系统，保证绿地面积。充分利用自然光、风、热，做到节能环保、人与自然共生。适度控制城市规模和建筑空间容量，以利于防灾与卫生，并发展山水城市；要有"历史文化"的理念，要整体保护历史文化名城，将保护、使用与发展结合起来，使历史文化更富有活力。尊重历史建筑文化，使新与旧、现代与传统建筑和合发展，新建筑继旧创新，同时发展地域建筑文化；要有"艺术美观"的理念，创造城市与建筑的标志，形成优美的城市立体轮廓线，控制建筑与道路的尺度和体量，协调城市肌理，掌握平衡变化的艺术规律，增强城市建筑整体完美；要有"便民交通"的理念，改变汽车主宰城市交通的格局，构建人行和自行车道路交通系统，并与城市公共交通连接成网，提升城市人气和市民出行便利与安全健康；要有"社会公正"的理念，落实城市居住、公共设施、文化教育、出行交通的社会公正，关怀广大居民及其弱势群体，缩小贫富差距，开放政府管理机构，贴近广大城市居民。

  我们相信，这个核心哲学观念及其6个具体理念是建筑科学文化现代化发展的方向，它会为提高中外城市、建筑和园林三位一体的规划设计做出突出贡献，并逐步实现建筑科学文化走向现代化发展的宏伟目标。

Part One: Concept of the Development of Chinese and Foreign Architectural Science

Foreword

Architectural science and culture is an integrated subject, which includes the plan and design of cities, buildings and landscapes. Its philosophy ideology, on the one hand, is the traditional Chinese natural world outlook, which conforms to the objective development rules of all things and follows the laws of nature; on the other hand, is a modern concept of ecological civilization, that is, the ecological balance of nature, economy, society and culture. Generally speaking, architectural science and culture is a comprehensive science and art, which is for the establishment of ecological civilization and living environment of human beings. This is the core philosophy concept of architectural science and culture and the essence of Chinese architectural science and culture. Cities and buildings all over the world have their own characteristics, but they are developing towards this direction as well.

We need to deepen our understanding of the integration of cities, buildings and landscapes. This is because they are an interrelated organic whole, indispensable to each other. The relationship between cities and landscapes is mainly for solving the problems of urban ecological environment, including natural mountains, rivers, flowers and trees in the outskirt of the city, and green space systems in the city and among buildings. The relationship between buildings and landscapes can be described as buildings cannot be separated from flowers and trees, "buildings are a part of the landscape and the landscape is a part of the buildings". This is what we need to improve the quality of living environment. The relationship between cities and buildings is that buildings are the main contents of cities, and they are the entities that urban residents need for life and production.

As for the integration of cities, buildings and landscapes, some famous senior architects are aware. They have basic knowledge of these three aspects, but they failed to put forward the concept of integrated subject. In the 1950s, after the founding of New China, there are "the south China representative - Mr. Yang Tingbao and the north China representative - Mr. Liang Sicheng" - the two persons were recognized as the leading figures of the construction industry. Mr. Yang Tingbao, obtained a master's degree in architecture from the University of Pennsylvania in 1924, and later served as the vice president and head of the architecture department of Southeast University. In addition to his outstanding architectural knowledge and works, he also has in-depth researches on urban planning and construction. He has made some comments on the city planning and construction of Lianyungang, Suzhou, Zhenjiang, Nantong, and Xiamen in his reports and speeches. At the same time, he also attached great importance to landscape architecture, and put forward his opinions on the landscape construction in Wuxi, the scenic area planning in Wuyi Mountain, and the renovation of Guyi Garden in Shanghai. Mr. Liang Sicheng, received a master's degree in architecture from the University of Pennsylvania in 1927, and later served as the head of architecture department of Tsinghua University. He has profound attainments in architecture and architectural history. He also invited Soviet experts to teach the principles of urban planning in Tsinghua. The graduation project of our "Class Five" (from 1951 to early 1956) was divided into three subjects for three groups: city, public building, and industrial building. Later, over 10 graduates from the "city group" were assigned to work in China Academy of Urban Planning and Design. Mr. Liang and Mr. Chen Zhanxiang jointly put forward the city master plan for the overall protection of the old Beijing.

In 1958, he made a report on the history and development plan of Qingdao at the National Urban Planning and Construction Conference hosted by Minister Liu Xiufeng of the Ministry of Works, and then published a book on this topic. In the early 1950s, he also arranged for landscape architecture students from Agricultural University to study and exchange in the architecture department of Tsinghua.

There are many examples in this respect. From 1929 to 1933, Mr. Cheng Shifu studied landscape architecture and urban planning at Harvard University and Cornell University. In the 1950s, he was responsible for the overall planning of the landscape in Shanghai and the reconstruction planning and design of Renmin Road block in the old city of Suzhou, and then participated in the overall planning of several provincial capitals. Professor Wu Liangyong from Tsinghua University is the academician of both the Chinese Academy of Sciences and the Chinese Academy of Engineering. He is a master of the planning and design of cities, buildings and landscapes. He has put forward great planning and design for cities such as Qufu, Xiamen, Shenzhen and Sanya. And from the city perspective, he changed the location of the original Confucius Institute in Qufu, and made it a garden-like building. Considering the original features of the old Beijing, he designed and rebuilt the quadrangle dwellings of Ju'er Hutong. Academician Mo Bozhi, the chief architect of Guangzhou Municipal Bureau for Urban Planning, is not only proficient in architecture and landscape, but also understands urban planning. From the 1950s to the beginning of this century, he has successively built 52 engineering projects, such as Guangzhou Beiyuan Restaurant and Guangzhou White Swan Hotel, which are all in line with the overall layout of the city and combine the design of buildings and landscapes together. Among the projects, the number of awards ranks first in the country. At the same time, he also put forward suggestions on the planning of Zhuhai, Hefei, Dunhuang, Hangzhou and other cities, which were published in the *Architectural Journal*. Among the well-known predecessors, there are others whose architectural works also meet the requirements of urban planning and landscape, such as Mr. She Junnan in Guangzhou, Mr. Feng Jizhong in Shanghai, and Mr. Xu Shangzhi in Chengdu. The famous architects of the later generation include Mr. Qi Kang, Ms. Zhang Jinqiu, and Academician He Jingtang.

In foreign countries, the "Radiant City" (unrealized) designed by the world famous architect Le Corbusier in 1931 and the Indian city "Chandigarh" planned and built in 1951 are examples of the "integration" of cities, buildings and landscapes. In recent period, Norman Foster, a world-famous architect, and his partners planned and designed Masdar, a new Eco-city, with a scale of 90,000 people in Abu Dhabi. They also transformed Duisburg, a city with the population of 500,000 in Germany. Both of the two cities have created a community concept of sustainable development. They designed and built many other buildings as well, such as Commerzbank Tower in Germany, Swiss Tower in London, and Terminal 3 of Beijing Capital International Airport. They added colors to cities and integrated landscape with indoor and outdoor space, which reflected their emphasis on the organic combination of cities, buildings and landscapes, and their target to create a beautiful ecological environment. The works of SWA Group from the United States, such as Tuen Mun New Town in Hong Kong, the Filinvest City Center in the Philippines and Hyatt Hotel in Arizona, all reflect the characteristics of cities, buildings and landscapes as a whole.

Looking back at history, it has been more than 2,000 years since we have practiced the integration of cities, buildings and landscapes. The urban planning and construction of Chang'an (now Xi'an) in Han Dynasty and Tang Dynasty, Luoyang in Tang Dynasty, Kaifeng in Northern Song Dynasty, Hangzhou in Southern Song Dynasty, Beijing in Yuan Dynasty, Ming Dynasty and Qing Dynasty, and the ancient capital of Nanjing, all reflect the ideological concept of integration. But no exact scientific name was put forward for this practice.

In June 1996, Mr. Qian Xuesen put forward the new concept of "architectural science", which is the "integration of cities, buildings and landscapes." Mr. Qian included it as one of 11 major disciplines, which are: natural science, social science, mathematical science, systematic science, military science, human science, thinking science, behavioral science, geographical science, architectural science and literary theory. This new concept is an innovative development of scientific theory, which conforms to the development and changes of history, the law of nature, and especially the needs of comprehensive development of cities, buildings and landscapes. It improves people's understanding of cities, buildings and landscapes, and plays a guiding role in the development of urban construction around the world. We believe that in the new stage of becoming a society of knowledge, the subject synthesis will also develop to the integrated science, so this theory will gradually be recognized by the world.

The highest philosophical guiding ideology of architectural science and culture put forward by Mr. Qian Xuesen is that the planning, design, construction and development of cities, buildings and landscapes included in the major discipline of "Architectural Science", should take the natural world review as the guiding ideology. This idea comes from Lao Zi in China. As early as 2500 years ago during the Spring and Autumn Period and the Warring States Period, Lao Zi put forward that "Tao is great; the heaven is great; the earth is great; the man is great. There are the great four in the universe, and the man is one of them. Man models himself after the earth; the earth models itself after heaven; the heaven models itself after Tao; Tao models itself after nature." Its core idea is "Tao models itself after nature", which holds that the evolution of all things in the universe (including heaven, earth and man) must obey the objective development rules, namely "Tao", and follow the law of nature. Otherwise, it will hinder development or be eliminated. This is the basic viewpoint in Lao Zi's *Tao Te Ching*. Later, Confucianism, Taoism and Buddhism also raised the idea of respecting nature (but the specific ideas are different). Therefore, Lao Zi's natural world outlook of "Tao" kept developing. It has always been the guiding ideology for the planning, design, construction and development of the integration of cities, buildings and landscapes in China.

Historically, this natural world view when reflected in cities, buildings and landscapes in China, is the science of geomancy. Those who master this science are called geomancers, and they understand the Tao of heaven and earth. That is they use astronomy, geography, natural science and their objective development rules to plan, design and construct cities, buildings and landscapes. As for our natural world view that cities, buildings and landscapes are integrated with nature, the British scholar Joseph Needham wrote in his Encyclopedia of the World about Chinese architectural spirit, "There is no other place that is as enthusiastic as Chinese to embody their great thought, that is, 'man cannot leave nature'. This man is not a person who can be separated from the society, but also include the major buildings such as palaces and temples. Houses in cities whether concentrated or scattered in grange, often embody a feeling of 'cosmic pattern' and have the symbolism as direction, season, wind direction and stars. This is the symbolism of the combination with nature and the feeling of Chinese people on the 'cosmic pattern'." This thought of respecting nature has also influenced Frank Lioyd Wright, who is a world-famous American architect in the 20th century. He has read Lao Zi's *Tao Te Ching* (translated from German to English) and deeply understood Lao Zi's natural world view of Tao, which he believes is not only the facade of things but also includes the internal spirit. His architectural features are the organic combination of architecture and nature. He put forward the "organic architecture theory". "Organic architecture" refers to natural buildings, which grow from the ground to the sunshine. Organic is the inherent essence of things, "organic architecture" comes from the natural essence of things, and the essence of architecture is nature. Wright has a deeper understanding of Lao Zi's natural world outlook, and believes that it is not only the representation of all things in nature, but also the inner spirit. There is also Ricardo E.Bofill, a world-famous Spanish architect, who said that he had a design career of 50 years, and only when he designed the Barcelona Port Hotel (completed in 2010) did he realize that "it shall change his attitude towards architecture, change his attitude towards nature, ecology and environment. Architecture should keep respecting nature".

About the guiding of natural world outlook on landscape design and construction, Mr. Sun Xiaoxiang, a well-known predecessor of Chinese landscape design, made a clear statement in the preface of my book, *Introduction to World Landscape Development*. He also made a comparison between the eastern and western landscape aesthetic concepts before the 18th century. He said: "the aesthetic theme of European geometric landscape is that 'man' is the master of the universe, and 'man' is the purpose of the universe. Nature must be transformed according to the order, rules, method and geometric patterns in the human mind. While the aesthetic theme of Chinese natural landscape is that the nature is the master and the purpose of the universe. Man is just one of the sentient beings in nature." He also pointed out that "China is the first country in the world that designs the landscape based on nature. Moreover, China's deep loves for nature, understanding of nature and sensitivity to natural beauty have all penetrated into all cultural fields such as philosophy, art, literature and painting." This content is an academic report made by Mr. Sun Xiaoxiang at the "International Conference on Landscape Planning Education" held by Harvard University in 1986, which was praised by Professor Steinitz, the host of the conference, and was selected as the most outstanding international education model in the discipline of landscape planning.

Mr. Qian Xuesen praised Chinese gardens many times for reflecting the natural world outlook. He said that "Chinese gardens are known as the 'mother of gardens', famous ones can be seen all over the country and are well-known to the world; "Landscape architecture in China is a pearl of national culture heritage"; "Chinese landscape art is a national treasure and has a glorious history of thousands of years". He also proposed many times that natural Chinese landscape art should be integrated into architecture and urban construction to create a natural Chinese city of mountains and rivers.

Lao Zi's natural world review points out the relationship between the development of human society and the objective development rules of all things, and mankind shall be in harmonious coexistence with all things. This is an important concept to solve the problem of "ecological environment crisis" in today's world. We should inherit and develop this concept, make it modern, and explore the ways of the organic integration of human beings, nature and society under new historical conditions. The concept expands to the ecological balance of nature, economy, society and culture will be explained in the following six concepts.

The establishment of ecological civilization and living environment is both a scientific and artistic issue. The culture mentioned in this book is "integrated culture". From this concept, it includes science, technology, culture and art. A cultural circle or the culture of a certain region is the special life style or life principle of the region, which includes all the artificial products, knowledge, beliefs, values and norms there, and comprehensively reflects the characteristics of society, economy, science and technology, concepts, customs and natural ecology. Therefore, architectural science belongs to the category of integrated culture, which contains culture, art, science and technology. What is science and technology? It is the exploration, research and understanding of nature and things with logical thinking and language, which is the matter of scientists, engineers and technical experts. What is art? It is the depiction, expression and promotion of nature and things by thinking and feelings in images, which is a matter for writers and artists. Architectural scientists, a combination of planners, architects and landscape architects, are professionals who give consideration to the above two kinds. This is determined by the attributes of architectural science. Therefore, we analyze and study Chinese and foreign architectural science and culture from the viewpoint of integrated culture, and draw the conclusion that architectural science is the science and art of establishing ecological civilization and living environment for human beings. It is good for promoting the development of Chinese and foreign architectural science and culture.

Through the development in the past 40 years after the reform and opening up, China's urban, architectural and landscape undertakings have made great progress and accumulated some good experiences. However, there are still some defects: the city scale is too large, the floor-area ratio is too high, and the one-sided pursuit of form is not in line with reality and environment; water, air, garbage and automobile pollution; demolition of historical buildings and destruction of the natural environment, lack of overall protection planning; pay no attention to the development of urban green space, waste resources, and unilaterally pursue immediate interests; cities have no characteristics and just disordered texture; some big cities blindly develop private cars, roads become parking lots, traffic is chaotic, and residents' lives and health are affected. Some of these defects also exist in foreign architectural science and culture. In view of these defects, combined with my investigation in China since 1956 and in many foreign cities since 1979, I have selected excellent examples at home and abroad with reference value. At the same time, I absorbed the reasonable viewpoints of foreign theories about "modernism", "post-modernism", "new urbanism" and "critical rigidity", and put forward six concepts under the guidance of the core philosophical idea of establishing ecological civilization and living environment for human beings, which are "innovative development", "natural environment", "history and culture", "artistic beauty", "convenient transportation" and "just society". These concepts can be regarded as the correct track and direction of the comprehensive development of cities, buildings and landscapes in the past, at present and in the future. If the development of cities both home and abroad deviates from the direction, then we will definitely go to a detour. If it develops along this direction, we will be able to embark on the road of ecological balance, scientific progress and sustainable development. Some excellent cities all over the world have gone through tortuous development, but finally they fully or basically embody these ideas. The following will explain these six concepts in detail, mainly citing urban examples, rarely rural contents.

Chapter 1 Innovative Development

Develop industrial economy, promote urban construction, attach importance to innovative development, and move to a higher level.

From the history of the dozens of cities at home and abroad, it is the industrial economy that drives the development of cities, buildings and landscapes. Therefore, developing industrial economy is the last word and economy is the foundation. Industrial economy includes primary, secondary and tertiary industries. During the feudal society, the leading industry was the primary industry - agricultural economy. In the period of capitalist society, the secondary industry - industrial economy was the main one. At present, it is changing to the society of knowledge and information. In all developed areas, the tertiary industry accounts for the largest proportion. This is the general trend of industrial economy development. Every city should actively develop its own industrial economy. For the content and proportion of the primary, secondary and tertiary industries in each stage, we should consider it from a wide range in order to achieve the ecological balance of economic development in ecological civilization.

1. Innovative development in China

At the beginning of this century, Chinese scholars put forward the theory of "second modernization". Mr. He Chuanqi, a researcher at the Chinese Academy of Sciences, believes that the first modernization is in the industrial age, and the priority is economic development; the second modernization is in the era of knowledge development, and the priority is the life quality. Knowledge and information expand spiritual life, and cultural life will be highly diversified. From industrialization-oriented to knowledge-oriented, knowledge economy has gradually surpassed material economy and the society of knowledge was initially formed. They also pointed out that different regions have different characteristics. China can be divided into three parts: first part - six developed regions of Beijing, Tianjin, Shanghai, Hong Kong, Macao and Taiwan; second part - 15 elementary developed regions such as Liaoning; third part - 13 underdeveloped regions such as Gansu. Developed areas should develop knowledge economy and build knowledge society through accelerated "industrial transfer" and "urban growth", and gradually realize the second modernization; elementary developed areas should coordinate the development of two modernizations; underdeveloped areas should continue to promote the first modernization and pay attention to the second modernization. By the middle of this century, China can basically realize the second modernization and catch up with the moderately developed countries in the world.

According to the time sequence after the founding of New China in 1949, I will explain the general situation of industrial economy development in some representative cities. The "First Five-Year Plan" started in the early 1950s, and combined with 156 key industrial projects aided by the Soviet Union. Beijing, Shanghai, Nanjing, Hangzhou, Suzhou, Qingdao, Shenyang, Kunming, Chengdu, Lanzhou and other provincial capitals mainly developed the secondary industry and part of the tertiary industry, which promoted the local cultural construction. After the 1970s, the international oil and dollar crisis appeared, and the world economy was in recession. Some developed countries invested in Hong Kong and Taiwan, which promoted the development of Hong Kong's finance, commerce and other industries, and also promoted the development of high-tech industries in Taipei and Hsinchu of Taiwan. Hong Kong and Taiwan of China, Republic of Korea and Singapore are called the Four Asian Dragons. After 1980s, four special economic zones were established in Xiamen, Zhuhai, Shantou and Shenzhen along the coast of China, which rapidly developed high-tech industry, electronic industry and processing industry, and promoted the fast development of urban construction and cultural construction. After 1990s, Shanghai developed Pudong New Area, which accelerated the development of finance and commerce. Beijing established Yizhuang and other new economic and technological development zones, and developed the CBD of the East Third Ring Road and the financial zone of Xicheng District. As Beijing prepared for the 2008 Olympic Games and Shanghai prepared for the 2010 World Expo, the industrial economy of these two cities developed rapidly, and the construction of architectural science and culture also got further progress.

The above is an overview of China's cities and their industrial development. Here are the details in three aspects.

(1) Industrial economy development of Deyang in Sichuan Province, which I participated in its planning and design during the "First Five-Year Plan" period, and Xi'an, which is now an elementary developed area.

The planning and design of Deyang began in March 1956, when I just graduated from the Construction Department of Tsinghua University and was assigned to work in the China Academy of Urban Planning and Design. I had the honor to participate in the overall planning and design of the city with my classmate Mr. Qi Ligen on behalf of the national urban construction part. At the beginning, we jointly surveyed and selected the sites of heavy machinery and electronic industry with the First Ministry of Machinery Industry, the Fourth Ministry of Machinery Industry and the Ministry of Geology of China. The experts from Soviet Union guided the overall planning of the city. At that time, Deyang had a population of only 10,000, which lived in the strip-type eastern district composed of north-south railways and rivers and the western district mainly in the open plain. The dominant wind direction was the north wind. Because the electronic industry has little pollution, it is arranged in the windward section of the northeast railway. The two electronic industry projects were mentioned by code names, and the specific contents were kept confidential. The heavy machinery factory, which covers a large area, generates pollution and requires soil bearing capacity, so it is in the downwind section of the southwest railway. The west district is the main part and the east district is the auxiliary part, which constitutes a unified living and residential area in Deyang. The heavy machinery factory is still producing the largest heavy machines in china so far. Deyang is one of the four major industrial cities in Sichuan. Like Chengdu, Mianyang and Chuanxi, it mainly develops heavy machinery, electronic industry and special steel. After more than 60 years of the development of secondary and tertiary industries and urban construction, it has become an elementary developed area.

Xi'an in Shaanxi Province, also belongs to the elementary developed area, which is one of the eight major cities planned and constructed in China's "First Five-Year Plan" period, and has now become one of the emerging industrial bases in China. Since the reform and opening up, Xi'an has adjusted its secondary industry development and expanded its electronic, high-tech and aerospace science and technology industries.

At the same time, combined with its own characteristics as an important historical and cultural city, Xi'an has vigorously developed its tertiary industries such as culture economy. It formulated the cultural revival plan of the old city of the "Imperial City of Tang Dynasty" and the regional plans with local cultural characteristics, including international tourism in Lintong, jade culture in Lantian, Chang'an ecological residence, farmer's painting in Hu County, Lao Zi Culture in Zhouyu and modern agriculture in Gaoling. Xi'an's approach is correct. It attaches importance to the development of urban industrial economy and develops the secondary industry and the tertiary industry like culture and service in a coordinated manner. It promotes the protection and development of Xi'an, the city of history and culture, through industrial economy progress.

Xi'an and Deyang should further coordinate the development of two modernizations to reach the level of developed areas as soon as possible.

(2) The industrial economy development of four special economic zones: Shenzhen, Zhuhai, Shantou and Xiamen.

In 1980, the state approved these four special economic zones, which are located in the southeast coastal area: Shenzhen, Zhuhai, Shantou and Xiamen, in order to accelerate the development of China's national economy. In 1960s and 1970s, it was not suitable for development because it was close to the coastal area of Taiwan. In the 1980s of reform and opening up, the international situation tended to be favorable for peaceful construction, introduction of foreign capital and foreign economic cooperation, and then China began to do the pilot by giving new policies to special economic zones. Practice has proved that this approach is correct and successful. At that time, I worked in the Architectural Society of China, and had many contacts with these four cities, deeply feeling their amazing achievements in construction and development.

The name of Shenzhen began in the eighth year of Emperor Yongle in Ming Dynasty. It had a small population and was a small town. The city was established in 1979 and was designated as the first special economic zone in 1980. Most of the residents moved from other places. But its development speed is unprecedented, creating a new record of urbanization and industrial modernization, and now it has become the fourth biggest city in China. Shenzhen's industries are mainly export processing, including machinery, electronics, textiles, light industry, building materials, medicine, clothing, food, etc. In recent years, it has vigorously developed the tertiary industries such as information, bioengineering, finance, commerce, tourism, service, etc., and also provided agricultural and sideline products for Hong Kong.

Shantou is an important port city in China that opens to the outside world. Its commerce and trade economy are developed. Since the establishment of the special economic zone, it has developed high-tech industrial development zone, Zhuchi industrial zone, Taiwanese investment zone and marine industrial zone. It is near Chaozhou in the north, which is a city with a long history and culture and the birthplace of Chaoshan culture. It has Chaoshan music, song and dance, and cuisine. This profound traditional cultural industry has driven the development of tourism industry in Shantou.

Zhuhai, originally a small town focusing on developing primary industry, agriculture and fishery, was designated as one of the first special economic zones in 1980. There are 164 islands under its jurisdiction, a wide range, and it has developed aquaculture and convenient foreign exchanges. Now its aquaculture and export processing industries are developing rapidly. The International Aviation and Aerospace Exhibition is held every two years, which has deepened its influence on foreign countries. In June 1984, the Architectural

Society of China was invited to organize famous experts from all over the country to study the overall planning of Zhuhai special economic zone and put forward suggestions. The leaders of the city attached importance to and adopted the opinions of the experts, and now Zhuhai has developed into a city with advanced economy, seaside garden style, suitable for living and traveling and vacation.

Xiamen is also one of the first special economic zones. In the 18th century, it was already one of the "five trade ports" in the southeast coast. It used to be an island. In 1955, the seawall of Yingtan - Xiamen railway was built, and then Haicang Bridge was built, connecting Xiamen Island with the mainland. In 1958, the China Academy of Urban Planning and Design was invited to organize a working group with members of complete types of work and came to Xiamen to propose a planning scheme for its urban development. At that time, due to the tense relationship with Jinmen Island at opposite, the city was not suitable for large-scale development. It was suggested that only small-scale construction and infrastructure improvement should be carried out, and attention should be paid to protecting the famous Gulangyu Island. Meanwhile, Xinglin and other areas opposite Xiamen were also been investigated, and there are deep-water ports that can carry ships above 10,000 tons. Therefore, it was suggested that Xiamen can develop into an important port city for foreign economy and foreign trade in the future. The above suggestions have been gradually realized in the process of economic development for decades. Not only the light industries like export processing, high-tech, information, commerce and tourism industries, have been vigorously developed, but also an open port city has been built. With the rapid development of economy, the city construction has been promoted, and the beautiful Yundang downtown area and the landscaped new public residential area have been built, which won the United Nations Habitat Award. Gulangyu Island, known as "Sea Garden" and "Music Island", has also been protected and developed, and it was listed as "World Cultural Heritage" in July 2017.

These four special economic zones all attach importance to "innovative development" and the protection of the "natural environment", so they are all listed as "national environmental protection models", "national garden cities", "national health cities" and "excellent tourist cities in China". The GDP of the four special economic zones is increasing very fast, and Shenzhen has exceeded the level of developed urban areas such as Tianjin proposed by researcher Mr. He Chuanqi. It can be seen from this that Shenzhen, Shantou, Zhuhai and Xiamen have developed to or been close to the standards of developed regions, and should further develop knowledge economy and promote the second modernization.

(3) The industrial economy development of six developed areas: Beijing, Tianjin, Shanghai, Hong Kong, Macao and Taiwan.

Beijing, the capital of the People's Republic of China, is the center of national politics, culture and international exchanges, as well as a famous ancient capital and the center of China's financial decision-making and management. Beijing's economy is diversified, with the coordinated development of the secondary and tertiary industries. From traditional industries such as iron and steel, machinery, electronics, chemical industry, light industry, textile and printing, it has gradually shifted to modern high-tech industries such as microelectronics, new biomedicine, automobiles, petrochemical materials, and integration of optics, mechanics and electronics. The latter part has become a new pillar. The industrial areas in the west and east of the original urban area have been transferred outward. In recent years, electronic information, IT network, tourism culture, services and other industries have developed rapidly, making the tertiary industry account for the largest proportion. The city's GDP is more than 3 trillion RMB.

Culturally, Beijing's traditional culture is rich and varied. Many of them have been listed as World Cultural Heritage (Forbidden City, Great Wall, Summer Palace, Temple of Heaven, Tonghui River Section of Grand Canal, Peking Man Site in Zhoukoudian, etc.), and Beijing is the city with the most cultural heritages in the world. Its folk culture includes Hutong, quadrangle dwellings, Peking Opera, local cuisine, arts and crafts (cloisonne, jade, carving), etc. These traditional cultures are precious wealth, which will definitely promote the development of tourism industry and the related aspects like transportation, hotel, restaurant, and featured commodity.

Tianjin is a central municipality and an important industrial, commercial and port city in China. After the reform and opening up, it has developed rapidly and has become one of the important industrial bases in China, including textile, automobile, electronics, marine, petrochemical, sea salt and other national key industries. Tianjin Binhai Economic New Area has been established in the east part near the sea, including high-tech development industrial zone, advanced manufacturing industrial zone, coastal chemical industry zone, central business industrial zone, seaport logistics zone, seaport leisure tourism zone and Tianjin port free trade zone. Among them, Tianjin port free trade zone plays an important role in the development of international trade, modern logistics and export processing industry. Tianjin port is the largest international port in north China, focusing on import and export trade. Tianjin is also a famous national city of history and culture, with many places of interest. Its traditional folk culture includes the famous Tianjin crosstalk, Yangliuqing New Year pictures, Clay Figurine Zhang and catering culture. These contents promote the development of tourism in Tianjin, and its regional GDP is above 2 trillion RMB.

Shanghai is one of the largest comprehensive industrial bases in China, with many industrial categories and large scale. It has metallurgical enterprises with an annual output of more than 10 million tons and other heavy industries such as shipbuilding, machinery, automobile and petrochemical industry. There are also high-tech industries such as electronics, aerospace, computers, communications, medicine and bioengineering. There are consumer goods manufacturing industries as well such as cotton textile, woolen cloth, silk, plastics, household appliances, clothing, leather shoes, cosmetics, clocks and toys. Pudong District was developed in 1990s with the establishment of Pudong foreign economic development zone and Lujiazui Shanghai financial center. Shanghai pilot free trade zone was established in August 2013, which greatly promoted the development of Shanghai's economy and industry. At the same time, the transportation industry of the port, aviation, railway and highway in Shanghai has also developed rapidly. Shanghai port has become the third largest port in the world, with more than 60 berths of 10,000 tons. Shanghai airport has been listed as one of the three major international airports, and railways in the city connect all parts of the country. The tertiary industry in Shanghai, such as transportation, finance, information network, insurance and real estate, accounts for the largest proportion. Its GDP is moving forward to 4 trillion RMB. The Yangtze River Delta City Cluster centered on Shanghai has become one of the six recognized world-class city clusters (the other five are the Green Lakes of North America, New York, London, Paris and Tokyo), with an annual economic aggregate of more than 15 trillion RMB, accounting for 15% of China's economy. This kind of city cluster is an inevitable trend of urban development, which shall be paid attention to.

On July 1, 1997, China resumed the exercise of sovereignty over Hong Kong and established the Hong Kong Special Administrative Region. Hong Kong's economy mainly focuses on export processing, international trade, logistics, finance, tourism, real estate and service industries. Export processing industry includes textiles, clothing, electrical appliances, electronics, clocks, toys and so on. Its transportation industry is very developed. Because it is a free trade port, it has developed into the second largest airport in Asia and the main aviation center in Asia and beyond. The sea port has also become the second largest container port in the world. There are 20 maritime routes leading to more than 100 countries in the world, and railways of Beijing - Kowloon and Guangzhou - Kowloon connecting all parts of the country. The Hong Kong-Zhuhai-Macao Bridge has been completed and opened to traffic, bringing smoother land traffic.

On December 20, 1999, China resumed the exercise of sovereignty over Macao and established the Macao Special Administrative Region. Macao is a free port. In the past, its economy was dominated by the world-famous gambling industry. Since 1960s, its economic contents began to change, and now it has formed a diversified economic structure system such as tourism gambling, foreign processing, foreign trade, finance and real estate construction. Its foreign processing industry is mainly small and medium-sized enterprise, including knitting, clothing, electronics, toys, etc. These products are mainly sold to Europe and America.

Japan occupied Taiwan in 1895 and returned it to China after the victory of the Anti-Japanese War in 1945. Its provincial capital is Taipei. Taiwan's industry is largely export processing. Before 1970s, it mainly developed light industry, like textile and food processing. After 1970s, it focused on steel, machinery, petrochemical industry, airplanes, ships, automobiles, electronics, electrical appliances, computers, sporting goods, clothing, shoes, toys, handicrafts and so on. For these industrial products, it paid attention to the continuous adopting of high and new technologies and upgrading, that their export products meet the international standards of various periods. Taiwan's transportation industry is developed. There is an electrified railway network from Keelung to Kaohsiung. In terms of highways, there are expressways and roundabout road networks. It has a natural port network centered on Keelung port in the north, Kaohsiung port and Hualien port in the south. In terms of civil aviation, there are Taoyuan airport in the north, Kaohsiung airport in the south and more than 10 other airports, which connected to Hong Kong, Southeast Asia, Japan and the United States.

All these cities attach importance to the development of inncvative industrial economy according to the characteristics of their own conditions, thus driving the development of architectural science and culture.

2. Innovative development abroad

Let's take a look at the development of industrial economy in some foreign cities I have visited. For example, in Seattle of the United States, Boeing Aircraft Corporation and its four subsidiaries were established in 1916, and its 80,000 employees accounted for half of the total number of employees in the city. It became a free port in 1949 and the largest port city in the northwestern United States. With the improvement of urban environment, the shipbuilding and medical equipment manufacturing industry have developed, and now become an important aircraft and shipbuilding center in the United States. The economic development of these industries has driven the construction of architectural science and culture in this city. Boston, the eastern city of the United States, is the largest port city in New England with a long cultural history. In the last century, the machinery industry, electronic industry and financial industry were developed, and the National Aviation and Aerospace Research Center was established. The rapid economic development of these industries has led to the cultural construction of this old cultural city.

Another example is Europe. I have visited France many times, which is a developed capitalist industrial country with the fifth largest GDP in the world. Its industries mainly include: nuclear energy, which ranks third in the world in terms of nuclear power generation; automobiles, which ranks third in the world regarding export; machinery, shipbuilding, chemistry, petrochemical industry, textile industry, as well as rapidly developing aerospace and marine exploration. Its capital, Paris, is the manufacturing center of France, and one fifth of the country's industrial production is in the Greater Paris area. Paris has a developed transportation, with railways and highways leading to all parts of Europe and aviation leading to all continents, which is extremely convenient. In May 2001, I took a day trip by high-speed railway to and from Paris in northern France to Marseille on the Mediterranean coast in the southernmost part of France. Paris has developed tourism, many historical and cultural places of interest, and is the center of modern art, and has become a famous tourist city in the world.

Switzerland was the first country I visited abroad in 1979. It is also a developed capitalist industrial country. Apart from metallurgy, machinery and chemical industry, its industries such as clocks and watches, precision instruments, textiles and food are outstanding, and it is known as the "kingdom of clocks and watches". Bern, the capital, vigorously develops the watch industry and is famous for its watches. Basel also focuses on the development of watch and medicine industry. After the impact of the advent of electronic watches, Swiss watch industry is now developing mechanical automatic gold watches with more advanced technology. This high-grade product enjoys a good reputation and sells well all over the world. Bern, Zurich and other cities with international organizations, such as Geneva and Lausanne, are very prosperous in tourism.

I visited Italy in 1995, which is also a developed capitalist industrial country. Foreign trade is its main pillar, and exports are largely automobile, steel, chemical industry, machinery and equipment, textile and others. The country's tourism industry is very developed, with a long history and culture and numerous places of interest. Rome, the capital preserves many historical sites in its entirety; Florence, the birthplace of the Renaissance; Venice, the water city, are all famous tourist cities all over the world. Tourism accounts for a large proportion of its economy and industry. Among them, the handicrafts in Florence mainly focus on artworks, including leather, jewelry, textiles, pottery, and silverware.

In 1996, I visited Belgium, which is also a developed capitalist industrial country. Its industries include iron and steel, machinery, chemical industry and non-ferrous metal industry. Brussels, the capital, is the site of NATO Secretariat and European Union headquarters, and is also known as the "European Capital". It is a famous tourist city in Europe. Antwerp, the second largest port city in Europe, is the processing and trade center of diamond. It has 250 diamond processing factories and 7 diamond schools, and its processing capacity accounts for half of the world total.

In 1985, I visited Egypt in Africa, which is one of the ancient civilizations in the world and one of the countries with relatively developed industries in Africa. Its traditional industries are mainly light industries such as textiles and food processing. In recent years, petroleum, steel, machinery, chemical industry and electric power have developed rapidly, especially the petroleum industry, which helped Egypt become the fourth largest oil producer in Africa. Its capital, Cairo, where one third of the country's industries are located, preserves Giza Pyramid Group. It is one of the world seven wonders and promotes the development of tourism in the city. Located in the northern part of Egypt, near the Mediterranean Sea, Alexandria is the largest port city in Egypt, where 90% of the country's import and export goods are handled. It is also a world-famous cotton market and summer resort.

Finally, Japan, the neighbor in Asia I visited in 1992, is the third largest economy in the world, with highly developed industries, including iron and steel, machinery, automobiles, electrical machinery, electronics, petrochemical industry, shipbuilding, construction, nuclear energy, etc. Among them, nuclear energy is more prominent, and over 50 nuclear power stations have been built. The shipbuilding industry is also in the forefront of the world. In addition, light industrial products such as paper making, printing, cameras, food and ceramics also occupy a certain market position in the world. Many raw materials are imported, and the import and export trade is developed. Tokyo, Nara, Kyoto and other cities have advanced tourism. Kyoto was the capital of Japan from

794 to 1869, and the religious and cultural center as well. It was designated as an international cultural sightseeing city as early as 1950. The imperial cities in the center were built in imitation of Chang'an and Luoyang in Tang Dynasty of China. Kyoto has the largest number of historical relics and historic sites, and 21 buildings, gardens, paintings and sculptures are listed as national cultural relics. In the example part of this book, it introduces the Heian Shrine, Kinkakuji Temple, Ginkakuji Temple and Stone Yard of Ryoanji Temple. The city's industrial economy is mainly based on cultural tourism. In 1974, Kyoto became a sister city with Xi'an, China.

The preceding part introduces the industrial and economic development of Chinese and foreign cities with their own characteristics, and the development promoted the construction and progress of cities.

Regarding the industrial economic innovation and development in Chinese and foreign cities: in 1980s, the United States took the lead in shifting from traditional industries dominated by steel, automobile, textile and construction to the development and utilization of high-tech industries such as information, integrated circuits, software, emerging materials and bioengineering. In 1979, the producers of the tertiary industry such as information and services accounted for more than 70% of the total. Japan and European countries with good modern science and technology foundation, such as Britain, France and Germany, are also catching up. China and developing countries in Asia, Africa and Latin America have begun to attach importance to this new trend of high-tech development and strive for innovative development as well. During this period, the developed cities both at home and abroad are accelerating the development of knowledge economy, widely adopting high technologies such as the Internet, big data, digitalization and artificial intelligence, and strengthening the construction of infrastructure software and hardware. Many cities have also developed comprehensive science parks and tried to develop digitalization. All have achieved good results in promoting development and the overall innovation of production and lifestyle. so as to gradually complete the second modernization and realize the society of knowledge.

It can also be seen that the industrial economics of the above-mentioned Chinese and foreign cities are constantly innovating, and innovation is productivity. Only by attaching importance to and implementing innovative development can they move to a higher level. Innovation is the driving force for the development of human society. If we do not attach importance to innovation and adopt high technology, we will lag behind the times. Therefore, we must do innovative development.

Chapter 2 Natural Environment

To achieve environmental governance enhancement, coexistence between man and nature, moderate space capacity, livability, disaster prevention and hygiene.

1. The natural environment is a place that everyone loves and yearns for, but many cities around the world have destroyed the natural environment, so it is necessary to strengthen environmental governance and improve the living environment

The development of cities, buildings and landscapes shall gradually reduce and eliminate the pollution to water and air, and treat the garbage harmlessly. For more than a century, all over the world have been doing this work, because factories and cities are emitting harmful gases, sewage and garbage at any time, and motor vehicles are also emitting tail gas. The environmental treatment of these four aspects (sewage, harmful gases, garbage and automobile tail gas) requires a lot of funds, new technologies and a long time. All advanced and livable cities pay attention to solving this environmental problem.

At present, many cities in China are gradually strengthening environmental governance. For example, Xi'an not only retains the Ming Dynasty city wall, but also retains the moat water system around the city, forming a green belt around the city, which greatly improves the environment in this area. In the master plan of Xi'an in 2020, they plan to restore the "Chang'an Eight Waters" in history, that is, Ba River and Chan River in the east; Wei River and the Jing River in the north; Feng River and Lao River in the west; Yu River and Hao River in the south. At the same time, they will invest billions to treat the environment. A staff of Xi'an Planning Bureau once told me that in order to treat the overall environment, it needs the cooperation with Gansu Province by managing the water in the upper reaches of Wei River at the same time. Jinhua of Zhejiang Province attached importance to urban environmental governance and developed the ecological city long before. On September 30, 2006, there were 5,045 biogas purification ponds built in the urban area of the city, with a total capacity of 126,900 m^3, which can purify domestic sewage discharged by about 600,000 people. As a pretreatment system, it can reduce the load and pressure of 80,000 m^3 sewage treatment plant. In addition, by the end of 2005, there were 170 ecological public toilets and 475 ecological buildings built in Jinhua, with a building area of 1.839 millionm^2 and a roof greening area of 146,400m^2.[①]

Recently, Beijing began to attach importance to the development of natural green space. From 2000 to the end of 2003, a total of 158,000hm^2 forests were planted. Since 2000, the ecological project of sandstorm source control in Beijing and Tianjin has been implemented. By 2010, the national investment reached 50 billion RMB, and 600 small green areas above 3000m^2 were added in the urban area, so that residents can enjoy natural green space and exercise within 500 m of travel. Beijing has made outstanding achievements in harnessing the environment of rivers and lakes in urban areas. There are 18 key rivers with a total length of 180 km. In 2007, the treatment of 120 km has been completed, and their sewage outlets have been successfully shut off. The sewage treatment rate of the central city has increased from 39.4% to 70%, and the water quality of rivers and lakes has increased from the past IV and V to II and III. According to the plan, five sewage treatment plants will be built in the central city and 15 sewage treatment plants will be built in the satellite city. In 2008, the urban rivers and lakes were basically cleared, and the smelly water no longer flowed through the urban areas, thus improving the living environment of residents in the capital.

In recent years, Nanjing has focused on improving the environment of Qinhuai River and the actions were completed in 2010, involving municipal administration, water conservancy, environmental protection, transportation, landscape, tourism, community, commerce and so on. Now it has revealed a harmonious ecological environment, and won the UN Habitat Special Honor Award in 2008. Shaoxing in Zhejiang Province also attaches great importance to the overall rectification of the environment. It has comprehensively rectified 18 rivers in the old city, built underground sewage pipelines and sewage treatment plants, closed and relocated more than 120 polluted factories, and introduced the living water outside into the rivers of the old city to improve the water quality and restore the original water town style. At the same time, it reduced the population

---

① The above information is provided by Mr. Yu Yigeng, former leader of Jinhua City.

density of streets in the old city, protected cultural attractions and cultural relics, coordinated architectural styles, and won the UN Habitat Honor Award in 2008. Suzhou is also paying attention to river regulation to restore the original style of the old water town.

In the future, China's "Fourteenth Five-Year Plan" should continue to focus on pollutant emission reduction, environmental governance, pollution source prevention and control, and also vigorously promote ecological protection and restoration, expand ecological space and capacity, and promote industrial structure adjustment, so as to lay a good foundation for establishing an ecological and civilized living environment.

Environmental governance in some cities of developed countries is half or even a century earlier than that in China. In 1993, I heard an introduction from the local planning commission in Seattle. More than half a century ago, Seattle had serious environmental pollution, traffic congestion and unstable residents' lives. In order to change this bad phenomenon, they reflected. First, they treated Washington Lake, which is located in the northeast of the city center, monitored and discharged the sewage in the whole area, made the lake clear, treated the whole city's water system, and strictly managed the sewer system. At the same time, it improved transportation, developed the road system of walking and public transportation, and developed free urban public transportation, which restored the original natural beauty of Seattle. The improvement of the city's environment has promoted the economic development, which in turn promoted the further improvement of the environment, and has embarked on the road of urban scientific development, becoming a "livable city" recognized by people.

When we visited Switzerland in 1979, we saw that the cities and villages in this country were clean and sanitary, which met the environmental requirements. The water of the Ale River, a tributary of the Rhine River passing through the capital Bern, is crystal clear; along the Lake of Geneva via Lausanne to Geneva, the water is clear and blue; Zurich Lake, on which Zurich depends, has been treated for a long time, and the lake has gradually become clear. This environmental effect is the result of strengthening environmental governance in Switzerland. In terms of garbage disposal, Swiss cities have already adopted the method of grading treatment, which can not only utilize waste, but also help clean the environment. The government encourages the use of solar energy in buildings and gives preferential policies to save energy and protect the environment. Switzerland is one of the countries with the largest number of solar energy buildings in Europe.

In 1996, we came to Bruges in northwest Belgium, which is called "Venice of the North" and really deserves its reputation. The rivers in the old city are netted and crystal clear and the whole city is neat and orderly. We have learned that urban management departments have a set of standards and management methods that meet the environmental protection requirements for river water quality, garbage disposal and flue gas emission, so as to ensure the environmental quality for living and tourism.

2. To create the natural environment, we should also ensure the green area and the coexistence between man and nature. By using the light energy and wind energy to save energy, protect the environment and live a comfortable life

Ensuring the area of green space is the foundation of building a natural ecological environment suitable for living and working, and conducive to disaster prevention and health. Therefore, the green space of natural scenic spots, parks, streets, residential areas and public buildings is an important part of the city, and it is the basic factor to ensure the coexistence between man and nature and create a good living environment. Therefore, we should pay attention to the construction of green space, including the large ecological environment outside the city, the ecological environment in the city itself (green space system) and the small ecological environment inside and outside buildings.

With regard to the green space construction of large, medium and small ecological environments, there are outstanding examples of cities in China. Hefei in Anhui Province was one of the cities, which was mentioned many times when I held the City Exhibition of the 10th Anniversary of National Day in 1959. Hefei was a small city with an old city area of 5.3 km^2 and a population of 50,000 in 1949. Now it is the capital of Anhui Province. In 1992, it became one of the first three cities in China that won the national title of "Garden City". At present, an annular green belt with a circumference of 8.3 km has been built around the old city, and four parks with more than 30hm^2 have been developed at the four corners of the green belt in combination with historical sites. A forest park of 550hm^2 has been opened on Dashu Mountain in the west, which is 9 km away from the city center. In Chaohu Lake, a southeast suburb 17 km away from the city center, it is planned to gradually develop a distinctive scenic tourist area. It is proposed to build a wind-induced forest belt in the southeast direction from Chaohu Lake to the urban area for introducing fresh air into the central area of the city along the dominant wind direction of the city. Surrounding the inner and outer edges of the two large green areas in the west and southeast, it is planned to gradually build the second and third ring green belts to connect the green space in the city and form the natural ecological green space system in Hefei. Since 2006, a lot of green space has been added. This natural green space system is very valuable, and it plays an important role in improving the urban ecological environment. It is an important part in the transition to the knowledge age and improving people's quality of life. After the reform and opening up, I found that the planning and construction of Zhongshan in Guangdong Province also reflected this idea. In 1996, the city won the title of national "garden city", and it has a more complete planning of natural green space system, which is gradually being realized. In 2003, the conceptual planning of the core area of Kunming, which was made by China Academy of Urban Planning and Design, won the international bidding, and three others participated, including Kisho Kurokawa from Japan, SASAKI from the United States and PTW from Australia. The spatial model adopted by China is "paralleled clustering and leapfrogging linear development pattern of Dian Lake", which separates and connects the natural ecological green space with rivers as the main trunk among the clusters, forming the characteristics of the spring city of Kunming, where urban areas coexist with the nature. The green space among the clusters and those inside (no less than 30%), makes the green space coverage rate of Kunming exceed 50%. The coverage rates of green space in Hefei and Zhongshan have also reached more than 50%, which makes the buildings in the whole city integrate into the green space.

Many ethnic minority villages, such as Ma'an village in Sanjiang of Guangxi, left a deep impression on me for being integrated with nature. It is a village of the Dong nationality, backed by green hills and surrounded by streams in the front. On the left and right of the village, there are the famous Chengyang Bridge and Pingyan Bridge lying across the water, used for communicating with outside world. Jinzhu village in Longsheng of Guangxi is a stilt-style residential village of the Zhuang nationality, which is built on the mountainside, shaded by bamboo trees, flowing by streams under the mountain, and integrated with nature as a whole. Other examples are Xizhou and Zhoucun of the Bai nationality in Dali of Yunnan Province, which are located between Cang Mountain and Erhai Lake. The climate is warm and humid, and the trees are lush. Their residential buildings are completely integrated into the beautiful environment of nature. Xishuangbanna, the southernmost part of Yunnan, has a tropical northern marginal climate, which is rich in bamboo, wood and water resources. In the Dai villages here, each household occupies a small square land, and many tall coconut trees shade it. Villages are generally built by ditches or rivers. The big green tree at the entrance of the village is a symbol of Dai people's worship. The whole village is surrounded by bamboo forests, coconut groves and clear water. The green coverage rate of these ethnic minority villages far exceeds that of cities, and truly reaches the coexistence of living place, man and nature. Cities should follow the spirit of attaching importance to the combination with nature.

Paying attention to the use of natural lighting, natural ventilation, solar energy and geothermal energy in architectural creation is an important part of saving non-renewable resources, protecting the environment and creating comfortable surroundings. With the development of architectural technology, people rely too much on air conditioning and other electrical facility, which not only increases the consumption of energy and resources, but also is not good for health. Mr. Mo Bozhi, a famous architect in China, has long known this problem. In the buildings he designed and built, natural light and ventilation are mainly used, and natural flowers, trees and landscapes are combined with indoor and outdoor space of buildings. He is the guide of Chinese architecture towards nature. His early works include Guangzhou Beiyuan Restaurant, Panxi Restaurant and Baiyun Mountain Villa. Guangzhou Mineral Water Guest House built during 1960s to 1970s is a typical example. There is a water court in front of the main building, a small garden with a small stream flowing behind the building, and a pillar layer at the bottom of the main building, which has excellent natural ventilation. All guest rooms have natural lighting and ventilation. Facing the natural water court and garden, they are energy-saving, environment-friendly, natural and comfortable. However, Mr. Mo's idea that the inside and outside of buildings should be integrated with trees and flowers has not attracted the attention of Chinese architects so far.

The Hong Kong and Shanghai Banking Corporation (HSBC), built in the mid-1980s, has a transparent lower floor, with the main supporting structure and vertical transportation system on both sides and the middle space is transparent in the north-south direction, highlighting the natural lighting and ventilation. The designer, Norman Foster, a famous British architect, attached great importance to the use of sunlight, and installed lenses at the top of the atrium to refract sunlight to the square. At the beginning of the 21st century in Beijing, the office building of the Architectural Design and Research Institute of Tsinghua University built by the University itsecf changed the conventional practice. It set up a sun visor and a court side of flower and wood facing south, made a solid wall facing west and left a gap, and set up a high skylight with east-west wind on both of the north and south sides, thus forming a system of natural lighting, natural ventilation and summer sun protection. It has obvious effects of energy saving, environmental protection and comfort. This concept of utilizing nature, attaching importance to ecology, saving energy and protecting environment should be the correct direction of architectural development in the 21st century.

With regard to the green space construction of large, medium and small ecological environments, Islamabad in Pakistan is one of the examples I have seen. Islamabad, the capital of Pakistan, was founded in 1961 and basically completed in 1970, with a population of over 200,000. The city is bordered by Margalla Hill to the north, Laval Lake to the east and Shakar Parian Mountain to the south. The city has a long strip and checkerboard layout from east to west, which is accompanied by landscapes and flowers. There are administrative office areas and public utility areas in the east, residential communities in the west, and light industrial areas, colleges and universities in the southwest. There are no high-rise buildings in these areas. The residential communities are 1-storey or 2-storey houses, and the floor-area ratio is below 1. All buildings in each area are integrated into green space. Road network and greenbelts connect green lands, gardens and parks, together with the mountains, lakesides and landscape around the city as a whole. Green space inside and outside the city is connected, and the coverage rate of green space within the built-up area reaches 70%, making it a city of forest garden.

Barcelona in Spain is a coastal city with mountains on its back and sea on its front. The jungle stretching on the flat mountain on the back constitutes a natural green barrier, which is parallel to the seashore of rich and lush trees. Montjuïc Hill stands in the west of the coast, which is adjacent to the old city to the east. After hundreds of years of construction, there are famous castles, museums, exhibition halls, botanical gardens and the main stadium for hosting the Olympic Games in 1992. These new and old buildings are hidden in the green sea of forests. In December, 1995, we visited more than 10 scenic spots in this area, and also investigated the famous gardens and parks in the northeast at the foot of the mountain. There are the natural Güell Park designed by Gaudi, and the famous historical garden of Laberint built at the end of the 18th century. In July 1996, we participated in the activities in Barcelona held in the gardens and parks in the northwest at the foot of the mountain during the conference of International Union of Architects. There are more than 10 green areas of various types. After the conference, we visited more than 10 parks in the southeast of the sea, and walked along the boulevard of new city. These east-west, north-south and diagonal radial green belts connect more than 60 green patches in the above four directions into a whole, forming a green space system with regional characteristics. It is embedded between the green mountains and blue seas of nature, creating a good ecological environment for residents. The green space coverage rate of the city has reached 50%.

In 1989, we visited Washington, the capital of the United States. Its central location is the intersection of Potomac River and Anacostia River, surrounded by large green space along the river. The inner central area of the city is composed of the 3.5km long central axis from Capitol to Lincoln Memorial, the green areas on both sides and low buildings scattered around. It highlights the natural green space environment, is a garden-like area, and reflects the political and cultural temperament of the capital. There are many large forests in the periphery of the central area, and the residential areas are mostly low-rise houses with low density. There are no high-rise buildings in the city. The regular road network and green belts connect these forests, parks and gardens together, making the buildings in the city melt into green space. The green area coverage rate reaches 60%.

In addition, Tuileries Garden in Paris of France, St Stephen's Park in Dublin of Ireland, Central Green Park in Boston, Grant Park in Chicago, Leisure Island near the lakeside in downtown Toronto of Canada, Pearl Continental Hotel in Lahore of Pakistan, etc., all of which have considerable green areas, creating a living environment in which man and nature coexist.

Give another example of a high-rise building that creates natural environment. In October, 2014, the world's first pair of "forest towers" was built in Isola District in the center of Milan, Italy. Trees and flowers were planted on the platforms of the towers, which improved the living environment of the towers. In the early 1990s, Mr. Qian Xuesen also advocated that in the development of landscape cities, vegetation should be planted on the platform of high-rise buildings to create "high-rise buildings with the sky garden" where man and nature coexist. This idea and practice are worth popularizing. Another example of integrating trees and flowers inside and outside the building is the Aventura Apple Store in Miami, Florida. The designer is Norman Foster and his partners. The indoor trees and tables extend to the double-floor hall and connect with the outdoor garden for tourists to rest. This kind of "garden building" which creates natural environment deserves Chinese architects' attention.

The use of natural lighting and natural ventilation is emphasized in architectural creation. Some world-famous architects have long known this matter. Most of the buildings they designed and built mainly use natural light and natural wind. For example, the houses designed by Frank Lloyd Wright, a world famous architect, including Chicago houses and studios of his own, are all "natural buildings". Natural lighting and ventilation are combined with the outdoor natural environment as a whole, and even its plane layout reflects nature and flexibility. As for the details of windows, he opposed sliding windows, and adopted casement windows to facilitate natural ventilation. Another example is Minoru Yamasaki, a famous American-born Japanese architect, who advocates natural light and water. The exhibition hall in northern Seattle and other low-rise buildings designed by him all reflect brightness, whiteness and reflection. The buildings have natural lighting and natural ventilation, accompanied by a pool fountain, which are of clear reflection, novelty, eye-catching scene, energy saving and environmental protection. In 1979, we visited Basel in the north, Bern in the middle and Fribourg in the south of Switzerland. Their houses, gymnasiums and indoor swimming pools all use natural lighting and ventilation, which shows that this country has long attached importance to energy conservation, environmental protection and comfortable experience. In 1990, we visited Montreal in Canada. The architectural decoration here is simple and practical, such as the Desjardins building in Montreal, which is located in the east of the downtown area. Its underground floor connects the buildings between the north and south subway stations in the east area to form an underground space network. Flowers and trees, pools and fountains are arranged in the basement and the upper floor of this building, creating a natural and comfortable environment. More ingeniously, natural light is introduced into the underground floor, and the hall on the upper floor is open and bright, making full use of natural light and natural ventilation, and achieving the effects of energy saving, environmental protection and comfort.

3. Moderate control of urban space capacity and moderate development of urban scale according to the load of urban land, water and other resources are also very important for livability, disaster prevention, sanitation and improving natural ecological environment in cities

At present, the floor-area ratio is too high and the population is too dense in many cities in China, which is an important reason for the unsatisfactory quality of natural ecological environment in cities. It not only brings chaos and contradictions, but also reduces the usage and economic value of cities and buildings. When faced with infectious diseases like SARS in 2003 and COVID-19 in 2020, areas with excessive population density will be very passive. In normal times, the environmental sanitation in these areas is also negatively affected.

Saving land is correct, but considering the basic requirements of urban sustainable development, disaster prevention and sanitation, we should find out the reasonable limits of urban space capacity, and within a reasonable range, divide different grades based on different limits.

For example, Hupo villa residential community built in Hefei in 1990s has a land area of $11.4 hm^2$, a total construction area of $117,000 m^2$ and a floor-area ratio of more than 1. Gongbei garden new community in Zhuhai has a land area of $6.92 hm^2$ and a total construction area of $110,000 m^2$. Chunyuan community in Kunming has a land area of $15.34 hm^2$, a total construction area of $18.66 m^2$ and green space of 42.4%. Sanlinyuan community in Shanghai has a gross density of $13,200 m^2/hm^2$, a residential building density of

12,000m^2/hm^2 and green space of 37%. Kangle new village and other residential areas in Xiamen have a building density of about 14,000m^2/hm^2. The floor-area ratio of these newly-built residential communities is mostly below 1.5, and their space capacity is relatively moderate, which is conducive to disaster prevention and hygiene.

As for the floor-area ratio, from the average level of Chinese cities and towns, we think that the ratio of residential area should be around 1.5, and the higher level ones should be around 1 or lower. The central business area is 2 to 4, and only some sections or big cities are above 4. At present, China's floor-area ratio is generally high, and its environmental quality is not that good. Most buildings in Xiamen, Zhuhai, Changzhou, Wuxi, Hefei, Yangzhou and other cities have suitable floor-area ratio. For example, the Yundang central area built in Xiamen in 1990s has created a better living environment. With regard to population density, in terms of a city as a whole, the residential land will be developed to 200 people /hm^2 in the future, with an area of about 50m^2 per person, and the further development should be 100 people /hm^2, with an area of 100m^2 per person. The residential land mentioned here does not include industrial, warehouse, railway, airport and military land. These analyses emphasize that we should have the idea of "attaching importance to disaster prevention, health and moderate space capacity", and we should not blindly expand the scale of cities and towns and improve the floor-area ratio of buildings.

On the issue of moderate control of urban space capacity, here are some foreign examples I have visited. Some cities are divided into different grades due to different limits. For example, Seattle is divided into four grades: the living density of Urban Center Villages is 15 to 50 living units per acre, and that of Hub Urban Villages is 15 to 20 living units per acre. Residential Urban Villages have 10 to 15 living units per acre, and Neighborhood Villages have 8 to 10 living units per acre. In order to ensure the environmental quality of the city, the city adopts the concept of urban village. Of course, Seattle has a high standard, while we should determine the appropriate space capacity according to the actual situation of small land and large population in China, but its concept and practice have reference value.

In the old city center of some foreign historical and cultural cities, the population density is relatively high. In order to improve the environment, subtraction was adopted to reduce the floor-area ratio. Paris adjusted the space capacity moderately in the core part of its old city, and reduced the floor-area ratio of residential buildings from 4 or 5 to below 2.7. It also reduced the floor-area ratio of Districts 10 and 11 that operate various activities to 2. In the old business district, it is not allowed to increase the floor-area ratio, and some of them should be reduced to less than 2, so as to retain the urban structure of Districts 8, 14 and 17 formed in the 19th century.

Another example is Los Angeles on the west coast of the United States. Its urban layout is decentralized, with some high-rise buildings built intensively only in the central urban area. In other areas, the floor-area ratio is not high, and most of the houses are 1 to 2 floors made of wood, with low space capacity and comfortable living environment, which is conducive to disaster prevention and health. The standard of this example is high, but its decentralized layout mode has reference value. Florence in Italy is an ancient city of history and culture and the birthplace of European Renaissance. It has always controlled the space capacity of the city. Only the Florence Cathedral (Basilica di Santa Maria del Fiore) is 107 m high, the bell tower next to it is 84 m high, and the old palace tower in Signoria Square is 95 m high. The rest are all low-rise buildings, with no new high-rise buildings. The city scale is moderate, so is the urban space capacity, and the environment is comfortable.

In this chapter, it describes strengthening environmental treatment, creating natural environment, ensuring green space area, using natural light and wind, and constructing moderate urban space capacity, and the purpose is to achieve ecological balance of nature in ecological civilization.

Chapter 3 History and Culture

Protect famous cities as a whole, make rational use and development, inherit the old and create the new, and develop architectural culture.

1. The history and culture are the roots of urban development, that we must have the idea of overall protection and rational use and development

In China, there are cities that have done a good job in protecting historical and cultural old cities as a whole, such as Hefei in Anhui Province, which has been mentioned in the previous chapter. The skeleton of 5.3 km^2 of the old city is completely preserved. The water system of the outskirt was fully preserved, forming a ring-shaped green belt in the city. It has developed into a big landscape city with the old city as the center. Another example is Xi'an in Shaanxi Province, which has a 9 km^2 old city of Ming Dynasty. Its north-south, east-west central axis and the relevant buildings are basically intact. All the city walls and moats of the old city were protected and utilized, and became a green sky garden on the wall around the city accompanied by water This ecological landscape around the city has a circumference of 11.9 km, which is a rare excellent example in famous historical and cultural cities of China. In addition, Chang'an city of Tang and Han Dynasty, Jianzhang Palace Garden and the ruins of Shanglin Garden have been preserved as a whole. There are also a few smaller historical and cultural old cities that have been preserved overall, such as Ancient City of Pingyao in Shanxi Province. In general, many famous historical and cultural cities in China were destroyed by war in the first half of the 20th century, and many were demolished by construction in the second half of the 20th century. After 1990s, the idea of "protecting the history and culture of famous cities as the priority" was gradually established, and the historical and cultural blocks and buildings of famous cities were rarely demolished. The methods of "transferring centers", "subtracting", "micro-circulation" and preserving old houses and time-honored brands were adopted, which achieved good results. Here are some examples I have seen.

When China's famous historical and cultural cities develop to medium scale, most of them should consider the multi-center layout, and move out the administrative, economic, commercial, scientific and educational centers to the new urban areas, except for the situation that the old cities have no history and culture, or the old cities have become historical and cultural centers. The overall planning of Xi'an has made a decentralized and multi-center arrangement, which transfers the administrative center to the northern area outside the old city for driving the development of the new district. The transfer of multi-center can also improve the natural ecological environment quality of the old city, alleviate the contradictions and pressures there, and ensure the restoration of the historical and cultural blocks in Xi'an. The overall planning and construction of Lijiang in Yunnan Province is an excellent example. In order to completely protect the historical and cultural panorama of the ancient city of Lijiang, other centers are arranged outside the ancient city. The administrative agencies of the municipal government are moved to Bahe district in the south of the new district, and economic, trade and other centers are arranged in several new districts in the west, thus ensuring that the historical culture of the famous city of Lijiang will not be damaged by the new construction and the ancient city of Lijiang will be protected as a whole.

Large and tall new buildings have been built in the central areas of old cities in some famous historical and cultural cities in China, which has destroyed the overall harmony. Therefore, "subtraction" should be actively implemented.

Beijing takes the lead in adopting this practice, which is worthy of publicity. Beijing has reduced two floors of the Di'anmen department store on the right side of the Drum Tower on the central axis, and removed the office building of the Municipal Housing Management Bureau on the east side of Nanchizi by three floors, so as to ensure the harmonious scale of the buildings in these two zones. For the disharmonious large buildings between Jingshan Park and Beihai Park, it is proposed to demolish and rectify them. There are two large buildings standing in the southwest corner and northeast corner of Xi'an Bell Tower Square. Subtraction should be carried out to highlight the image of Bell Tower and Drum Tower and maintain the overall artistic appearance of the core old city. The old city of Nanjing has made a plan for protection and renewal, emphasizing the protection and development of the old city culture. The general principle is good. However, modern culture appears in the central area of Xinjiekou, and many high-rise buildings have been built, and it is proposed to build super-high-rise buildings, which is not proper. "Subtraction" should be implemented in this core area, so as to make the overall scale of the buildings in old city harmonious and reduce the pressure on the population, traffic and facilities in the central area of the old city.

It is a good method to improve the blocks of historical and cultural cities by adopting the "micro-circulation method". The improvement of Zhongshan Road in the central area of Quanzhou in Fujian Province is a good example. It has no major demolition and renovation, no new tall buildings, no widening of large roads, only focusing on improving infrastructure, related buildings and environment, and basically maintaining the size of the original streets. This "microcirculation method" can ensure the original historical appearance, lifestyle and cultural customs of the old city, improve the living environment and meet the requirements of building a well-off society. It is worthy of emulation by local governments. In 2005, Beijing put forward the protection plan for the buffer zone of the Forbidden City, adopting the methods of "micro-circulation" and "organic renewal" to strictly protect the Hutongs and quadrangle dwellings in the buffer zone. In principle, there will not be demolishment in batch, the main streets and lanes will not be widened, and the infrastructure in the area will be actively improved instead of being completely abolished and rebuilt. In addition, the old city of Shanghai retains the residence in lanes and alleys, the old city of Kunming retains the "seal-shaped dwelling", and the old city of Kashgar retains the Kashgar residence.

These projects to protect old city blocks or buildings should be fully utilized and developed to make the cities more energetic. Some increase green space and living content, and become cultural leisure areas, such as the Big Wild Goose Pagoda in Xi'an, which opens up a modern urban green space square in the north. Taking the pagoda as the end of the central axis of the square, the pagoda, Ci'en Temple and the culture of Tang Dynasty they represent are highlighted, and there are also pools and fountains covering the central axis, all of which provide an excellent cultural and leisure place for residents in Xi'an tourists both home and abroad. Another example is Huaqing Pool in Li Mountain of Lintong in Xi'an. Huaqing Palace, a historic site with beautiful scenery, is the place where Emperor Xuanzong spent the winter and bathed with Lady

Yang. On the eve of the 10th anniversary of the founding of the People's Republic of China in 1959, the "Nine Dragon Pool" green garden of 5,300m² was built in the northwest of the original "female bathhouse", including Feishuang Hall, Chenxu Pavilion, Sunset Pavilion and traditional courtyard-style hotels. In 1990, the "Imperial Bathhouse Ruin Museum of Tang Dynasty" was built based on newly excavated cultural relics. The whole scale is 11 times larger than the original, and it has become a cultural scenic spot. Others have added the content of cultural studies and become cultural research areas, such as Dunhuang Mogao Grottoes. A large museum has been built at the opposite of the 1,600-meter-long (north to south) grottoes, and the existing Dunhuang Cultural Relics Research Institute has been expanded, making it a base for studying Dunhuang culture. In the old city of Jingzhou in Hubei, Jingzhou Museum and Three Kingdoms Green Park were built next to Kaiyuan Taoist Temple in the west; Binyang Tower was restored in the east, and a forest of steles was built there, which eventually became the cultural research area of Jingzhou old city. Some have increased the content of social activities and become the areas for cultural and social public welfare activities, such as Southern Putuo Temple in Xiamen of Fujian Province. Buddhist education and social public welfare undertakings have been developed within the scope of the temple. A college of Buddhist education has been set up, which is divided into two departments for men and women, and the male department is located in the temple. At the same time, charitable foundations and free medical institutions have been set up to serve the public and are welcomed by the public. Another example is Lugang Longshan Temple, which is the largest Longshan Temple in Taiwan Province. Besides offering sacrifices to gods and spirits, it is also a social relief agency and is popular among the general public.

Other good examples of making full use of and developing historical and cultural buildings are: Emperor Qin's Terracotta Warriors in Xi'an, Xuanwu Lake (park) in Nanjing, Yu Garden in Shanghai, Scenic Area of West Lake in Hangzhou (including Three Pools Mirroring the Moon, Xiling Society of Seal Arts and Guo's Villa), Green Vine Study in Shaoxing, Gulangyu Island in Xiamen, Yuelu Academy in Changsha, Du Fu's Thatched Cottage in Chengdu, Id kah Mosque in Kashgar, Macau City Hall Plaza, etc. The green space or museums make the places more lively and vibrant.

There are many ancient cities in Europe, and there are the most examples of overall protection. The city I saw which is the best practice is Rome, capital of Italy, which has a history of more than 2,000 years. From the ruins of the central square of ancient Rome in BC to the Monument to Victor Emanuele II built for the founding king of Italy in the early 20th century, all have been completely preserved. This famous historical and cultural city has become a huge historical museum without any damage. The new capital was built 7km south of the old city. This practice of separating the old and new cities has protected the famous historical and cultural cities as a whole.

Another good example of protecting historical and cultural cities as a whole is Paris. It is the capital of France. Instead of moving out the administrative center and building a new city, it adopts the overall protection principle of "protecting in groups and treating at different levels", which was put forward in the reconstruction plan of the old urban area of Paris adopted by the Paris Parliament in 1977. The old urban area of Paris refers to the area within the city wall built in 1845, including two forest parks in the east and west with an area of 105km². This overall protection principle is to divide the old urban area into two circles. The elliptical inner circle is the core part of the tradition protection and belongs to the first-class protection area. Its renovation has three specific principles: a) Protect the old residential area, and strengthen the living function to improve its comfort level; b) Preserve various functions and improve public activity space, including the improvement of cultural, entertainment, commercial and other facilities; c) Maintain the unity of architecture in 19th century. Why take the inner circle as the core part of keeping tradition? This is because Paris originated and developed in this area, and the famous central axis of the city runs through this area. The city's traditional culture, commercial districts, streets, famous ancient buildings and residential neighborhoods are also in this area. The outer circle belongs to the second-class protected area, which is also divided into three categories. The specific principles of each type of treatment are: a) First category, there are neighborhoods with preservation value and shall keep the characteristics of the original built-up area; b) Second category, maintain the original living function and properly rebuild existing buildings; c) Third category, develop the function of residence and re-control the existing buildings by changing the original characteristics properly. The density and height of buildings are strictly controlled. The standard of floor-area ratio has been partially explained in the previous section. The height of the "roof" in the core part of the inner circle is limited at 25 m, the highest building in the outer circle is limited below 37 m, and the neighborhoods, squares and some scenic areas that need to be preserved in the outer circle are also limited at 25 m. This principle of "protecting in groups and treating at different levels" in the planning of the old city has indeed maintained the historical and cultural characteristics of Paris.

In Barcelona of Spain, the earliest old city lies to the west of the seashore, which was completely preserved when the city expanded in the 19th century. The newly opened main road is connected along the east side and southwest side of the old city. Rambles Road and the north-south central road of the old city, are connected with the north-south main road of the new city through Catalonia Square. In the new urban areas developed in the 19th and 20th centuries, the modern buildings, such as the church and Meera House designed by Gaudi, have been completely preserved and stood out on important streets. Similar to the example of Barcelona's overall protection of historical and cultural old cities, it is common in Spain's capital Madrid, Switzerland's Bern and Geneva, Belgium's Antwerp, Italy's Florence and Venice and other European countries.

These projects to protect old city blocks or buildings should be fully utilized and developed. Some have added life content, such as Piazza San Marco in Venice. There are various shops and restaurants in the bottom of the municipal building and the small street in the north of the square. Some outdoor coffee seats are set in the east of the square, which is convenient for life and makes tourists feel that it is a "beautiful living room". Asakusa Temple in Tokyo of Japan has many festivals in one year. Around this protected historical building, shopping malls, game rooms and other facilities have been gradually built, and the flower market has been expanded, forming a commercial tourist area that preserves the customs and history of old Tokyo, and has now become a must-see for foreign tourists visiting Tokyo. Others, such as the Mayor Square in Madrid of Spain, the coastal area in downtown Seattle of the United States, and the Roman Square in Frankfurt of Germany, all add living facilities in and around the protected historical buildings, so that these places have acquired life interest and vitality. Others add green gardens, such as Ueno Park in Tokyo of Japan, where green space and zoos are built next to the original museum buildings, which are combined into a large park covering an area of $52.5hm^2$ and become a green place where people can enjoy history, culture and animals as well as relax and refresh themselves. Huntington Cultural Park in Los Angeles was originally the residence of Mr. Huntington, with a library of precious books and an art collection. Later, it developed a 130-acre botanical garden and built 12 flower gardens. Now it has become a national cultural and educational center in Los Angeles, which can be called Huntington Cultural Park. In this park, you can read precious manuscripts, enjoy historical and artistic treasures, and visit or rest in different types of gardens. Others, such as El Escorial Palace in Madrid, Plaza de Espana in Madrid, Montjuic Mountain in Barcelona, Germany Pavilion in Barcelona International Exhibition, Rubens Museum in Antwerp of Belgium, Castles, Badshahi Mosque and Shalimar Garden in Lahore of

Pakistan, Todaiji Temple and Deer Garden in Nara, Heian Shriane, Kinkakuji Temple, Ginkakuji Temple and Ryoan-ji Temple in Kyoto, all add green gardens to the original historical buildings to make them more attractive and lively.

2. Respect historical buildings, deal with the relationship with historical buildings when develops new ones; inherit the old and create the new and develop architectural culture

Xi'an in Shaanxi Province has made a lot of exploration on the development of regional culture by combining the old with the new. Among them, Academician Ms. Zhang Jinqiu, chief architect of China Northwest Architecture Design and Research Institute Limited Company, has achieved the best planning and design results. For example, the Bell and Drum Tower Square in Xi'an built in 1998. It strives to highlight the images of the two ancient buildings of 14th century, extending in both ancient and modern directions along the theme of "morning bell and evening drum", and arranging green squares, sunken plaza, sunken commercial streets, underground shopping malls and commercial buildings between the Bell and Drum Tower. It not only echoes the Bell and Drum Tower, but also protects the historical sites well, and solves the living needs of the development of the old city, so that this central area has many functions such as sightseeing, rest, shopping, dining, etc., and becomes a veritable citizen square, providing an "urban living room" with regional cultural characteristics for the old city of Xi'an. The National Art Museum of China, built in Beijing in 1962, is located to the east of the Palace Museum, Jingshan Park and Beihai Park. Next to these royal palaces, how to combine with them and develop new buildings? The designer, academician Mr. Dai Nianci, adopted new structures and materials, but made a large roof with yellow glazed tiles of double eaves at the center and a gable and hip roof at the entrance. It was partly spread on the east and west sides, and the bottom and top floors were made into an empty corridor, small bottom and big top, with a scale similar to that of the traditional buildings in the west. The overall shape was stretched with orderly contrast between the actual and the virtual, and the colors are bright and beautiful, with pine, bamboo and flowers. This kind of treatment is in harmony with the imperial buildings of the Forbidden City, so that the old and new buildings can be combined together. The Macao Museum, built in 1998 on the east hill of the Ruins of St. Paul, was next to the fortress (built in 1616 to repel Dutch invaders). The designer made full use of the terrain and placed part of the building in the fortress. At the entrance of the museum, there was a retaining wall built 400 years ago. The museum has three floors, and the exit of the exhibition hall on the third floor is the top of the fortress. The panoramic view of Macao can be seen there, which makes the Macao Museum, the fortress and the Ruins of St. Paul a harmonious whole.

When creating new buildings, we should consider inheriting excellent traditions and developing regional architectural culture. From 1968 to 1972, Taipei Sun Yat-sen Memorial Hall, designed and built by Mr. Wang Dahong, a famous architect from Taiwan, adopted the variant Chinese traditional yellow glazed tile roof form, which is an important example of exploring "Chinese modern architectural style" of inheriting the old merits and making innovations. This project has been selected into the East Asian Volume of the *World Architectural Masterpieces Collection in the 20th Century*. Similar to Sun Yat-sen Memorial Hall, there is the Taipei Palace Museum. In 1980s, Xinjiang Architectural Design and Research Institute took the lead in attaching importance to the inheritance and development of regional architectural culture. For example, the Xinjiang People's Hall in

Urumqi absorbed the traditional combination of square and circle, taking the main building as the square and the auxiliary building as the circle. A circular tower was built at the four corners of the main building as the pipe room, the top of the tower was made of stainless steel as a small dome, the outer wall columns were decorated with continuous arched parts, and the eaves were made of deep golden glazed tiles. These contents all reflect the style of traditional Islamic architecture in Xinjiang, but they are also new structures and of new features, which make us feel the regional cultural characteristics of Xinjiang architecture. Another example is the Xinjiang Guest House in Urumqi, which was also designed by the institute and built in 1985. It combines function, structure and regional traditional style together, uses traditional pointed arches in many places, and transforms the water tower structure into an image that expresses the beauty of Uygur culture. Huanglong Hotel, which was built in Hangzhou in the late 1980s, is located in the north of Stone Mountain in West Lake Scenic Area. The designer, Master Cheng Taining, adopts the garden-style general layout of breaking up the whole into parts, and has the traditional characteristics of China in the big picture. The combination of flowers, trees, bridges, stones, pools and pavilions in the central garden is completely southern China style, which has the characteristics of regional garden. The lobby at the entrance faces the real garden space which is like the long scroll of Chinese landscape. This embodies the layout characteristics of the main hall of Chinese traditional garden facing the main scene. In the garden, you can see the Baochu Pagoda on the distant Stone Mountain, which is combined with the external environment, reflecting the arrangement of Chinese traditional gardens by borrowing and facing the scenery. In the south lobby, the north restaurant, the east multi-function hall and the west gym can feel the changing spatial sequence when you wander around this different spatial environment from north to south. The walls of each single building are light-colored, with the top of dark green tiles, similar to the pink walls and tiles of southern China dwellings. However, this group of buildings adopts new structures, new materials and new facilities, and its various architectural spaces are spacious, novel and concise according to the new demands, so this is an example of inheriting the tradition, making innovations and developing regional architectural culture. The White Swan Hotel located in Shamian of Guangzhou, designed by academician Mr. She Junnan and academician Mr. Mo Bozhi, is a new building which is inheriting the past and making innovations and of regional cultural characteristics. It first pursues the natural artistic conception with Chinese characteristics, combines architecture with natural landscape, and further points out the artistic conception of "night moon in the goose pond", one of the eight scenic spots in the Ram City. In addition, a white drum-shaped main building with a height of 100m was built, which was like a white swan playing in the water. There is also an atrium with Ling-nan garden style, surrounded by open corridors, which are connected with the lounge and restaurant. Clover is planted all over the corridor, and the garden is composed of rocks, pools, pavilions, bridges and flowers, which are poetic and picturesque. The blue sky can be seen in the upper part of the atrium, and the top is a glass ceiling of caisson type. The sunlight is sprinkled into the garden, which is very lively. Especially, the stone under the high pavilion near the top is engraved with the inscription "Hometown Water", and the waterfall rushes down from the pavilion. This inscription and the sound of water flow will cause overseas travelers to miss their motherland and hometown. Walking through the atrium corridor, you can reach the Chinese restaurant on the third floor, and at the entrance, you can see a ring corridor and pool-style Ling-nan garden overlooking the sky. The large and small restaurants next to the garden are all Chinese decoration and Chinese tables and chairs, which are simple and elegant. The decorative hanging screen and some furniture of the main building rooms are also Chinese. By adopting new technology and innovating space combination, the interior and exterior of buildings are integrated with gardens, and the architectural culture of Ling-nan is developed.

Taiwan Taipei Jiantan Youth Activity Center, Taichung Tunghai University Campus and Luce Memorial

Chapel, Shanghai Sports Center (including the gymnasium built in 1975 and the stadium built in 1997), Xiamen Gaoqi International Airport Terminal 3 built in 1996 and Terminal 4 put into use in December 2014, China Pavilion of Shanghai World Expo built in 2010 and Terminals of Beijing Daxing International Airport built in 2019 are all good examples of "innovation with Chinese characteristics" by adopting new technologies, developing space with high efficiency, low energy consumption and artistic style.

Out of respect for historical buildings, how can foreign countries develop new buildings with important historical buildings? Trinity College in Dublin of Ireland left a deep impression on me. Located in the downtown area of Dublin, the college was founded in the 17th century and is the most famous institution of higher learning in Ireland. There are two courtyards with a high-rise clock tower in the middle, which separates the Parliament Square from the Library Square. A new square is built on the side of the Library Square, and on the side of the new square, there is an old library and a new library. And there is also a new art gallery on the opposite side. The height, volume, color and materials of these new buildings are consistent with those of the old ones, but new technologies are adopted to make the division and decoration of wall doors and windows simpler. The center of the square is paved with lawns and a new sculpture is erected, forming a complete and harmonious courtyard, where students sit on the floor, do some reading and some talking, which is harmonious and lively. Yale University in New Haven of the United States was founded in 1701. It is the third oldest institution of higher learning in the United States, and the campus is built with classical renaissance courtyard buildings in Greek and Roman style. In the second half of the 20th century, some new buildings were added, such as Yale Centre for British Art designed by world famous architect Louis Kahn, which was built in 1977. It is completely new material and new structure, but the designer coordinated the height of the building and the division of exterior wall lines with the surrounding buildings, so that the old and new buildings could be combined together. There is also an expanded new library with white marble walls to control the height of the building and coordinate it with the surrounding masonry buildings. The tallest building in the whole campus is the Gothic Harkness Tower, which was built in 1917, with a height of 67 m. It dominates the space of Yale University complex and harmonizes the old with the new. There is also the Palais des Beaux-Arts de Lille in France. The original museum was built in 1895 and is a classical renaissance building. It was expanded in 1970s. First, the original museum was cleaned and restored, and a new museum with new technology was built on the opposite side, plus a thin piece building. The surface of the thin piece is transparent glass, and the small lens in the middle is made into a square grid. The mirror reflects the old art museum, combining the new with the old. Others, such as Le Fresnoy National Contemporary Art Center designed by French famous architect Bernard Tschumi, the East Building of National Gallery of Art in Washington designed by Chinese-American world famous architect Ieoh Ming Pei, and the new Boston Public Library designed by American famous architect Philip Johnson, all respect the original historical buildings, coordinate with them and combine them into a harmonious integration of old and new.

When creating new buildings, there are many excellent foreign examples of inheriting traditional spirit and developing regional architectural culture, such as Faisal Mosque in Islamabad of Pakistan, which was built from 1970 to 1986. The designer was Turkish architect Vedat Dalokay, who was selected through international design competition. This group of buildings includes the chapel, the courtyard of the bathing area and an international Islamic university. He broke through the traditional concrete mode of arch, dome and Islamic patterns, but applied new materials and structures to embody the spiritual meaning of traditional mosques. He formed the roof of the chapel into a huge tent with reinforced concrete space frame. There were four towering Minarets in the four corners of the lobby, and a large courtyard was arranged in front of the chapel. The "Mecca pilgrimage wall" and "holy shrine" were located on the main axis of the hall and courtyard. This building innovated and developed the architectural culture of the mosque. Another example is the Garden Grove Community Church in the southern suburbs of Los Angeles, which was built in 1980 and designed by Philip Johnson, a famous American architect. This design completely makes innovations on the architectural layout and spatial environment of the church, but only inherits the spirit that the church should echo the sky and the practice of natural ventilation and lighting. In order to make each seat face and approach the altar, the plane of the classical Greek cruciform church is changed into a quadrangular star shape. The altar is in the center of the long side, with a rectangular seat pool in front, east and west halls and a pedestal pool in the south. It adopts a new structure of white steel grid and new material reflecting glass, which only allows 8% of light to penetrate, forming a clear and transparent feeling of quiet space corresponding to the sky without roof and wall. There is no air conditioning facility inside, and natural ventilation and natural lighting are used completely. The innovation of this building has developed the architectural culture of the church. In addition, buildings such as the National Library of France in Paris (La Bibliothèque Nationale de France), the Boston City Hall, the Washington Monument, and the Canadian Centre for Architecture in Montreal, all adopt new technologies and consider combining with the traditional spirit of the original buildings, creating new buildings suitable for their locations and developing the architectural culture of their regions.

There are also outstanding innovations in new buildings, including the use of the latest technology, new space layout and new content, which play a pioneering role. Early examples like Carson Pirie Scott Department Store in Chicago, which was built in 1904 and creatively embodies the characteristics of modern frame structure. Another example is the Disneyland in Los Angeles, which was built in 1955. It is a world-famous cartoon amusement park, and many countries have imitated it since then. The Atomiun designed and built in 1958 to hold the World Expo in Brussels, Belgium, expressed the beautiful vision of mankind for the peaceful use of atomic energy. The National Art and Cultural Center of Georges Pompidou (Le Centre National d'art et de Culture Georges-Pompidou) in Paris was built in 1977. The outside part looks like a factory painted in different colors to represent different functions, and the interior is a large column network, which can flexibly change the use functions. The Illinois Government Building in Chicago, which was built and used in 1985, is based on high technology, using classical dome and arranging open offices around the tall atrium used by the public to express the public meaning of the modern state government building.

The well-known historical buildings and landscapes scenic spots under complete protection in China include Forbidden City, Temple of Heaven, Xiyuan, Summer Palace and Great Wall (Badaling) in Beijing; Chengde Mountain Resort; Nanjing City Wall; Humble Administrator's Garden, Lingering Garden and Tiger Hill in Suzhou; Jichang Garden and Li Garden in Wuxi; Slender West Lake in Yangzhou; Orchid Pavilion, Mr. Lu Xun's Former Residence and Sanwei Study in Shaoxing; Yellow Mountain Scenic Area in Anhui; Sanya Ends Area; Li River and Reed Flute Cave in Guilin; Dujiang Dam, Fulong Temple and Leshan Giant Buddha in Sichuan; the Dragon Gate of West Hill, the Five Hundred Arhat Statues in Qiongzhu Temple and Lunan Stone Forest in Yunnan; Gaochang Old City and Jiaohe Old City in Xinjiang, etc. While in foreign countries, there are Giza Pyramid Group and Luxor Sun Temple in Egypt; Iraq Babylon City; Athens Acropolis in Greece; Roman Forum, Roman Colosseum, Column of Trajan, Arch of Constantine, Piazza Spagna, Emmanuel II Monument, Trevi Fountain, St. Peter's Cathedral, Piazza Della Signoria, the public buildings and residential areas in Pompeii Ancient City and Villa Lante in Italy; Vaux-le-Vicomte Garden, Versailles Palace, Louvre Palace, Concorde Square and the Arc of Triumph in France; Madrid Transport Department Building and Alhambra Palace in Granada City in Spain; Parliament Building in London; Brussels Central Square in Belgium; Moscow

Red Square in Russia; Snow Scenery in Swiss Alps; and Niagara Falls Scenic Area in Canada, etc.

The architectural science of cities, buildings and landscapes mentioned above has the characteristics of artistic beauty.

This chapter focuses on the overall protection of famous cities, rational use and development. When creating new buildings, we should respect historical buildings, inherit traditions and make innovations, and also develop regional architectural culture in order to achieve the ecological balance of culture in ecological civilization.

Chapter 4 Artistic Beauty

Create city symbols, lead the space environment, master the artistic rules, and coordinate urban architecture.

The examples related to history and culture mentioned in the previous chapter all contain the contents of artistic beauty. In this chapter, we will expound and emphasize three aspects of how architectural science and culture create artistic beauty.

1. Create city symbols and lead the space environment

Every city has its own characteristics, so it is necessary to create a landmark building that conforms to its own characteristics, and form a beautiful three-dimensional contour line in its center and surrounding areas, so that people can know which city it is at a glance. These landmark buildings play a leading role in urban space environment and make people feel beautiful.

Shanghai Oriental Pearl TV Tower, built in 1995, is located at the bend of Huangpu River in Pudong, surrounded by river water on three sides, looking towards the Bund on the west bank across the river. It is facing the bustling Nanjing Road, and it is also the focus of the important streets such as Beijing Road, Fuzhou Road, Yan'an Road and Siping Road. The tower is 468 m high, and the frame of the tower body is three cylinders, with 11 spheres of different sizes as connectors, forming a positive upward, spherical cylinder space modeling. This tall and magnificent TV tower has become a new landmark building in Shanghai. The Building of Bank of China, located in Central Garden Road of Hong Kong, is 368 m high, and its rising shape is outstanding, which controls the spatial environment in Central Hong Kong and becomes a landmark building there. Located in Harbour Road, Wanchai, the Central Plaza Building, with a height of 375 m, is a triangular structure, which controls the spatial environment in the eastern part of Hong Kong Peninsula and is also a landmark building. The landmark buildings on the east and west sides of Hong Kong Island are not high enough, and a building of about 600 m is in the centre, just like a screw-shaped residential building with a height of 667 m planned to be built behind the Grand Park in the central area of Chicago, which will dominate the spatial environment of the whole area and highlight the three-dimensional outline of the city in the central area.

In the center of the well-preserved old city of Xi'an, which was built in Ming Dynasty, stands the Bell Tower, the landmark building. The city is shaped as the Chinese character "Field" with ancient crossing roads and surrounding city walls. The main axis in north-south direction and the small axis in east-west direction lead to the four gates of Guancheng in east, west, south and north. The Bell Tower is located in the center of the whole city where the two axes cross. The Bell Tower was built in the 17th year of the reign of the first emperor in Ming Dynasty (1384). The lower part is a brick pier of 35.5 m square and 8.6 m high with four round holes directly facing for Guancheng in four directions. A tall wooden tower of pyramidal roof with three eaves is built on the brick pier, and a big clock is hung on the upper layer. With a total height of 36 m, it is the tallest building in the old city and the most beautiful building in terms of volume, shape and color. The Drum Tower is located in its northwest. During Ming and Qing Dynasties, the buildings in the roads of the whole city were all low-rise buildings with a height of 5 m to 8 m. The Bell Tower and Drum Tower, along with the surrounding city walls with a height of 12 m and the gates of Guangcheng, formed a rhythmic three-dimensional corridor in the old city of Xi'an, and the central Bell Tower dominated the spatial appearance of the whole city. Although some buildings along the street and inside have increased the number of floors, the Bell Tower can still control the spatial environment of the old city (only two tall buildings beside the Bell Tower need to lower the height).

Look at the preserved old city in Beijing. It was built in the 18th year of Emperor Yongle in Ming Dynasty (1420). There are three city walls from inside to outside. The inner palace is called the Forbidden City, the middle is called the Imperial City, and the outer one is called the Capital City or Inner City. In the 32nd year of Emperor Jiajing in Ming Dynasty (1553), an outer city was built to form a convex shape. There are 9 gates in the Inner City and 5 gates in the outer city, with gate buildings and arrow buildings more than 30 m high. In the Inner City, there are the white pagoda in Xiyuan and Wanchun Pavilion in Jingshan Park, both over 60 m, and the white pagoda of 50.86 m in Miaoying Temple. There is a large residential area of one floor or two floors in the city with a height of 5 m to 8 m. These three most prominent high points, together with the 30-meter-high main palace and gate towers, constitute a rhythmic three-dimensional outline of the old city of Beijing, which is picturesque. The ratio of ordinary buildings to high-rise buildings is 1:5 to 1:8, which highlights the high landmark buildings and dominates the spatial environment of the old city. Seen from the north-south central axis of the old city of Beijing, the southernmost part is Yongdingmen, which has recently rebuilt the gate tower with a height of nearly 40 m. Going north, it enters Zhengyangmen Tower and Arrow Tower with a height of 40 m, Chairman Mao Memorial Hall, Monument to the People's Heroes and the Tiananmen Gate Tower with a height of nearly 40 m. Then pass the Hall of Supreme Harmony, which is 33 m high in the center of the Forbidden City, followed by Wanchun Pavilion, the highest of the five pavilions in Jingshan Park at the opposite. To the north of this pavilion are the Drum Tower with a height of more than 40 m at the northernmost end of the central axis and the Bell Tower with a height of 47.95 m at a distance of 100 m. These buildings are all symbolic buildings, which still play a role in controlling the spatial environment in the central area of Beijing. People can know where they are when they see one of the symbolic buildings. These symbolic buildings are characterized by high height and beautiful shape. Along the north-south central axis of the old city, you can reach the newly-built sports center area by extending northward. A national stadium, for the opening and closing ceremonies of the 29th Olympic Games in 2008 and track and field events, was built on the east side of the extended central axis. The project was selected through the international architectural design scheme competition. The building is oval, and its external shape is like the surface of the branch structure with soft objects filled in by a bird for building a home, commonly known as the Bird's Nest. The oval building is large in size, spectacular and novel. A national swimming center of a square shape is also built on the west side of the extended central axis. These two large-scale new buildings, one round and one square, one rigid and one soft, with sharp contrast, have become the landmark buildings that lead the space environment of the sports center area in the north of Beijing's extended central axis.

The Statue of Fisher Girl in Zhuhai of Guangdong Province is located in the sea water to the north of Binhai Park between Xiangzhou and Gongbei District, which is connected with Lovers Road by a bridge. It is of simple shape, friendly manner and pleasant scale, with beautiful natural landscape as the background, and has become the focus of attention in this central area. This exquisitely made and touching statue is naturally a landmark work that can control the space environment between Xiangzhou and Gongbei area. It also has a moving fairy tale, which can be read in the following explanations. There is also the Drum Tower in Ma'an Village of Sanjiang in Guangxi, which is a landmark building of the Dong ethnic group, with large volume and located in the center of the whole village. The plane figure is square, and the center is a public hall. There are four main load-bearing columns more than 10 m high supporting the drum pavilion on the roof, and the outer contour is a square cone with seven layers of dense eaves. There is a big square in front of the Drum Tower, which is a place where the residents of the whole village meet and communicate. People gather in the Drum Tower and its square for deliberation, ceremonial celebrations, welcome guests, rest and entertainment, love affairs and so on. Dong people call it "the gut of the village", which means the soul of the village. This social activity center of the whole village residents dominates the spatial environment of the whole village living and residential area.

There are also many foreign examples where landmark buildings dominate the urban spatial environment, such as the TV tower and lakeside buildings in the central area of Toronto in Canada. There is a Canadian National Television Tower with a height of 553 m along the west side of Ontario Lake, beside which there is a large stadium with an operable roof. This group of contrasting buildings leads the high-rise buildings on the lakeside, forming a three-dimensional contour with rhythm and beautiful space balance, which is also the symbol of Toronto city. Another example is the high-rise buildings in the central area of Chicago, which are arranged orderly behind the green space of the strip park on the lakeside of Michigan. The south is the tallest dark Sears Tower, the middle is the second tallest light-colored Oil Building, and the north is the third tallest dark Hancock Building. Other lower buildings are arranged between these three landmark high-rise buildings, forming a high and low architectural complex. Looking from east to west from the lakeshore, a three-dimensional outline of a beautiful building with musical rhythm and poetic rhythm is presented in front of us, which has become the symbol of Chicago. The layout of Chicago's high-rise buildings is planned. We saw this plan when we visited SOM Architects in June 1993. The Institute has been responsible for this work for decades, and they planned and built it from the aspects of land balance, use function and improvement of environment. This practice is worthy of reference by big cities in various countries. There is also another example of the seaside complex in the central area of Seattle, which is divided into three sections from north to south: the northern section is a low-rise complex, but it has the highest landmark building "Space Needle" in Seattle with a height of 185 m; the middle section preserves the original low-rise public buildings along the lakeside, followed by new high-rise buildings, including financial district, business center, communication and city hall, etc.; the southern section is a low-rise area, including Pioneer Square and historical reserve area, and a new low-rise but huge circular dome landmark building Kingdome for entertainment, followed by the international area. North, middle and south constitute the three-dimensional outline of the coastal city in Seattle's central area with rhythm and hierarchy. The buildings along the coast are low, and the high-rise buildings are arranged backward according to the mountain line. This layout not only has a three-dimensional outline of artistic beauty, but also preserves historical and cultural buildings, and achieves reasonable functional zoning and balanced urban land use.

The old palace in Signoria Square of Florence is now the City Hall. On the cornice stands a tower with a total height of 95 m, which is characterized by a solemn and solid castle in the Middle Ages. It controls the downtown area of Florence, including the open-air sculpture museum in front of it and on its left and right sides, and the space environment along the Arno River on its left. It is a prominent landmark building.

Notre Dame de Paris in France is another prominent landmark building. It is a world-famous Catholic church. It was built from 1163 to 1345, and is located on Cite Island on the Seine River in the center of Paris. The inner hall is 32.5 m high, which can accommodate 9,000 people for religious activities. It is a Gothic church structure, but it has been greatly improved, which not only saves materials, but also forms a light style. The 13-ton clock is hung on the top of the main hall, and the 90-meter-high spire and a pair of towers in front of it become the focus of people's attention. It stands out as a landmark building, not only because its building is tall and magnificent, and the famous French writer wrote the novel Notre Dame de Paris, but also because it has become a political and social activity center. In the new area of Barcelona in Spain, the Basílica i Temple Expiatori de la Sagrada Família, designed by the world famous architect Gaudi, was built in the early 20th century. It is a typical landmark building with eight towering circular towers, which embodies the "natural architecture", composed of curves and curved surfaces created by Gaudi. People call it the Sagrada Família Museum. The Parliament Building in London is also a typical landmark building. It stands on the Thames River. The facade of the building is 280 m long. There is a Victoria Tower with a height of 104 m in the southwest corner and a square clock tower with a height of 98 m in the northeast corner, which is the world-famous Westminster Bell Tower, with an extra-large clock of 21 tons on the top. In 1923, the BBC sent the bell tone to the whole world. Notre Dame de Paris, the Basílica i Temple Expiatori de la Sagrada Família, and the British Parliament Building all play a leading role in the spatial environment of their respective downtown areas.

Through the analysis of Chinese and foreign architecture, we can draw the conclusion that urban landmark buildings should have the conditions of location, height, volume, artistic modeling, sight gathering and so on.

2. Control scale and volume and coordinate urban texture

This matter has been involved in the third chapter of history and culture. Here, we only emphasize that architects should pay attention to mastering the scale and volume of buildings, and control different scales and volumes according to the different locations of new buildings, so as to make the whole harmonious beauty in the city texture.

There are many good examples, like Queli Hotel in Qufu of Shandong Province. The designer, academician Mr. Dai Nianci, paid great attention to the relationship between architectural scale and volume, and basically positioned the height of guest house as a decentralized two-story building. Rooms are courtyard-style, decorated with pools and flowers and trees, and they are fastidious in terms of facade division, door and window size, etc. This guest house complex is in harmony with Confucius Mansion and Confucius Temple, which keeps the historical and cultural features of the original buildings in this central area and the original texture of the city. In a similar way, there is Zhuhai Hotel, which is located next to Stone Scene Hill Scenic

Spot in Zhuhai. The scenic spot is full of granite egg-shaped stones, and there is a large pool of white lotus. The designer, academician Mr. Mo Bozhi, also adopts a low-rise scattered courtyard-style layout. Its scale, volume and garden are completely coordinated and integrated with the scenic spot, enriching the landscape here.

Another example is the Shanghai Museum, which was built in the 1990s. It is located at the southern end of the central axis of People's Square in Puxi central area of Shanghai, far opposite to the north city government building, with a total construction area of 38,500m^2. The designer is Mr. Xing Tonghe, a famous architect. In order to adapt to this open square space and echo the surrounding new high-rise buildings, the designer attaches importance to controlling the scale and volume of the building itself, and adopts a horizontally stretched and flat body shape, which is both stretched and stable, avoiding depression. The lower part of the building is made into a square with two steps, which looks like a pedestal, and the top part is made into a round shape, which looks like a Chinese bronze mirror, implying the "round sky and square earth" in the ancient Chinese natural view theory. With complete internal functions and advanced technology, this project has been selected into the East Asian Volume of the *World Architectural Masterpieces Collection in the 20th Century*, which is an innovative building with appropriate scale, volume, regional cultural characteristics and promoting the harmony of the city. On the west side of the square, Shanghai Grand Theatre built in 1998 has a total construction area of 68,000m^2. The design units are French Arte Charpentier and East China Architectural Design and Research Institute. The whole building attaches importance to mastering moderate scale and volume and adopts a multi-layer pedestal with dispersed volume and partially extending into the ground. It arranges a moderate audience hall and stage on the pedestal, and unfolds to the sky on the top roof, just like the cornucopia of the Chinese nation, which inherits the grace and wisdom of mankind from the universe, symbolizes Shanghai's enthusiastic pursuit of world culture and art. The building makes full use of modern high-tech means and new materials to enrich and create its own light, transparent, elegant and magnificent image.

Among foreign famous historical and cultural cities, there are good examples of mastering architectural scale and volume, such as the Opéra de la Bastille, which is located in the famous Place de la Bastille, the symbol of French revolution. The design scheme was selected through the international design competition, which paid attention to volume division and comparison, so as to adapt to the original scale of blocks and squares and establish a concept of overall harmony and beauty, so it was selected. Another example is the expansion project of the Louvre in Paris. In the 1980s, French President Mitterrand decided to move the Ministry of Finance out of the Louvre because of the need to expand its scale. Finally, he chose the pyramid expansion design scheme of Ieoh Ming Pei, a world-famous Chinese-American architect. The expanded building area was completely placed in the underground space. The scale and volume of the entrance building of the glass pyramid on the ground were suitable for the original building space, while maintaining the original urban texture. Similar practices include the Cultural Center of Africa, Near East and Asian of Smithsonian in Washington, which was completed in 1986 and covers an area of 668,000 square feet. It was built on a 4.2-acre section between the headquarters of the Victorian Smithsonian Society built in 1849 and its two historical and cultural museums. This location is located in the front left side of the Capitol in Washington, among many museums, which attracts great attention. The designer boldly adopted the underground scheme, only three entrance pavilions were built on the ground, and 96% of the area was placed underground. The total height of the entrance pavilions was 12 m, which kept the historical and cultural features of the original buildings on the ground and the harmonious beauty of the original texture of the city. It was well received by all walks of life in the United States and won the Tucker Award Excellence in 1988. Then there is the Manezhnaya Square Shopping Center in Moscow of Russia, which was completed in 1997. The square is located in the north of Red Square, adjacent to the Kremlin wall. In order to improve the living environment here, this square was reconstructed in 1990s, and the multifunctional shopping center with four floors underground was built. Because it was built underground, it kept the original urban texture.

There is also the newly-built East Hall of National Art Museum, located on the right side of the Capitol in Washington, which is close to the east of the old National Art Museum and designed by famous Chinese-American architect Ieoh Ming Pei. It attaches importance to mastering the scale, coordinates with the old museum in height and volume, and has simple and varied appearance. It is mainly made of solid walls, and the scale ratio of window opening is appropriate. Especially, the connection between the old and new museums is combined through the ground square, fountain, water bucket, skylight and underground passage waterscape. In the new museum, according to the new functional requirements, new structures, new materials and new triangular space are adopted for organization. After its completion in 1978, President Carter attended the opening ceremony and cut the ribbon. He praised the building as a symbolic building combining public life with art. There is also the Seattle Public Market Center, which is close to the Puget Strait water surface, and retains an old market that mainly supplies seafood for citizens. In the opposite and surrounding areas of this old market, attention is paid to controlling the scale and volume of buildings, and adopting the appropriate height and volume that are in harmony with the old market and connected with the high-rise buildings in the east of the center for administration, commerce and trade. Only buildings of 3 to 6 floors with moderate size were built, which achieved the overall harmonious effect of urban texture.

3. Grasp the artistic law of balance and change, and enhance the overall perfection of urban architecture

Art form is an important content of urban architecture, which can also be said as an important component, because the functions of architecture include the use function of physical space and spiritual function. The beauty of form has a great influence on the spirit of users. Form is included in functions, and it is an indispensable and unavoidable content of any building. Therefore, when we create urban and architectural design, we should consider the combination of function and form, content and form, and attach importance to the external and internal physical space of buildings and the form creation that in harmony with the city. This kind of attention is not to consider the form first, and then put the physical space carrying the function of materials into it. As a result, many spaces are not applicable, and it leads to the misunderstanding of one-sided pursuit of form. The relationships mentioned below all considered that the form is consistent with the content and function.

(1) Urban buildings have comparative spatial changes, which can achieve impressive overall perfection.

In the Summer Palace of Beijing, from the east entrance, people can gradually go to the courtyard centered on the Hall of Benevolence and Longevity. It used to be the place where the emperor went to court to handle government affairs. It is a solemn closed space, with luxurious and colorful displays. When passes through the transitional space to the west, you can see the extremely open landscapes such as Longevity Hill and Kunming Lake. There is not only a large space but also a turquoise tone as a whole. The contrast between small and large, solemn and natural, strong and elegant gives people a perfect feeling as a whole. After entering the scenic spots, there are many spatial changes of this contrast, such as the courtyard of Hall of Happiness and Longevity in the east at the foot of Longevity Hill, the courtyard of Hall for Listening to Orioles Singing in the west, Jingfu Pavilion in the east of the hillside, the scenic spot of Tour into the Picture, etc. This is the artistic and aesthetic law of planning and design. Going back to Xiyuan (Beihai Park), the local scenery you see from the main entrance in the south is quite different from the open environment of Beijing you see after crossing the bridge and climbing to the high land of the white pagoda in Qiongdao. This change in spatial contrast also exists in many places in Xiyuan. For example, after entering a group of garden attractions with closed space in Heart-East Study from the back door in the north, you can climb the Diecui Pavilion for viewing or walk out of the front door of Heart-East Study to enjoy the open panoramic view of Xiyuan or the landscape of Jingshan Park beyond.

In the West Lake Scenic Area of Hangzhou, you can go to the jungle in the north of Gu Hill via Bai Causeway, then go to the closed high platform at the back of Xiling Society of Seal Arts, and then walk into Sizhao Pavilion to look down to the south, and you can expect to see the panoramic view of the open West Lake. This sudden change in landscape space makes you feel relaxed and happy. And you also can travel westbound to the Guo's Villa Garden Scenic Spot in the north of the West Lake. First enter the closed main scenic spot of "One Mirror Opens the Sky", then enter the main building Jingsu Pavilion, and enjoy the wide and natural beauty of the West Lake from the second floor. This change in space contrast also makes you get the same perfect feeling. If you want to visit Dujiang Dam in Sichuan, you'd better go to Fulong Temple at the intersection Precious-Bottle-Neck of two waters. This temple is of a three-story terrace courtyard. You can enter the Iron Buddha Hall courtyard on the second floor from the Old King Hall courtyard on the first floor, and then climb to the Jade Emperor Hall courtyard on the highest third floor. Looking to the north and west, you can get a panoramic view of beautiful landscapes such as the broad Minjiang River, the spanning Anlan Rope Bridge, the Erwang Temple in the green jungle, Zhaogong Mountain and snowy mountains. This scenic spot is vast, natural, simple and quiet, and it is deeply impressed and unforgettable by the overall perfect spatial contrast changes.

There are also many places with spatial contrast changes in foreign countries. In Venice of Italy, I visited Piazza San Marco twice. I had to go through a long narrow commercial lane with a width of several meters before I could enter the trapezoidal square through a circular arch. The square is with a length of 175 m, a width of 90 m in the east and a width of 56 m in the west. It connects with a little square in the south. At the corner of two connected squares stands a bell tower with a height of 100 m, giving you a feeling of contrast and change between the two spaces. One contrast is the narrow lane turning into Piazza San Marco, and the other is between the 100-meter-high tower and the churches, government houses, old and new municipal buildings which are only three or 4-storey high. All of these enhanced my impression of the overall beauty of the world-famous Piazza San Marco. Montjuic Mountain in Barcelona of Spain is located in the west of the old city, with a castle built in the 17th century. In 1929, in order to hold an international exposition, an exposition art exhibition hall was built in the north of the hill. When looking down from the little closed space in the hall to the north, it is an open, rich and moving picture combining architecture with nature. Close-up is the green square before the strip-shaped Expo, and mid-range is the Plaza de Espana with six roads crossing. There is a beautiful triangular pavilion sculpture fountain in the center of the square. The northeast corner of the square is a large competitive bullring built in the 19th century with a capacity of 26,000 spectators. The long-term view is the landscape of horizontal green mountain forest as the city background. This impression of spatial contrast still exists in my mind. I have visited the church next to Hancock Building in the southwest of Boston's central district. The interior of the church is full of regular space and exquisite decoration. Then I came to the viewing hall at the top of Hancock Building, which is 240 m high. Looking east, I found a large green park and high-rise buildings in the center of the city as the front middle scene, and an open picture with low-rise buildings and water surface as the background in the distance. Looking further north, first, the front scene is the low-rise residential area interspersed with green belts and the Charles River, then the middle scene is the main building of Massachusetts Institute of Technology on the north bank of the Charles River, and other buildings, with large green areas. From the closed space environment of the church to the top of the building, I saw the wide green space of Boston, the contrast is extremely strong, which makes people remember forever.

(2) The methods of opposite scenery and borrowed scenery have enriched the landscape art of buildings and cities.

The Taiye Pool in Jianzhang Palace Garden in Chang'an City, first introduced in this book, is the origin of the garden form of "One Pool of Three Mountains". There are Penglai, Abbot and Yingzhou in the pool, which symbolize the three sacred mountains in the East Sea of Chinese myth. These three mountains are in a scenic relationship with the main buildings on the shore. Later, Hangzhou West Lake imitated Taiye Pool and adopted the practice of "One Pool of Three Mountains". There are three small islands in the lake, namely Three Pools Mirroring the Moon, Huxin Pavilion and Ruangong Islet. Later, the overall layout of the Summer Palace in Beijing completely imitated the pattern of Hangzhou West Lake Scenic Area. The West Dike and Branch Dike built in Kunming Lake divide the water surface into one big and two small parts, and there is one island in each of the three waters, symbolizing the three sacred mountains in the East Sea. These three islands are the opposite sceneries of the landscape in Pavilion of the Fragrance of Buddha on Longevity Hill in the north, and the three small islands in Hangzhou West Lake are the opposite sceneries of in Gu Hill and Stone Mountain. In this view, the main scenery on the mountain can be seen from the three islands in the lake, and the scenery of the three mountains in the lake can be expected from the main scenery on the mountain, which has the artistic effect of enriching the landscape. Another example is the white pagoda in Qiongdao of Xiyuan, which is the opposite scenery of the Five-dragon Pavilion in the north. In the Humble Administrator's Garden in Suzhou, the following scenic spots are all opposite sceneries of each other, like the main building Drifting Fragrance Hall to Fragrant Snow and Cloud Pavilion on the northern hills, the spot of Hidden Beauty to Bamboo Seclusion Pavilion, Yulan Hall to Mountain in View Tower, Xiangzhou to Pavilion of Lotus Breezes, etc., all of which are opposite to each other. Most of the opposite sceneries are about 50 m away from each other, and you can enjoy a clear landscape. Five-pavilion Bridge and the white pagoda in the middle of Slender West Lake in Yangzhou are opposite to the Chuitai Pavilion. The bright landscapes of the pagoda and bridge can be seen from the two round doors in the Chuitai Pavilion. From Five-dragon Bridge, you can also see multi-level pictures with the Chuitai Pavilion as the middle scene and the misty rain of the four bridges as the distant view. These scenes have enriched the landscape of scenic spots.

There are many examples of borrowed scenery in China. For example, the Summer Palace in Beijing borrows the pagoda on the Jade Fountain Hill and distant mountains into the park, and the scenery is very far-reaching. Another example is Jichang Garden in Wuxi, where Longguang pagoda on Xi Mountain is borrowed into the park. These two places have expanded the space environment in the park and enriched the artistic landscape of the two parks.

There are also many examples of opposite scenery in foreign countries, like the Trinity Church in Piazza Spagna, the fountain at its lower end and the narrow and long fashion street at opposite are all opposite sceneries of each other. In the Villa Lante of Rome, there is a statue fountain on the circular island in the center of the square pool at the bottom floor, which is opposite to the circular fountain on the second floor, and then the continuous water flow on the third floor is opposite to the central dolphin fountain on the fourth floor. The Washington Monument with a height of 169 m and a pyramid shape at the top in the center of Washington has an obvious relationship of opposite scenery with the Capitol in the east, the Lincoln Memorial in the west and the White House in the north. All of the above landscape layouts have played an artistic role in enriching architecture and urban landscape.

(3) The symmetrical layout of the building can obtain the perfect effect of solemnity, balance and unity.

Symmetric layout is widely used in famous historical and cultural cities and buildings at home and abroad. In ancient China, Chang'an in Tang Dynasty and Beijing in Ming and Qing Dynasties were symmetrical, such as the Inner City of Beijing in Ming Dynasty, in the front is the imperial court and in the back is the market, on the left is the Imperial Ancestral Temple and on the right is the Altar of Land and Grain. The Forbidden City is in the middle, and the central axis was prominent. Even the grid-shaped roads leading to the nine gates in the north, south, east and west were symmetrically arranged, with Chongrenmen (Dongzhimen) and Qihuamen (Chaoyangmen) on the left and Heyimen (Xizhimen) and Pingzemen (Fuchengmen) on the right. The front is the main entrance of Zhonglimen (Zhengyangmen), with Wenmingmen (Chongwenmen) and Shunchengmen (Xuanwumen) on both sides, followed by Deshengmen and Andingmen, with the names in Qing Dynasty and at present in brackets. Now the Forbidden City is centered on the Hall of Supreme Harmony (the core), the Hall of Central Harmony and the Hall of Perfect Harmony, with a symmetrical overall layout, reflecting the supreme dignity of the former emperors.

After the founding of New China, Beijing built Tiananmen Square in 1959, together with Zhengyangmen they are the center line. The Great Hall of the People was built on the west side of the square, the Chinese Revolution and History Museum was symmetrically built on the east side, and the Monument to People's Heroes was built in the center of the square. This newly opened rectangular square with symmetrical pattern is 40hm^2, which is used to celebrate the 10th anniversary of National Day. It is a symbol of New China, with magnificent momentum, unified architectural colors and harmonious old and new buildings, forming a complete space for architectural art activities. The famous Beijing quadrangle dwellings also have a neat and symmetrical layout. Generally, three main rooms are arranged on the main axis, with a lower wing on each side of the main room. The east and west wing rooms are symmetrically arranged in front of the main room, and form a central courtyard with the south wall and the hanging flower gate as the front yard or backyard. Many urban residential areas outside Beijing also adopt symmetrical courtyard or patio layout, which is formed by people's distinction between young and old, psychological balance and living habits. Most temples in China also choose a symmetrical pattern. For example, the Lugang Longshan Temple in Taiwan, the general layout of the building is a symmetrical three-courtyard style. The first courtyard is the front hall, and the main hall of the second courtyard stands out in the middle. It is a double-eaves and mountain-resting style, dedicated to the Lord God Guanyin. The front sides of the main hall are symmetrical corridors and two big lush banyan trees, which are spectacular. The third courtyard is the back hall. This rhythmic symmetrical spatial layout leaves me with a solemn, balanced, unified and perfect impression.

The symmetrical style abroad is most prominent in the palaces built in medieval Europe. For example, the Vaux Le Vicomte Garden in Paris of France was built in 1656 and designed by the famous gardener André Le Nôtre. He adopted a strict symmetrical layout of the central axis, with a length of 1,200 m from north to south and a width of 600 m from east to west. It is the main building on the middle platform, facing the symmetrical flower beds, pools, fountains and architectural sketches, and has a horizontal canal at the end. This obvious central axis attracts people's eyes and makes people feel the majesty of their masters. This design met the requirements of the owner Nicolas Fouquet, who was the Finance Minister of Louis XIII and XIV. He was dictatorial at that time and built this garden to show his authority. However, Louis XIV was invited to visit the Garden for dinner after the completion in 1661. He felt that Fouquet might usurp the power, so Fouquet was jailed for questioning and sentenced to life imprisonment. After that, Louis XIV, who was only 23 years old, called Le Nôtre, the designer of Vaux Le Vicomte Garden, Louis Le Vau, the architectural designer, and Le Brun, the interior designer, to the palace, and asked them to plan and design Château de Versailles, proposing that the palace should surpass El Escorial Palace in Madrid, and the garden should surpass Vaux Le Vicomte Garden that it shall never been seen in the world. After decades of construction, Château de Versailles really met Louis XIV's requirement of showing his king's authority. This grand palace with an area of 800 hectares, a building area of 110,000m^2 and a central garden of more than 100hm^2 is called "Le Nôtre-style palace", because of its balanced and symmetrical layout, which highlights the grand momentum of the Grand Canal and the longitudinal axis. This pattern has influenced European countries for a whole century. In Germany, Britain, Austria, Spain, Sweden, and as far away as Russia, symmetrical palaces with prominent major axes embodying the majesty of kings have been built. The symmetrical pattern of Château de Versailles's major axis is that three rays outside the palace focus on the central main building, two large pools and statues are symmetrically arranged on the first floor behind the palace, and the north park and the south park are symmetrically arranged left and right. Downward along the central axis is the Latona Statue Pool Fountain, and further west along the central axis is the Royal Avenue with long lawn green space. Both sides of the avenue are symmetrically covered with statues and trees, and there are symmetrical gardens behind the trees on both sides. Next to the Royal Avenue are the Appolo Statue Pool Fountain and the 1.56-km-long Cross Grand Canal. Symmetric radiation is set beside the pool at the east end of the Grand Canal, which leads to the Grand Trianon and Petit Trianon in the northwest and the zoo in the southwest. This magnificent palace with far-reaching scenery and strict style reflects the intention of showing off the authority of the monarch.

Again, the world's largest cathedral, Basilica di San Pietro in Rome, was built from 1506 to 1626, which lasted for 120 years, and its overall layout is strictly symmetrical. The central dome is 41.9 m in diameter, 123.4 m high at its inner apex and 138 m high at its outer apex, which is the highest point in Rome. It is supported by four symmetrical piers. The front of the church is composed of a symmetrical oval and trapezoidal colonnade with a column height of 19 m, which is very imposing. There is an obelisk in the center of the square, with a pool fountain symmetrically arranged on both sides, which is particularly spectacular as a whole.

Judging from the solemn, balanced, unified and perfect effect of the above symmetrical pattern approach, it can be improved and applied according to actual needs in the future.

(4) The change of architectural shape and proportion can enhance coordination or perception of artistic conception.

Hall of Prayer for Good Harvests in the Temple of Heaven in Beijing is a symbolic building. Its base is a circular white marble platform with a height of 6 m and a diameter of 90 m. The circular hall with a diameter of 30 m stands on the base and has a height of 38 m. The original upper, middle and lower eaves are blue, yellow and green, respectively. When it was rebuilt in the 16th year of Emperor Qianlong (1351), it was changed into a blue glazed tile with three eaves. The round shape and blue color are more prominent symbols of the sky. It is the highest, largest and most perfect building at the northern end of the central axis of the Temple of Heaven. Another example is the terminal building of Xiamen Gaoqi International Airport, which was built in 1996. It adopts a new reinforced concrete roof structure, creating a unique shape with a large curved roof in southern Fujian, which is as light as a bird spreading its wings and flying. The upper side windows are several layers high, with natural lighting and natural ventilation. The atrium-style large space can be changed as needed. This is an innovative building of architectural modeling, energy saving and high efficiency.

As for the proportional relationship between distance and height of buildings, it is advisable to a comfortable horizontal perspective in gardens or other buildings. For example, the distance from the Drifting Fragrance Hall, the main scenic spot in Suzhou Humble Administrator's Garden, to Fragrant Snow and Cloud Pavilion is 50 m, and the pavilion as well as its background trees is 20 m high. So the horizontal view from Drifting Fragrance Hall is 18° to 27°, which is the best perspective, and the distance is 2.5 times the height. The proportion between residential buildings should be lower, which should be 1.7 to 2 times, so that the sunshine shines into the window of the bottom floor of the north house at 30° from the south during the winter solstice. In parks or squares, the proportion can be 6 to 8 times, and a better visual experience can be obtained. After mastering this proportion, harmonious and comfortable visual and use effects can be achieved.

Abroad, there is the Giza Pyramids in Cairo of Egypt, and its gold triangle means that Pharaoh incarnates the sacred meaning of eternal soul and upward sky. The triangular gable wall at the top of the Parthenon in Athens represents the intention of respecting heaven and striving upward, and its facade is divided into three rhythmic sections. The bottom step width, column and lintel beam gable wall have a certain proportional relationship with the column diameter, which is very harmonious. The stigma uses Doric style to represent men. It is a classical building with the intention of respecting heaven and being strong and perfect in shape.

Seattle built a 185 m-high Space Needle for the "21st Century Exhibition" held here in 1962, and made the top of the tower look like a flying saucer hanging in space. Its creative idea implied that it would become a symbol of space development in the 21st century. When I saw this light, towering and beautiful space needle tower, I really felt its symbolic intention. The Stone Yard of Ryoanji Temple in Kyoto built in the 15th century, has 15 selected stones on the white sand ground, which are arranged in five groups according to the number of 5, 2, 3, 2 and 3 in turn, symbolizing five island groups, and arranged according to the composition principle of triangle, achieving balanced and perfect effect. The white sand at the bottom is raked into water stripes to symbolize the sea. This group of abstract sculpture stone yard has reached the natural landscape symbolizing the sea and islands. The change of image and proportion has really enhanced people's perception of artistic conception.

(5) The spatial sequence change of architecture and city can enhance the harmony and perfection of the whole.

The sequence arrangement of Humble Administrator's Garden in Suzhou is of great reference value. The entrance gate to the rockery is a leading small space, which is mainly composed of stone-mountain barriers. Then it crosses a small bridge and goes around the rockery to enter a semi-open small space with landscape, and then enters the main building Drifting Fragrance Hall to enjoy the large space with open natural landscape. From Drifting Fragrance Hall to the west, you can see the small courtyard space with small flying rainbow as the main part, and then enter Mountain in View Tower from Willow Shades Road, and it is an open large space from Mountain in View Tower to the flower path in the north, which is a small space with the characteristics of a water town, and then come to Fragrant Snow and Cloud Pavilion, where you can enjoy the whole garden space centered on Drifting Fragrance Hall. After going down the mountain, you can see the small space with loquat and begonia as the theme. This spatial sequence can be simplified as small, small, large, small, large, small and closed, semi-open, open, semi-open, open, closed, open and closed, which is similar to the prosody of the flat tone in poetry, and is similar to the introduction, description, climax, turning and ending of an article, which constitutes a rhythmic flowing space.

We came to the Ma'an Village in Sanjiang of Guangxi. First is the famous Chengyang Bridge with magnificent architecture. Through the long passage of this bridge, it enters the open village space surrounded by before going to the back mountain, where the tall Drum Tower near the back mountain and its square in front control the spatial environment of the whole village. From the east of the central village along the river bank, we can see the undulating landscape and the semi-open space of the building. Finally, through the smaller Pingyanfengshui Bridge to the natural mountain forest outside the village, the spatial sequence is closed, wide open, semi-open, closed and open, which also makes people feel the harmony and overall perfection of the spatial sequence.

The first foreign architecture was Luxor Sun Temple in Egypt built in 14th century BC, which is the most spectacular temple in Egypt and has obvious spatial sequence. The entrance is from a narrow doorway. The front yard is a large open space with Ramesses II statue standing on the side. It enters the column hall consisting of 16 rows of 134 tall and dense stone pillars. Among them, the 12 pillars in the middle are the most tall and prominent, forming a towering and large-scale but not open hall space. Behind the column hall is a relatively open space with obelisks. The spatial sequence of this temple is closed, open, closed, open and small, large, extra large and small, which still makes me feel that its spatial environment is rhythmic and overall perfect. Then there is Alhambra Palace Garden in Granada of Spain, which was built from 1238 to 1358. Its architecture and garden are Spanish Islamic gardens, which are called Patio style. First enter the medium-sized Myrtle Courtyard. The main hall of this courtyard is where the emperor met ambassadors of various countries to hold a ceremony. There is a pool in front of the temple, and myrtle hedges are planted on the west side. From this courtyard, it enters the Lion House eastward. This courtyard is the residence of the empresses. The courtyard is decorated with cross-shaped canals. There are 12 fine stone lion statues at the center of the canal intersection, and there are round basin-shaped pool fountains on it. It is surrounded by 124 slender column arch arcades. Here is the most beautiful medium-sized courtyard in the palace. To the north of this courtyard is a small Patio-style garden, and then to the east is the open Pato Garden, which is not the Patio-style anymore and not surrounded by buildings neither. Its spatial sequence is medium, medium, small, large, semi-open, semi-open, closed and open, which enhances the overall harmony and perfection. Finally, there is the downtown area of Paris with urban sequence space. The starting point is the main building Louvre, which faces the east-west sinking long Tuileries Garden. This garden was rebuilt and expanded by the famous gardener André Le Nôtre during Louis XIV in the 17th century. It is a semi-open garden space, which goes west along its downtown axis, and is connected with an open Place de la Concorde extending horizontally from north to south. In the center of the square stands an Egyptian obelisk, flanked by a pair of fountains in front of St. Peter's Cathedral in Rome. Walk into the east-west strip Champs Elysees Commercial Street along the central axis of this square and at the west end of this street is the famous Arc of Triumph in Paris and its circular plaza, where all the radiation roads around it are focused, forming an open space environment radiating outward from the center. The four continuous spatial sequences along the central axis of Paris are semi-open garden space, open rectanyle square space, non-open street space and open radial square space. I have watched this rhythmic sequence space many times and deeply felt the harmony and perfection of its overall spatial changes.

This chapter explains that cities, buildings and landscapes included in architectural science have beautiful art and reflect people's various thoughts and feelings. This artistic beauty also belongs to the content of achieving cultural and ecological balance in ecological civilization, and also reflects the nature of architectural science as a comprehensive art to establish ecological civilization and living environment for human beings.

Chapter 5 Convenient Transportation

Change the mode of automobile-domination, build pedestrian system, build bicycle system and connect public transportation.

In order to facilitate the travel of urban residents, urban traffic cannot be dominated by automobiles, so it is necessary to build pedestrian, bicycle and public transportation systems, mainly networking the three, and change the concept that automobiles dominate urban traffic.

In the central area of a city or other important living and cultural areas, especially in the central area of the old city, a pedestrian street system should be gradually built, so that the street space environment serving the general public's life is not disturbed by automobile traffic, ensuring personal safety and people flow, and truly becoming a bustling street that meets the needs of urban citizens' life. This is an important measure for urban prosperity and development, as well as an important measure for promoting consumption growth and economic development. It reflects the level of urban modernization, urban development and the degree of caring for urban residents. With the development of urban cars, especially the rapid growth of private cars, urban traffic is dominated by cars and road traffic is congested. Cities in developed countries have reversed this chaotic situation, turned to control cars in the central area of the city, built a humanized walking road system, and connected with bus network and underground railway traffic network, which enriched the activities there and improved the artistic level of space appearance, facilitated and attracted more residents, and promoted the vitality of the city.

The following are some methods in different aspects.

1. Limit the number of cars driving in urban areas

　　Chinese and foreign cities have adopted various methods in this respect. In some downtown areas, there is a certain range, in which cars are not allowed to enter, and only pedestrians and public transportation are allowed. For example, in 1979, when I visited some cities in Switzerland, there were parking lots outside the designated range, and the original pollution-free trams and buses were still kept in the range. Some places of historic interest and scenic beauty in the downtown area are designated as pedestrian areas, and no vehicles are allowed to drive. For example, Florence opened a pedestrian area in the center of Signoria Square and its surrounding areas, from Basilica di Santa Maria and Bell Tower in the north to Arno River and Old Bridge in the south. Other places of historic interest in the city were also designated as pedestrian areas, and no cars were allowed, which was convenient for tourists. Others changed the square roads in front of famous buildings in the downtown area into pedestrian streets, such as Trafalgar Square in the central area of London, England. In 2003, the renovation project was completed, which closed the vehicle road in front of the National Art Museum, which is the central building of the square. The large step of the sunken plaza in front of the building was transformed into a large step at the entrance of the museum, connecting the building with the square as a whole, and forming a pedestrian road system on both sides and in front of the square, separated with car flows. Through this transformation, the number of people coming to the square for activities increased by 13 times compared with before, which promoted the popularity of this area and facilitated the residents. This area is no longer dominated by cars. In addition, on November 14, 2007, London Mayor Ken Livingstone said at the London Council that he planned to transform several bustling streets and parks in central London, such as Regent's Park, Oxford Street and shopping area, into pedestrian streets, allowing only pedestrians and bicycles to pass, and hoped that these streets in this area would be tree-lined roads like Rambles Pedestrian Street in Barcelona of Spain, and the reconstruction project would be completed in stages. Rambles Pedestrian Street, a famous pedestrian street in Europe and a flower market, is called "Flower Market Street", which has become a must-see for tourists from all over the world. The entrances and exits of this pedestrian street are all connected with the urban underground railway network, which is convenient for sightseeing. Others have been changed to pedestrian streets at fixed time, such as Ginza in Tokyo of Japan, which is the most prosperous business district in Tokyo. Since August 1970, vehicles have been forbidden to pass on Sunday afternoons and holiday afternoons, and they have been changed to pedestrian streets. There is also a coffee bar in the street center. At this time the place is safer and more active with more tourists. Ginza Street starts from Kyobashi in the north and reaches Shinbashi in the south, and there are underground railway network connected with it, which is convenient for people to visit here.

　　Chinese cities have also begun to pay attention to this problem. For example, Beijing opened Wangfujing Pedestrian Street in 1990s, covering a section from the south exit of Wangfujing to Bamiancao, which should be expanded in the future and connected with the northern underground railway network. Dashilan Street outside Qianmen has long been a pedestrian street. Now all the important streets outside Qianmen have been changed into pedestrian streets, and many streets and Hutongs around have also been changed into pedestrian areas, forming a pedestrian area network. Nanjing Road Commercial Street, which is famous in Shanghai, also became a pedestrian street in 1990s, and public transportation and motor vehicles pass through on the streets at both sides. Binjiang Road, beside Quanyechang in the central area of Tianjin, has been opened as a pedestrian street for a long time, and has been well received by the general public. These pedestrian streets in the central area are not allowed to pass by cars, which partly limits the number of cars in the central area.

2. Establish a three-dimensional urban road traffic system and develop a large-scale sidewalk system

Three-dimensional urban road traffic is to connect sidewalks, bicycle lanes, automobile lanes, underground railways and overhead lanes at different traffic speeds with buildings, and finally form a complete three-dimensional road traffic system. Nowadays, most Chinese and foreign cities have attached importance to connection, but only a few can form a complete system. The first city to attach importance to connecting into a system is Philadelphia in the United States. The planning of downtown area of Philadelphia is to gradually develop a three-dimensional sidewalk system in a planned way, connecting the original historic buildings, commercial service buildings, cultural and recreational buildings with sidewalks, and forming a connected whole in 1963. In 1973, it was further adjusted and connected with subway and bus lanes and developed into a complete system. The designer was Edmund N. Bacon, a famous American planner and architect. The central black block in the attached plan is an independent building with traditional history, and the gray grid is the activity area with the highest density in the central area. The black line indicates the pedestrian system under the street, and the white line indicates the pedestrian system on the street, which connects the shops in various parts. There is an independent building in the middle of the elevation expansion plan, and an open sunken plaza in front. The left side of the square is connected with the service building, and the right side of the square is connected with the market street, with a three-story building with a glass roof. The bottom floors of these buildings are connected together to form a walking system. The second floor of the building is connected with the street where cars pass. Under the central sinking square, an underground road passage is set up. This arrangement makes it very convenient for people to go to and from the central area. When they move in the central area, they have a sense of security without being disturbed by cars. The design of the city street in the central area also emphasizes different spatial treatments. In the walking system, the lower part of the street is arranged as a garden-style activity space, which is connected with the subway. For details, see the attached street view of the central area of Philadelphia. The bus terminal and parking lot are set at the end of the walk, which is convenient for transportation. Different levels of space can be seen from the street, with rich and hierarchical landscape.

A simple example of connecting into a three-dimensional transportation system includes the newly built Eaton Center in downtown Toronto of Canada, which is a large shopping mall in the form of an indoor pedestrian street with a length of 274 m and 300 shops in three floors. Indoor commercial streets are connected with surrounding buildings through overpasses and tunnels, and a large-scale pedestrian system is constructed, which is connected with underground railways and above-ground bus roads. There are also examples of large-scale connection becoming a three-dimensional road transportation system, such as downtown Montreal of Canada, which started to develop this system from the underground space of Vile Marie Tower in 1960. There are three connected complex buildings in the center, namely, Desjardins, Place Ville Marie and Les Cour Mont-Royal, which are connected with the subway ring network in the central area and intersect with Catherine and Rene-Levesque, the main roads in the central area. After 28 years of construction, the city's three-dimensional road traffic system has been basically completed, and nearly 300,000 people use it every day. According to the statistics of six subway stations in three complex buildings, they are connected with 1.7 millionm^2 of office space, 1,400 fashion shops, 3 concert halls, 2 department stores and hotels with 3,800 rooms in total. A large number of street entrances of shops have been converted into indoor entrances, increasing from 2.7% to 36% in 1961. Shoe stores increase from 0 to 48%, and women's clothing stores from 1.8% to 67%. The development of indoor pedestrian streets and squares has promoted the improvement of commercial and cultural facilities on the street, adding electronic games, fast food, bars, etc., making the famous outdoor commercial streets connected with indoor pedestrian streets Catherine and Mai-Sonneuve for common prosperity.

In China, Beijing has basically formed this three-dimensional road traffic system. Tiananmen Square, the core area of the center, has been connected with two underground subway stations, Tiananmen East and Tiananmen West. The commercial and cultural zone of Qianmenwai Street in front of Tiananmen Square has formed a pedestrian zone network, which is connected with Qianmen and Zhushikou subway station in parallel. The Gulou South Street Commercial and Cultural District behind the central core area is connected with two subway stations, Beihai North and Nanluoguxiang. In the Wangfujing Commercial and Cultural District in the east of the central core area, Wangfujing Street has been changed into a pedestrian street. Dong'an Market in the east of the street has connected the streets in the east, and it is connected with Wangfujing subway station at the southern end of this street, but the northern end of the street is far from the entrance of Dengshikou subway station in Dongdan North Street. Xidan North Street Commercial and Cultural District to the west of the central core area is connected by two subway stations, Xidan and Lingjing Hutong. The above five important areas are all connected with the above-ground bus transportation network, and connected with the three horizontal and three vertical grid networks of Chang'an Avenue, Qianwai Street, Gulou South Street, Wangfujing Street and Xidan North Street, which basically constitute the three-dimensional system of road traffic in Beijing. In the future, by adding an subway line from Yongdingmen to Di'anmen, improving the sidewalks of Gulou South Street and Xidan North Street which are not pedestrian streets, and strengthening the connection between various traffic lanes and buildings, Beijing can realize a complete three-dimensional transportation system, which is more convenient for the citizens and Chinese and foreign tourists.

This mode of connecting into a three-dimensional transportation system can really solve several major problems.

Avoid the contradiction between people and vehicles, so that residents living in the central area have a sense of security.

Meet the requirements of multi-function, like business, culture, entertainment, and even can connect office and hotel, which is convenient to use.

Provide a good environment, with natural light or good artificial lighting, green space, fountains, seats, etc., so as to facilitate activities, like meeting friends or rest.

Improve traffic, road network, subway network, sidewalk network, connect important buildings with public facilities, etc., and make it convenient to come and go.

Combined with civil defense engineering, the system of underground streets, underground squares and underground passages is good civil defense engineering. For example, the Moscow subway played a great role in the Great Patriotic War.

3. Try to develop subway traffic and expand the scope of underground buildings around the subway

Traffic tension and congestion in the central areas of big cities and megacities at home and abroad are one of the main problems in these cities. After continuous improvement in recent decades, the traffic in these cities has improved. Most of the big cities and mega-cities where the traffic problems in the central area been alleviated have developed well-organized subways. As described in the previous section, the subways have developed from connection to system. To sum up, the development of subway can be divided into three situations.

The first situation is thay the subway transportation network in the central area is consistent with the road network of the above-ground chessboard system or circular radial system, except that the density of the network is sparse, and the distance between the two lines is about 1km. There are entrances and exits of subway stations in the main locations of the central area, and the transportation to and from the central area is very convenient. Specific examples are Paris, London, Moscow, New York, Beijing, etc.

The second situation isthatin the central area, the core part forms the ring network of underground railway, and several urban trunk subway lines pass through the central ring network. Washington, for example, has five lines that cross the central area from four directions and form the core network of the center. The subway line in the central area of Chicago is combined with the traffic line on the viaduct to form a ring network in the central area, which leads to the north, northwest, west and south directions. The subway ring network in the central area of Toronto is relatively small, and several subway lines are connected to this central ring network.

The third situation is that thert are only one or two subway lines passing through the central area, which are mainly bus and automobile traffic lanes organized on the ground, such as San Francisco of the United States.

In these three cases, the first practice has the best effect and smooth traffic; although the second practice is not as convenient as the first one, it is basically convenient to directly reach the ring network in the downtown area from the main area through the subway, but some areas have to walk for a while to enter the subway network; the third practice is not as effective as the first two practices.

The most complete example of adopting the first practice is Paris subway network, and it is also the most convenient to connect with the buildings on the ground. When I first arrived in Paris in 1982, someone told me that if you need to go to work or hold meetings somewhere on time, you'd better take the subway. It is not that certain if you choose the ground transportation. The east-west main axis of Paris subway network has the densest flow of people, but it is very popular because of the high-speed cars, which shortens the time. In line with the first practice, the Beijing subway network started to build the first subway along the extension line of Chang'an Street along the east-west main axis in 1960s. Then, a subway on the Second Ring Road of the old city was built, followed by several ring-shaped subways outside the First Ring Road and the Second Ring Road, and subways radiating from all directions. Northeast to Shunyi, southeast to Yizhuang, northwest to Xishan in Changping, and southwest to Liangxiang and Suzhuang, it has formed an integrated subway system, which is convenient for the whole city's travel activities. In China, there are less land but more people and large cities and megacities with high population density and high floor-area ratio, so Shanghai, Tianjin, Guangzhou, Xi'an, Wuhan and other cities should develop the first practice.

With regard to expanding the connection between underground buildings and above-ground buildings around the subway, the central area of Montreal in Canada is a good example. From the underground space system diagram in downtown Montreal, we can see that there are more underground buildings around the black subway, which make full use of the underground space and reduce the pressure on the ground construction land. At the same time, the underground buildings are also connected with the enlarged buildings on the ground, which is more convenient for residents' lives. The central street of Nagoya in Japan is the main axis of the south-north direction of the city, with a green belt of a width of 100 m. An underground commercial street has been opened under this central street, and there are wide underground commercial and cultural square buildings at the intersections with east-west roads. At the same time, there are entrances and exits of subway stations, which communicate with the buildings on the ground, making it very convenient for citizens and tourists to come and go. This pattern of expanding the service scope of developing underground commercial culture and forming a large area of underground space buildings has played a beneficial role in saving land, facilitating life, avoiding the troubles of wind, sun, rain and snow, and increasing pedestrian facilities. China should pay attention to this practice, and engineers engaged in urban underground railway should study and develop underground railways and underground buildings together with planning architects, so as to expand the scope of underground buildings and achieve good results in many aspects.

4. Restore and develop bicycle lane system and control fast-moving vehicles

From 1950s to 1980s, the road traffic in cities of China was basically normal. There were many three-block road modes in various places, with motor vehicles driving in the middle and green isolation belts on both sides. Bicycle lanes were next to this belt, and many people used convenient bicycles to travel within 5-6 km. In the past 20 years after the reform and opening up, private cars have developed rapidly, and these three roads in big cities and megacities are gradually occupied by cars. Beijing is a typical example. Some of the original bicycle lanes are designated as turning lanes for cars, and a large part of them are used as parking lots for cars. Bicycle lanes have even been cancelled in some areas. This practice of reducing the area of bicycle lanes and temporarily meeting the needs of cars but exceeding the capacity is an important issue of Beijing's road traffic improvement. Speeding motorcycles and electric bicycles on bicycle lanes threaten the safety of cyclists, and the management departments should control these vehicles. Urban road traffic serves the general public, and it is the basic right of residents to use it reasonably. Many cities in Northern Europe and Germany have been developing bicycle lane systems, and the proportion of bicycle trips in the whole city traffic volume is about 30%. For example, the city of Feldberg in western Germany offers incentives for commuting, shopping and relaxing by bike. Thomas Herzog, a famous German architect, told us that he has two cars, which are only used when he goes to the suburbs or other cities. He rides a bicycle when he goes to work in the downtown area. He thinks that this is beneficial to the city, which can reduce traffic congestion, not pollute the air, save energy and exercise his body. Denmark, which is known as the kingdom of bicycles has set up bicycle lanes in the Danish Pavilion at the Shanghai World Expo and promoted this healthy concept. Beijing and all cities in China should re-recognize this concept, restore and develop bicycle lane system, and control speeding vehicles. To restore bicycle lanes, it is necessary to build a certain number of car parking lots, so that increasing the number of cars in cities can adapt to the development of urban road traffic and parking lots. Since developing a private car, the city needs to increase the area of roads and parking lots by many square meters. We should not only look at how much national economic output can be increased by producing a car, but how much urban area and how much investment and construction time of urban infrastructure should be increased accordingly. Urban planning, construction and transportation departments should put forward these construction funds that need corresponding land use, parking lot construction and air pollution control. The National Development and Reform Commission and local city leaders should have a comprehensive concept, seriously solve this problem, and can't ignore the needs of urban construction and unilaterally pursue the immediate interests of developing a large number of cars.

In addition to the above aspects, urban planning and design should also pay attention to the development of urban mixed communities, which not only have the residential function, but also can arrange some pollution-free tertiary industry institutions, which can facilitate residents' lives and reduce the pressure of road traffic and the traffic cost of people commuting to and from work. There is also the need to pay attention to strengthening the management of urban road traffic. In 1979 in Switzerland and 1982 in Paris, our friends drove us to travel, and once due to parking issue and once due to slow turning, they were quickly affixed with fines. I was surprised at that time since I didn't see traffic police around me. Now I recall that they used cameras to manage traffic and management personnel. Forty years later, on Zengguang Road on the south side of Beijing Construction Institute, I often saw illegally parked cars, not to mention that for a while, they were left unattended for several days, and on the bicycle lane, electric delivery trucks were running at a gallop, and still no one to manage that. Therefore, it is also an important aspect to strengthen urban traffic management.

We should pay attention to the concept of convenient transportation, because it is an important content of making cities livable and advanced, and it is related to the ecological balance of nature, economy, society and culture in ecological civilization.

Chapter 6 Just Society

Implement social justice, care for vulnerable groups, open urban space and serve urban residents.

When reflects social justice in the architectural science and culture, it means caring for the urban residents and vulnerable groups, and justly narrowing the gap between the rich and the poor in the urban living and working environment. Government management departments should give care and preferential treatment to the vulnerable people from the aspects of living, public buildings, road traffic, green space and so on. We should not widen the gap between the rich and vulnerable groups for the sake of economic interests. Also, it is necessary to continuously expand the open space of the city, so that the urban residents can equally enjoy various social and cultural life service facilities in the city, and make the government management departments more close to and serve the urban residents.

These contents are the basic ideas and ways to solve the problem of social justice. China has achieved poverty alleviation for all at the end of 2020 and reached the standard of a moderately prosperous society in all respects. It is of great significance for the world to improve the living standard of people at the bottom. The construction of urban green space both home and abroad has been explained in the previous Chapter 2, and the road traffic has been introduced in Chapter 5. Here, the work done for the public and some excellent examples in terms of housing and public buildings are briefly introduced as follows, which can be used for learning and reference.

1. In terms of urban living, we should try our best to improve the housing of residents

Xiamen in Fujian Province, considered the needs of low-income people and vulnerable groups in newly-built residential areas, and built six or seven-story residential areas with comfortable and livable environment, and gave preferential treatment and attention to housing prices. In 2004, it won the UN Habitat Scroll of Honor Award, being especially praised for solving the housing problem for the vulnerable groups. Since 2007, cities all over China have been increasing the construction of affordable housing to meet the needs of vulnerable groups with lower incomes. At present, the government department in charge of housing put forward that the housing problem of citizens who live in difficulties, rural migrant workers and newly employed college graduates will be solved by developing rental housing.

Another excellent example of solving the housing problem for the citizens is the reconstruction project of Beijing Xiaohoucang Dangerous House completed before 1990. Xiaohoucang Hutong is located in Xizhimen, Beijing, and covers an area of $1.5hm^2$. There are 298 households with 1,100 people, of a population density of nearly 800 people per hectare. The original houses are simple bungalows, and the residents are mostly low-income and vulnerable groups. The designer is Ms. Huang Hui, a well-known architect of Beijing Architectural Design and Research Institute. She made a careful investigation of these 298 households, talked with the residents and listened to their demands. Finally, the original residents moved back to the original place to live, which ensured the smooth progress of the demolition work. The residents' living standards were improved, and the area was enlarged, the orientation was adjusted, the functions were improved, the storage area was increased, and the green space was added, which improved the living environment in this area, while maintaining the original appearance of Beijing and meeting the living requirements of the public. The designer told me that after the completion of this district, an old lady grabbed her hands and said excitedly and repeatedly, "I used to be poor. I want to thank the government for letting me live in a sunny house, which is the sunshine and warmth given to me by the Party. Please help to express my appreciation to the government." This shows that caring for the disadvantaged people to solve the housing problem is very popular. This reconstruction project was published in the 7th issue of *Journal of Architecture* in 1991. I once told the designer that you designed for the people, and all of your works were praised by people, so you can be called "People's design master".

The solution to the housing problem in Hong Kong is that the government invests in the construction of public housing and Home Ownership Scheme (HOS). Public housing is a low-standard mode in which many households share toilets and kitchens and the houses are rent to vulnerable groups at low rents. HOS is like the affordable housing in the mainland, but its standard is much lower, and the government sells it to low-income people at preferential prices.

In terms of social justice, some good practices in foreign cities are that old residents are retained in the old urban areas, such as Paris, Belgium, Bruges, Barcelona, and Seattle, all of which retained and improved the original residential buildings, and most of the old residents are still there, instead of demolishing old buildings to build new high-rise buildings, moving the old residents and vulnerable groups to the suburbs, and turning the old city into the residence of rich people. This practice of improving or transforming the housing in the old city and retaining the old residents is worthy of recognition. Because this is not only an economic problem, but also a social problem, it preserves the historical and cultural blocks of the city. With the presence of old residents, its cultural life characteristics are preserved, and because of the government's care, it maintains social justice and promotes social stability. The concrete example is a residential area reconstruction project in London of England. The designer is the famous British architect Hackney. He takes care of these low-income residents, makes full use of the original building materials, improves the space environment, and organizes the construction and reconstruction by himself. The cost is not much, but it has achieved good results and is praised by the public. Therefore, he was elected as the president of both Royal Institute of British Architects and International Union of Architects. There is also a native residential area designed and built by the famous Egyptian architect Hassan Fathy, which conforms to the local living habits and appearance style of Egypt. It has complete space and functions for comfortable life and low cost, and has effectively solved the housing problem for the poor, and won the gold medal of the 15th Congress of the International Association of Architects.

Here, some foreign residential buildings are introduced, and the information is provided by architect, Ms. Huang Hui. The slogan of "World Habitat Day" in 1987 was "Housing for the Homeless". Although this slogan is unrealistic in some countries, it still aroused great repercussions. The issue of "housing for the public" has become a worldwide concern. Different countries have different political, historical, economic and natural conditions, so the ways and means to solve the housing problem are different. 1) Many countries pay more attention to housing construction. The government focuses on low-income households and builds "public housing". Private real estate developers focus on high-income households. The investment in building high-grade public housing comes from the government. After completion, it will be rented (or sold) to low-income citizens at very favorable rent. If the total household income of households rises above the quota of subsidized qualifications, it will stop preferential treatment that they need to move out, or pay high rent, or buy the original housing at a high price. This is the practice in France and other countries. 2) Apply Loans from the World Bank or domestic banks and governments are responsible for the construction. The land is sometimes provided by the government and sometimes by private individuals. One third of the population in Quezon, outside the Philippine capital, has a low income. In 1987, with the assistance of the Housing and Urban Development Association, a residential area was built on 1.8hm^2 of land through a loan from the World Bank. In 1990, 40 residential areas were planned in Craigland Village, all of which were loaned by the World Bank. There are more than 3,000 houses in Tangsonghong Community, 20km away from the center of Bangkok, Thailand, of which 50% of the construction funds are borrowed from the World Bank, 43% are domestic loans, 7% are drawn from the government's public welfare funds, and the loans are paid off by the residents within 20 years. 3) Households apply for low-interest loans from the state and local development organizations are responsible for the development of residential areas. For example, the Iremaya Community in Sri Lanka was partially built in 1985, and the main building, roads and supporting facilities were built by loans. The equipment part was self-raised and borrowed by the households, and the households should pay off the loans within 15 years. If the municipal facilities and destitute families are unable to repay, they should be supported by the government. 4) Large enterprises invest in building residential areas for employees, and sell or rent them to low-income employees at preferential prices. For example, a steel workers' area in Pakistan is built near a steel factory, which reduces fatigue of workers on the way to and from work, and is conducive to improving production efficiency of factories. 5) Private fund-raising and cooperative housing construction. In some countries, residents are organized to build houses in cooperation, and some funds are obtained from bank loans. Residents elect housing committees to be responsible for the implementation of construction plans and they are participated in the design. The Netherlands has a relatively complete working system in organizing residents to participate in design. There are great differences in the area standards and facilities of "public housing" in different countries. The kitchen and toilet facilities in developed countries are better, while many households in developing countries share a public toilet.

Although the area and facilities standards of "public housing" built in developed countries are high, there are slums of different races in the cities of developed countries such as in Europe and America. From the COVID-19 in 2020, it can be seen that these mixed slums have the highest infection rate and mortality due to poor environmental conditions and lack of medical care. Therefore, Manuel Franco, a famous Spanish medical expert, called for attention to social inequality in 2020. The infected cases and death toll of poor residents in various countries are many times higher than those of rich people. In the future, more attention should be paid to vulnerable groups in housing and other aspects.

2. In terms of public building service facilities, we should first consider serving the public

In terms of public buildings and cultural life, Hefei, Xiamen, Zhuhai, Shanghai and other cities all attach importance to the construction of various public facilities to meet the needs of residents' life. In the newly developed and old communities, besides the food and clothing shops serving the daily life of residents, there are kindergartens, nurseries, primary and secondary schools for teenagers to study, medical clinics for residents to see doctors, and other cultural, sports and repair service facilities. Some places have done a good job in caring for the residents of the whole city. For example, the Outline of Planning and Layout of Shanghai Vegetable Market and Public Toilets compiled by Shanghai Planning Bureau is very good. It is mentioned that in 2020, more than 200 vegetable markets and more than 800 public toilets will be added in the central city of the whole city, with the service radius of 500 m for the vegetable market and 300m for the public toilets. This service idea which is convenient for people's life is worth learning and emulating.

In foreign countries, social justice is reflected in the arrangement of public service facilities. There are examples of proper arrangement of public facilities for food, clothing and daily life in cities, such as the seafood market serving the citizens still preserved by the seaside in the central area of Seattle, and its buildings are original low-rise houses, but they are neat and orderly. Boston, a modern city on the east coast of the United States, still retains the old shopping malls serving the general public behind the city government building in the central area, which is welcomed by residents. They chose not to demolish the old shopping malls and build new commercial buildings, to avoid the local government and developers to damage the public interests and obtain their own money. These good practices are worth learning and affirming. Convenient for public life is also reflected in opening up more small squares for outdoor public activities in cities, such as sunken plaza in Rockefeller Center of New York, Leisure Square between high-rise buildings in central Seattle, Civic Center Square in Chicago, and Life Square in Brighton Residential Quarter of UK, etc. In these small square spaces, there are cafes, sculptures of flowers and trees, ice rinks and dancing grounds, which become public places for citizens to relax and play.

Open urban space, another point is that government agencies are open to the public. This is a new trend of urban development, which makes these urban spaces closer to residents. In 1990s, the architectural layout and management of Guangzhou Planning Bureau was open, and architect Mr. Lin Zhaozhang, then the deputy director, who had a better understanding of this development trend. In the 1990s, the design scheme of Shanghai

Architectural Design and Research Institute was finally selected for Xiamen Great Hall in the city center. One of the important points is that it is an open style. This building has already been built and used, and there are exhibitions of urban construction inside. It is often open to the public and is very popular among residents.

Open urban space, another point worthy of reference is that the municipal government office buildings, provincial or state government buildings are open to the public. For example, in the office building of Boston City Government, citizens can enter this office building at any time, and they all know that the prominent part of this building is the mayor's office. This approach narrows the distance between the city government and citizens. Another example is the Illinois State Government Building in Chicago, whose layout is public and approachable. In the middle is the atrium for public activities, while the lower part around the atrium is a shop for public use, and the government office floor is above it, which is open to the atrium and is an open office. There are many tourists and citizens visiting here every day, which gives the impression that the government staff is close to the urban residents. When we arrived in Philadelphia in 1989, we saw the city hall tower in the center of the city, which was also open to the public.

3. Social justice is one of the important concepts guiding architectural science development

At the end of this chapter, I want to further analyze Lao Zi's natural philosophy of "Tao models itself after nature". This world outlook is a correct concept of dialectical materialism, which means that the progress and development of everything in the world must conform to the objective law of natural development, which I call the law of nature. It is the guiding ideology of the development of architectural science and human society. Only when human society develops according to this objective law can it finally reach the ideal country of social justice. This goal is not affected by man's subjective consciousness. Leaders of all countries in the world can promote social progress if they follow this objective law, and those who do not will inevitably hinder social development. Therefore, it will take a tortuous road to achieve this goal. Lao Zi's idea of "Tao models itself after nature" is consistent with the lofty goal of the ideal country of communist society put forward by Marxism. We believe that it will take a long time to achieve this goal, and it can't be said how many centuries it will take. Only when the productive forces develop to a very high stage will there be a society in which there is no distinction between rich and poor people in production relations. The article named "To Establish a Scientific System of Social Forms of Consciousness" of Mr. Qian Xuesen published in the 9th issue of *Qiushi* magazine in 1988, mentioned that "Comrade Xia Yan once talked about two '70 years', from Marx and Engels writing *the Communist Manifesto* in 1847 to the victory of October Revolution in October 1917 is the first '70 years'; the second '70 years' is from the October Revolution in 1917 to the 13th National Congress of the Communist Party of China in 1987, when the theory of the primary stage of socialism was put forward. When adds another '70 years', that is, by 2057, to see if we can complete the tasks of the primary stage of socialism." This shows that the time to reach the communist society of the ideal country is quite long. Therefore, we must develop according to the objective laws of society and work to achieve a just society. To pursue the development of architectural science, we should have the idea of social justice, and leaders of various countries should take this as the guiding ideology to promote the progress and development of society. In short, we should serve the public. This is the case in China, and foreign countries should also attach importance to developing towards a just society. Lao Zi called social leaders as saints 2,500 years ago, because the legendary emperors of his time were leaders who promoted the development and progress of social civilization. In his book Tao Te Ching, he pointed out many times that saints should treat people with "inaction" and "action". "Inaction" means there can be no selfishness or lusts, while "action" is to seek benefits for the people and promote social development. The philosophical view of "Tao models itself after nature" advocated by Lao Zi is a dialectical and unified philosophical view composed of two philosophical elements: the "nameless" and the "named". The two symbolize the opposite relationship contained in everything, and they are unified to form everything. Foreign countries attach great importance to Lao Zi's philosophical thought of Tao Te Ching, which was translated several centuries ago, and the German and English versions in 19th and early 20th centuries were popular in Europe and America. The "organic architecture theory" of world famous architect Frank Lioyd Wright was put forward after being inspired by the philosophy of Lao Zi. In the second half of the 20th century, Reagan, then president of the United States in 1988, quoted the phrase "Rule a big country as you would fry small fish" in Lao Zi's *Tao Te Ching* in his State of the Union address. "Frying small fish" means keeping it delicious according to its own characteristics. "Ruling a big country" means knowing the times and social background, and adopting policies that lead the progress and development of society and people. Its central meaning is that no matter doing small things or managing big things, we must follow the objective laws of the development of things. In 2011, UN Secretary-General Ban Ki-moon also quoted the contents of "with all the sharpness of the way of heaven, it injures not; with all the doing in the way of the sage he does not strive" in *Tao Te Ching*. This is the last chapter, which means that sages should conform to the historical development and lead the society to make continuous progress, instead of leading others with their own subjective that would deviate from objective laws. Ban Ki-moon took this opportunity to explain that we should conscientiously implement the Charter of the United Nations for the consistent development of all countries. Because these leaders attached great importance to Lao Zi's *Tao Te Ching*, many American publishing houses published the English versions at the end of the 20th century, which had a wide influence.

These facts show that the creation and development of everything in the world must follow the objective laws of nature, and the ultimate goal is to establish a just society, which is also an important idea in the guiding ideology of the development of architectural science.

Social justice, caring for vulnerable groups, opening up urban space, retaining residents of the old city, arranging public facilities and facilitating people's lives are all social problems mentioned in this chapter, which are related to social security and stability. It is hoped that the social and ecological balance in ecological civilization can be achieved through the input of relevant Chinese and foreign administrative departments in thought, energy and economy.

Epilogue

The idea running through this book is that the core philosophical idea of establishing ecological civilization and living environment for human beings and its six specific ideas are the guiding ideology for the development of the architectural science, which originates from Lao Zi's natural world outlook and develops with the philosophical idea of ecological civilization and living environment modernity.

Through the analysis of the previous six chapters combined with examples, our understanding of the development of Chinese and foreign architectural science and culture is: under the guidance of the core philosophical concept of establishing ecological civilization and living environment for mankind, firstly it is necessary to have the concept of "innovative development", attach importance to the development of industrial economy with its own characteristics, so as to drive urban construction with special focus on digital economy, attach importance to scientific innovation and development, and promote architectural science and culture to a higher level. Secondly it is necessary to have the concept of "natural environment", strengthen the environmental governance of water, air, garbage and traffic, build urban green space system, ensure the green space area, make full use of natural light, wind and heat energy, ensure energy conservation and environmental, and ensure the symbiosis between man and nature, and moderately control the city scale and building space capacity in order to facilitate disaster prevention and sanitation. Thirdly it is necessary to have the concept of "history and culture", protect historical and cultural cities as a whole, not destroy history and culture, and combine protection, use and development to make history and culture more dynamic. We should respect the architectural culture of history, coordinate the development of new and old buildings, modern and traditional buildings, and develop regional architectural culture by inheriting traditions and innovating new buildings. Fourthly it is necessary to have the concept of "artistic beauty", create city and architectural signs, form beautiful urban three-dimensional frame, control the scale and volume of buildings and roads, coordinate urban texture, master artistic laws, and enhance the overall perfection of urban buildings. Fifthly it is necessary to have the concept of "convenient transportation", change the car-dominated urban transportation, build a transportation system of sidewalks and bicycle roads, and connect with urban public transportation to form a network to enhance the popularity of the city and the convenience, safety and health of citizens. Sixthly it is necessary to have the concept of "just society", implement social justice in urban housing, public facilities, culture, education, travel and transportation, care for the majority of residents and the vulnerable groups, narrow the gap between the rich and the poor and open government management institutions, so that they can be close to urban residents. We believe that this core philosophical idea and its six concrete ideas are the development direction of the modernization of architectural science and culture, which will make certain contributions to improving the planning, design and construction level of Chinese and foreign cities, buildings and landscapes, and gradually achieve the goal of establishing a scientific and artistic ecological civilization and living environment for mankind. This is also the aim of the modernization of architectural science and culture.

中篇：
中国建筑科学文化精品图说

Part Two:
Illustration of Chinese Architectural Science and Culture

一、陕　西

1. 西　安

西安地区位于陕西省关中平原"八百里秦川"的中部，这里平原舒展，远处群山环绕，东有"西岳"华山、骊山，西为太白山，南侧是著名的秦岭山脉，北为连绵的阴山山系，内部有号称"长安八水"流过，其中最大的河流是渭河，古称长安的西安3000多年来就在这渭河边发展着。

从公元前11世纪建立丰镐二京起，先后有西周、秦、西汉、前赵、前秦、后秦、西魏、北周、隋、唐等王朝在这里建都，时间长达1000多年。西安在全国著名六大古都中，其建都的王朝最多，时间也最长。西安还是历史上的国际大都市，它在西汉、盛唐时期开辟并发展"丝绸之路"，促进了汉唐与中亚、西亚和欧洲各国的经济文化交流，使中国的文化名扬中外。

1949年5月，西安解放，翻开了西安历史新的一页。西安作为陕西省的省会，是全省政治、经济、文化中心，已建成为我国新兴工业基地之一，也是我国西北最大的中心城市。西安还是国家历史文化名城。

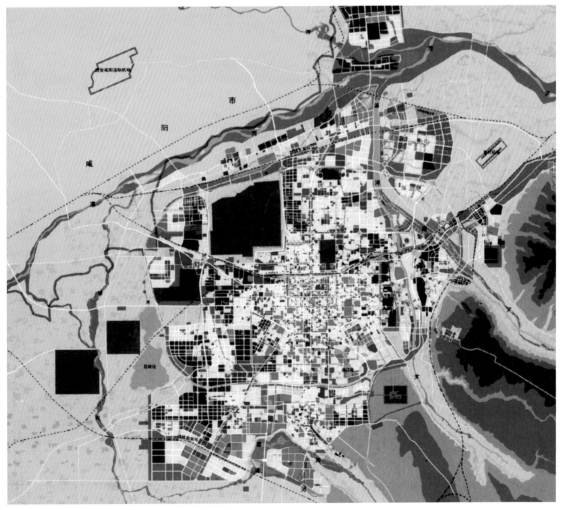

西安市主城区用地规划
Xi'an Urban Land use plan of the main urban area

实例1　西安秦始皇陵兵马俑

　　秦始皇陵在今西安市所属临潼区东5km处，骊山北麓，渭河南岸的平原上。陵封土如覆斗形，底边长350m，高43m，全部用夯土筑成。现陵墓保存完好，并未被盗。秦始皇陵于1961年被列为第一批国家重点文物保护单位，也是我国最早列入世界文化遗产名录的项目。

　　1974—1976年，在秦始皇陵东1.5km处发现3个秦代陶俑坑，其中一号坑最大。1979年在一号兵马俑坑上还建成秦始皇兵马俑博物馆。在这些兵马俑中，有弓卒、步兵、骑兵、战车兵和将军，其体形高大，武士俑身高1.78m至1.87m，身着战袍甲，挽弓挟箭，手持剑、矛，姿态威武，栩栩如生。兵马俑文物具有极高的历史和艺术价值。

骑兵鞍马俑
Cavalry pommel horse figurines

1. Shanxi

1.1. Xi' an

　　Xi' an is located in the middle of Guanzhong Plain, surrounded by mountains, with "Eight Waters of Chang' an" flowing through it. Since the 11th century BC, several dynasties have established their capitals here, and the number of historical relics ranks first in China. Xi' an is also an international metropolis in history. Now Xi' an is the capital of Shaanxi Province, the political, economic and cultural center, and one of the new industrial bases in China.

Example 1　Emperor Qin' s Terra Cotta Warriors

　　Emperor Qin' s Mausoleum is located in Lintong County of Xi' an. It is sealed like a bucket, with a bottom edge of 350 m and a height of 43 m. In 1961, it was listed as the first batch of Key Cultural Relics Site under the State Protection, and it was the earliest project listed in the World Cultural Heritage in China. From 1974 to 1976, three pits of pottery figurines were discovered 1.5 km east of the Mausoleum, and the Terra Cotta Warriors Museum was built in 1979. The warriors are tall, dressed in battle robes, lifelike, and of high historical and artistic value.

秦始皇兵马俑博物馆一号坑局部内景
Interior of Pit No.1 of Terra Cotta Warriors Museum

秦始皇兵马俑博物馆入口
Entrance of Terra Cotta Warriors Museum

实例2　西安建章宫苑

该宫苑建于公元前2世纪，位于西安旧城外西北部，它是"一池三山"园林形式的起源。古书中记载，建章宫"其北治大池，渐台高二十余丈，名曰太液池，中有蓬莱、方丈、瀛洲，壶梁象海中神山、龟鱼之属"。这种形式一直为中国后世所仿效，并影响到日本。如中国的杭州西湖、北京的颐和园等都采用了这一模式。从景观来看，这种模式确实可丰富景色：从岸上观水面，增加了景色层次；从水中三山上还可看到依水而建的主题景色。所以说"一池三山"的形式是一种造园的手法。

Example 2　Jianzhang Palace

Built in the 2nd century BC, the palace is located in the northwest outside the old city of Xi'an, which is the origin of the landscape form of "One Pool and Three Mountains". According to the book, "the Taiye Pool is in the north of the palace with Penglai, Abbot and Yingzhou in it symbolizing the three sacred mountains in the East Sea". From the landscape point of view, this model can enrich the scenery, viewing the water from the bank, the three mountains add layers, and the theme scenery built by the water can also be seen from the three mountains in the water.

建章宫鸟瞰（原载《关中胜迹图志》）1-蓬莱山；2-太液池；3-瀛洲山；4-方壶山；5-承露盘
Aerial view of Jianzhang Palace (originally published in Guanzhong Photo Story of Famous Historical Site)
1-Penglai Mountain; 2-Taiye Pool; 3-Yingzhou Mountain; 4-Fanghu Mountain; 5-Dew Tray

实例3 西安慈恩寺大雁塔

位于唐长安城东南端。现明城缩小，坐落于西安城南郊。隋代此处为无漏寺，唐高宗李治为纪念其母文德皇后重建，后坍掉，武则天与王公再建。为楼阁式砖塔，平面方形，塔身10层，现仅存7层，高64m。其内部是木制梁枋、楼板隔层、木制楼梯上下，于明代重修时外面加了一层装饰，但其外形轮廓风格仍是原来面貌。现于塔北面开辟绿地广场，更衬托出大雁塔雄伟壮观的气势。此塔于1961年被列为第一批国家重点文物保护单位。

Example 3 Big Wild Goose Pagoda in Ci' en Temple

It is located at the southern suburb of Xi'an. In Sui Dynasty, it was Wulou Temple here. Emperor Gaozong rebuilt it in memory of his mother, then collapsed and was rebuilt in Emperor Wuzetian's reign. It is a pavilion-style brick tower of 10 floors and only 7 floors are left. The interior is made of wooden beams, floor partitions and wooden stairs. When restored in the Ming Dynasty, a layer of brick was added outside, but its outline style was still the same. Now, a green square is opened in the north of the tower, which sets off the magnificent and upright momentum of the Big Wild Goose Pagoda. This tower was listed as the first batch of Key Cultural Relics Site under the State Protection in 1961.

大雁塔外景
Exterior

外观近景　Close shot of appearance

实例4　西安城墙

西安是我国六大古都中唯一保存有完整古代城墙及其护城河的城市，但此时西安已失去作为都城的地位，它只是明代一般府城的规制格局。该城墙明初在唐长安城的皇城范围内建起的。城墙用黄土分层夯筑，至明隆庆二年（公元1568年）在城墙夯土外又加砌了一层青砖。城墙围合成长方形，东墙长2590m，西墙长2631m，南墙长3442m，北墙长3244m，墙高12m，底宽15—18m，顶宽12—14m。城门设在四面墙的中间。于崇祯末年，在四门外修四关城。此项目于1961年被列为第一批国家重点文物保护单位。

Example 4 City Wall

Xi'an is the only city with intact ancient city walls and moats among the six ancient capitals in China. The city wall was rammed in layers with loess. The wall is formed into a rectangle, with the east wall of 2,590 m long, west wall 2,631 m, south wall 3,442 m and north wall 3,244 m. It is 12 m high, 15 m to 18 m wide at bottom and 12 m to 14 m wide at top. The gates are located in the middle of the four walls. During Emperor Chongzhen's reign, the Guancheng was built outside the four gates. It was listed as the first batch of Key Cultural Relics Site under the State Protection in 1961.

关城近景 Close shot of Guancheng

城墙与关城 City wall and Guancheng

2. 临 潼

实例5 临潼骊山华清池

位于临潼城南骊山西北麓，这里有山和温泉水，风景优美，现是国家重点风景名胜区。唐贞观十八年（公元644年）在此建汤泉宫，天宝六年（747年）发展扩建，称为华清宫。唐玄宗每年同杨贵妃来此过冬，在这里沐浴。华清池水温43°，水中含有多种矿物质，适合沐浴疗养。1959年中华人民共和国成立10周年前夕，在原来的"女汤"西北面修建起5300m²的"九龙池"，在池北面建有飞霜殿，整个园区经多次修葺扩建，其规模比原来扩大了11倍。在20世纪80年代前后发掘出"九龙汤""海棠汤"的遗址后，于1990年建成"唐代御汤遗址博物馆"。中国建筑西北设计研究院总建筑师张锦秋在设计该遗址博物馆时特意保留其地形高差，并根据汤池的不同服务对象，分别采用不同的体量与造型，使其成为富有历史文化风情的景点。

1.2 Lintong

Example 5 Huaqing Pool in Li Mountain of Lintong

It is located at the northwest foot of Li Mountain in the south of Lintong. Tangquan Palace was built here in the 18th year of Emperor Zhenguan, and was developed and expanded in the 6th year of Emperor Xuanzong, and was called Huaqing Palace later. Every year Emperor Xuanzong spent the winter and bathed with Lady Yang here. Huaqing Pool has a water temperature of 43° and contains many minerals that is suitable for bathing and recuperation. On the eve of the 10th anniversary of the founding of New China in 1959, the "Nine Dragon Pool" green garden of 5,300 m² was built in the northwest of the original "female bathhouse". After many repairs and expansions, the whole park has expanded its scale by 11 times. In 1990, the Imperial Bathhouse Ruin Museum of Tang Dynasty was built. It kept the terrain height difference and adopted different sizes and shapes according to different clients, making it a scenic spot with rich historical and cultural customs.

骊山华清池九龙池景色
Scenery of Nine Dragon Pool

右为九龙汤馆，左为海棠汤馆　Nine Dragon Hot Spring House on the right and Haitang Hot Spring House on the left

二、北 京

实例6　北京总体规划与建设

北京建城已有3000多年的历史，从辽、金、元、明、清至今作为都城，亦有850多年的文化古都史，现存的故宫、长城、明十三陵、颐和园、天坛、北京猿人遗址等已列入世界文化遗产。北京是我国重要的历史文化名城，在国际上享有很高的声誉。我生长、生活在北京已80多年，亲身经历了它的风雨沧桑变化。新中国成立前，北京房屋建筑危旧，古建园林荒圮，道路泥泞不平，这种破败形象就是旧中国的缩影。

1949年1月北京解放，经过70多年的规划与建设，北京的面貌大为改观。1953年编制的北京城市总体规划方案，提出城市向四周扩展，工业区布置在东北、东南和西部边缘地区，特别是梁思成、陈占祥提出的将党政机构迁至西郊、保护古都原貌的方案未被采纳。1959年对北京城市总体规划重新进行了编制，提出了"分散集团式"的总体布局，这个重要理念奠定了以后多轮修改总体规划的思想基础。关于这个问题，我印象比较深刻。1962年国家计委城市规划局局长曹洪涛带队到北京市建委了解北京规划与建设的情况，我们参加了国家计委的工作组，当时北京出面接待工作组的是建委主任佟铮和郑祖武处长，他们介绍了"分散集团式"的规划格局和在长安街旁修建国务院八大部委的设想。改革开放之后的1982年、1992年分别提出了修改北京城市总体规划方案。进入21世纪，以中国加入世界贸易组织和北京成功申办2008年奥运会为契机，北京于2004年又提出了《北京城市总体规划（2004—2020年）》，2005年1月国务院批复同意这个修编规划，并作了12条批复，进一步明确北京是全国的政治中心、文化中心，是世界著名的古都和现代国际城市；北京中心城的建设，要以调整功能、改善环境为主，控制建设规模，加强通州等11个新城的规划……；同意《总体规划》确定的2020年北京实际居住人口控制在1800万人左右（其中，中心城控制在850万人左右），应着力于提高人口素质，防止人口规模盲目扩大等。从这70多年的历程可以看出，北京城市的规划与建设工作已走向成熟，并取得了显著的成就，当然还存在着执行规划不力的一些现象，诸如人口规模的控制等。

2. Beijing

Example 6 Master plan and construction of Beijing

Beijing has a history of more than 3,000 years and was the capital of many dynasties. The existing Forbidden City, Great Wall, Summer Palace and other historical sites have been listed as world Cultural Heritage. As an important historical and cultural city in China, it enjoys a high reputation in the world.

After the founding of New China, Beijing has gone through many years of planning and construction. In 1953, the master plan was to expand the city around, and the industrial areas were located in the northeast, southeast and western marginal areas. In 1959, it was re-compiled, and the general layout of "decentralized group" was put forward, which laid the ideological foundation for revising the general plan many times later. After the reform and opening-up, the revised Beijing urban master plan was put forward in 1982 and 1992 respectively. In the 21st century, taking China's entry into the World Trade Organization and Beijing's success to host the 2008 Olympic Games as opportunities, and according to the new development requirements, Beijing Urban Master Plan (2004-2020) was put forward, which further clarified that Beijing is a national political and cultural center and a modern international city. From the 70-year history, it can be seen that Beijing's urban planning and construction work has made remarkable achievements today.

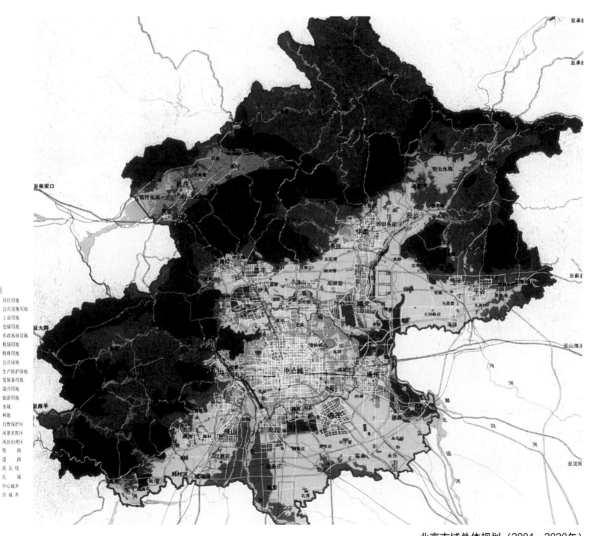

北京市域总体规划（2004－2020年）
Master plan of Beijing (2004－2020)

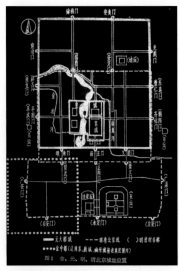

北京历代都城位置
Beijing capital locations of past dynasties

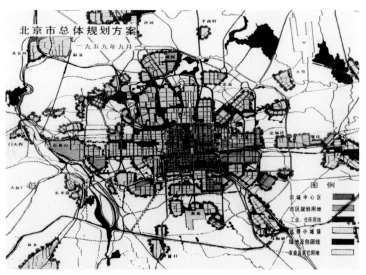

北京市总体规划方案（1959年9月）
Master plan of Beijing (Sep. 1959)

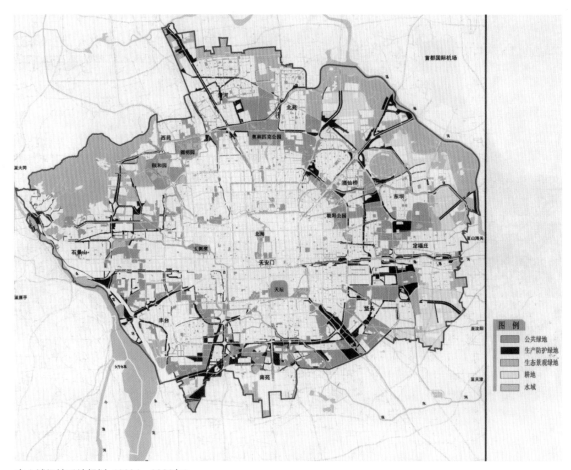

中心城绿地系统规划（2004–2020年）
Central City Green Space System Plan (2004–2020)

实例7 天安门广场

位于北京市区中心，南北纵向中轴线和东西横向轴线的交汇处，主体建筑天安门城楼坐落在广场北端的中轴线上，此城楼原为明清两代皇城的正门，创建于明永乐十五年（公元1417年），高33.7m，原名承天门，清顺治八年（公元1651年）改建后称天安门。天安门广场中心建有人民英雄纪念碑一座，于1958年4月落成。人民英雄纪念碑碑高37.94m，碑心正面镌刻毛泽东题词"人民英雄永垂不朽"8个鎏金大字，显得雄伟壮观、庄严肃穆。

在天安门广场两侧，1959年建成了人民大会堂，设计师是赵冬日、张镈建筑大师。人民大会堂建筑面积17.18万m²，面对天安门广场的正门矗立着12根高达25m的浅灰色大理石圆柱，它们组成了圆柱柱廊，人民大会堂建筑总高40多m，庄严绚丽。人民大会堂内设宽76m、深60m的万人大会场，是全国人民代表大会开会的地方，也是国家和人民群众开展重要活动的场所。在北边设有可容5000座位的大型宴会厅；南边是人大常务委员会办公楼。在天安门广场东侧，1959年建成了中国革命博物馆和中国历史博物馆，该两馆合成一座壮丽的建筑，设计师是张开济建筑大师。总建筑面积为6.5万m²，连接两馆的正门同样是12根巨型方柱组成的门廊，它同人民大会堂有所不同，其柱型分别为一圆一方，而廊型是一实一虚，两者在变化对比中取得和谐。中国革命博物馆和中国历史博物馆的色彩与人民大会堂一致，墙面为浅黄色。

沿南北中轴线往南，也就是人民英雄纪念碑的后面，1977年9月建起了毛主席纪念堂，再向南便是正阳门。正阳门是明清两代北京内城的正门，该城楼建于明永乐十九年（公元1421年），高42m，其后的箭楼则建于1439年。

1959年，在修建天安门广场两侧巨大建筑的同时，也开辟出规模为40公顷的长方形天安门广场，该广场为1959年10月1日热烈庆祝中华人民共和国成立10周年使用。天安门广场是新中国的象征，其气势磅礴，建筑高度从33m至42m不等，建筑色彩统一，新老建筑和谐，形成了一个完整的建筑艺术活动空间，它吸引着成千上万的中外旅游者前来观赏。

天安门广场中轴线上的天安门城楼、人民英雄纪念碑于1961年被列为第一批国家重点文物保护单位，北京正阳门于1988年被列为第三批国家重点文物保护单位。

Example 7 Tiananmen Square

It is located in the center of Beijing, at the intersection of vertical and horizontal axes. The main building, Tiananmen Gate Tower, is located on the central axis at the northern end of the square, and was originally the main entrance of imperial city in Ming and Qing Dynasties. In April 1958, Monument to the People's Heroes was set up in the center of the square and Chairman Mao's inscription "The people's heroes are immortal" is engraved on the front of the monument. On the west side of the square, the Great

Hall of the People was built in 1959 with a building area of 171,800 m^2. On the east side of the square, the Chinese Revolution and History Museum was built in 1959, with a total construction area of 65,000 m^2. Along the north-south central axis to the south, the Chairman Mao Memorial Hall was built in September 1977, and the Zhengyangmen is located to the south.

Tiananmen Square with an area of 40 hm^2 was built in 1959 and it is a symbol of New China. The

天安门
Tiananmen

天安门广场全景（中间是人民大会堂、人民英雄纪念碑）
Panorama of Tiananmen Square (Great Hall of People and Monument to the People's Heroes in the middle)

surrounding buildings are unified in color and the old and new are in harmony, forming a complete space for architectural art activities.

Tiananmen Square and Monument to the People's Heroes were listed as the first batch of Key Cultural Relics Site under the State Protection in 1961, and Zhengyangmen was listed as the third batch in 1988.

从西边东望人民英雄纪念碑、中国革命博物馆和历史博物馆以及毛主席纪念堂
Monument, Chinese Revolution and History Museum and Chairman Mao Memorial Hall

实例8 故宫(紫禁城)

故宫原为紫禁城,是明清两代的皇宫,是我国现存最大最完整的历史文化建筑群,建成于明永乐十八年(公元1420年),占地72公顷,建筑面积约15万m^2,它的布局是"外朝""内廷"坐落在中轴线及其两侧的对称位置上。"外朝"以太和、中和、保和三大殿为中心,这里是皇帝处理朝政的场所。"内廷"有乾清宫、交泰殿、坤宁宫和东西六宫等,是皇帝处理日常政务和后妃、皇子居住生活的地方。故宫南面是端门与天安门形成长条形前庭。故宫的正门是午门,北门是神武门,东门是东华门,西是西华门。乾清宫东西两边各布置6组院落,即东六宫和西六宫。内廷还建有3座花园,位于神武门南边的花园为御花园,位于宁寿宫西面的花园为乾隆花园,位于慈宁宫北面的花园是慈宁宫花园。故宫建筑群雄伟壮丽,它是中国古代建筑艺术的伟大成就,其内部还保存有历代的珍贵文物。故宫于1961年被列为第一批国家重点文物保护单位,并于1987年被列为"世界文化遗产"。

Example 8 Palace Museum (Forbidden City)

Palace Museum was the imperial palace of Ming and Qing Dynasties and is the largest and most complete historical and cultural complex in China. It was built in the 18th year of Emperor Yongle in Ming Dynasty (1420) and covers an area of 72 hectares with a building area of about 150,000 m^2. Its layout is "front imperial court and back palace". The "imperial court" is centered on the three Halls of Supreme Harmony, Central Harmony and Perfect Harmony, where the emperor handled the government affairs. The "back palace" includes Palace of Heavenly Purity, Hall of Union, the Palace of Earthly Tranquility, and imperial gardens etc. It is the place where the emperor handles daily affairs and the empresses and princes live.

It was listed as the first batch of Key Cultural Relics Site under the State Protection in 1961 and World Cultural Heritage in 1987.

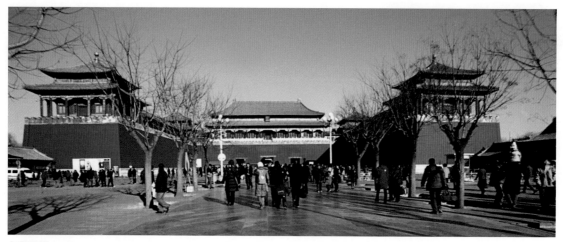

午门全景
Panorama of the Meridian Gate

从景山万春亭望故宫
Palace Museum from the view of Wanchun Pavilion in Jingshan Park

故宫全景鸟瞰（新华社稿）　Aerial view of the panorama of Palace Museum (from Xinhua News Agency)

故宫太和殿

故宫的主体建筑是太和殿,它是故宫三大殿中最重要的建筑,建于明永乐十八年(公元1420年),原名奉天殿,清顺治二年(1645年)始称太和殿。这里是皇帝召见群臣议政和举行各种大典活动的地方。现存建筑是康熙三十四年(1695年)重建的,为面阔11间重檐庑殿顶,进深5间,高35m,宽63m,建筑面积2377m²,并立于高2m的汉白玉石台基上。其前后各有3层石阶,中间石阶是一块巨大的石料雕刻艺术品,上面刻有以海浪、飞云为衬托的蟠龙,栩栩如生。建筑院落宽阔,红墙红柱黄瓦,色彩亮丽,使其成为故宫最壮观的建筑,也是全国最大的殿堂。

Hall of Supreme Harmony

It is the main building of the Forbidden City and is the place where the emperor summoned his ministers to discuss politics and hold various grand ceremonies. The hall has 11 rooms of double-hipped roof top with a building area of 2,377 m². It stands on a 2-meter-high white jade platform with three stone steps in front and back. The courtyard here is wide, with red walls, red columns and yellow tiles, all in bright colors.

太和殿前中间与三大殿后石阶石雕蟠龙
Carved stone dragons on the steps in front of the Hall of Supreme Harmony

石阶平台上铜鸟雕饰
Carved bronze bird on the stone platform

从太和门外高视点望太和殿全景
Panorama of the Hall of Supreme Harmony from the view of the Gate of Supreme Harmony

故宫乾隆花园

位于紫禁城东路、宁寿宫后寝的西侧，花园呈狭长方形，南北长160m、东西宽37m，做成纵轴南北向5个庭园空间，是为乾隆皇帝退位后之游园，故称之为乾隆花园。从南面衍祺门进入第一个庭园，中心建筑是北端的"古华轩"，西侧建有禊赏亭，外边凸出敞亭地面作一流觞曲水渠，取绍兴兰亭"曲水流觞"之意境；由古华轩后垂花门进入第二个三合院庭园，主体建筑为遂初堂，是一处雅致的休息之地；第三进是一封闭的山石空间，布满庭院的石假山上立一耸秀亭；第四进庭园的主体建筑是一个面阔、进深各五间的二层重檐的"符望阁"建筑；符望阁后身北面为第五进庭院，中心建筑是倦勤殿，西侧有一小院落"竹香院"。从这五进庭园的内容、空间的开敞、建筑的样式与高低大小、建筑同山石花木的配合等方面来看，都是富有韵律节奏和对比变化的，体现了中国造园艺术与工艺的最高水平。

Emperor Qianlong's Garden

It is located on the east road of Forbidden City and is a long and narrow square with a length of 160 m and a width of 37 m. It is a garden of five courtyards. The center building of the first courtyard is Guhua Pavillion, with a sightseeing pavilion on the west side and a meandering streamlet to take the artistic conception of drinking and poetry. The main building of the second courtyard is Suichu Hall, which is an elegant resting place. The third one is a closed space of rockery with a pavilion on the top. The main building of the fourth courtyard is a Fuwang Pavilion with two floors and double eaves. The central building of the fifth courtyard is Juanqin Hall and there is a small Bamboo Fragrance Pavilion on the west side.

石假山上耸秀亭
Songxiu Pavilion on the rockery

禊赏亭内流觞曲水
Meandering streamlet in Xishang Pavilion

"竹香馆"小院
Yard of Bamboo Fragrance Pavilion

实例9　北京西苑（今北海公园部分）

西苑是北海、中海、南海的合称，但此苑的起源部分是在北海，因而在这里仅着重介绍北海部分。辽代时这里是郊区，是一片沼泽地，适于挖池造园，随在此建起"瑶屿行宫"；金大定十九年（1153年）继续扩建为离宫别苑，现团城为其一部分；元至正八年（1348年）建成大都城中心皇城的禁苑，山称万岁山，水名太液池，山顶建广寒殿；清顺治八年（1651年）拆除山顶广寒殿，改建喇嘛白塔，山改名白塔山，至乾隆年间，题此山为"琼岛春荫"景，成为燕京八景之一。该园的特点是：

（1）"琼岛春荫"主景突出。北海的中心景物就是白塔山，即琼岛。岛上轴线与团城轴线呼应，构成了北海的中心。岛内立有乾隆皇帝所题"琼岛春荫"碑石。

（2）城市水系重要一环。自13世纪元代之后，这里已成为城市的中心地带，北京城的水系是自西北郊向东南郊方向连贯，北海太液池水是北京城水系的重要组成部分，起着连通的作用。

（3）西苑宫城相依相衬。元、明、清三代皇宫皆在今紫禁城位置，西苑北海、中海、南海与景山在其西面与北面，以拱形相依，相互依存、相互衬托，构成一个宫苑整体。

（4）城市立体轮廓标志。北京历史文化名城优美，还在于它具有韵律般的城市立体轮廓，近60m高的琼岛白塔是北京旧城立体轮廓的一个重要标志。

（5）园中之园相互联系。北海水面东、北两岸设置许多景点，游览路线将其联系贯通。

北海及其团城于1961年被列为第一批国家重点文物保护单位。

Example 9 Xiyuan (now part of Beihai Park)

Xiyuan is the collective name of Beihai, Zhonghai and Nanhai, but its origin is in Beihai. During Emperor Qianlong period, this mountain was named "Spring Shade on Qiong Island" and became one of the eight scenic spots in Yanjing. The park is characterized by:

(1) The main scene of "Spring Shade on Qiong Island" stands out.

(2) It is an important part of urban water system.

(3) It is echoing with the Forbidden City to form a whole palace.

(4) Part of the three-dimensional outline sign of the city.

(5) The scenic spots in the garden are interrelated.

Beihai Park and Round City were listed as the first batch of Key Cultural Relics Site under the State Protection in 1961.

琼岛白塔全景（新华社稿）
Panorama of White Pagoda on Qiong Island (from Xinhua News Agency)

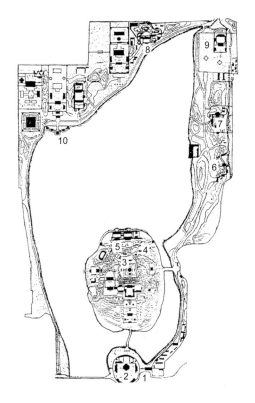

西苑平面　1-入口；2-团城；3-白塔；4-琼岛春阴碑；5-承露盘；6-濠濮间；7-画舫斋；8-静心斋；9-蚕坛；10-五龙亭

Plan　1-Entrance; 2-Round City; 3-White Pagoda; 4-Spring Shade on Qiong Island Monument; 5-Dew Tray; 6-Haopu Yard; 7-Painted Boat Studio; 8-Heart-East Study; 9-Silkworm Altar; 10-Five-dragon Pavilion

"琼岛春荫"碑（"燕京八景"之一）
Spring Shade on Qiong Island Monument (one of the eight scenic spots in Yanjing)

琼岛南面荷景（左部为团城）
Lotus scenery at the south of Qiong Island (Round City is on the left)

从五龙亭望琼岛
Qiong Island from the view of Five-dragon Pavilion

中篇：中国建筑科学文化精品图说/北京

125

北海团城

位于北海公园南门西侧，金时为御苑的一部分，元代在其上修建仪天殿，明时重修改名承光殿。高5m的城外围加砌砖墙，使之成为今日的团城模样，其面积有5760m^2。承光殿现存的方殿为清康熙二十九年（公元1690年）重建，平面为十字形，黄琉璃瓦绿剪边重檐歇山顶，四面单檐卷棚抱厦，飞檐翘角，宏丽庄重。殿内供奉由整块玉雕成的白玉坐佛一尊，高1.6m，清透亮润。乾隆十四年（1749年）又在承光殿前边建玉瓮亭，将黑玉酒瓮陈设亭中。此玉瓮为元初制作，高0.7m、长1.82m、宽1.35m，重约3500kg。玉质是青白中有黑，周身浮雕着海浪和海龙、海马、海鹿、海猪等，形态生动，原陈设在琼华岛广寒殿里，后几经转移，于乾隆十年在真武庙发现后，以重金购得置放承光殿中。这黑玉酒瓮是我国现存形体最大、时间最早的珍贵玉器文物。岛上古松甚多，大都为金代所植，其中一棵乾隆封它为"遮荫侯"，起因是有一年夏季白天闷热，乾隆来到团城坐在此树下，小风吹过，顿感凉爽，遂封为"遮荫侯"。这些古松亦可属于重要文物。还有一点值得介绍，就是在台地下面修建雨水收集工程，以利雨水排放和供林木生长，这是几百年前建造的技术工程，直至今日仍有参考价值。

Round City

It is located on the west side of the south gate of Beihai Park with an area of 5,760 m^2. The Chengguang Hall in the center is cross-shaped in plane, with multiple eaves of yellow glazed tiles and green trim. In the hall, a white jade carved from a whole piece is enshrined as a sitting Buddha with a height of 1.6 m. There is also a Jade Urn Pavilion with a black jade wine urn placed inside, which is a precious cultural relic as the largest and earliest existing jade shape in China. There are many ancient pine trees here, one of which was named "Duke of Shade" by Emperor Qianlong. Brick works of rainwater collection culverts are built under the platform to facilitate rainwater drainage and forest growth.

玉瓮雕刻
Carved jade urn

遮荫侯
Duke of Shade

承光殿与玉瓮亭
Chengguang Hall and Jade Urn Pavilion

北海静心斋

位于北海北岸，建于清乾隆二十三年（公元1758年），原名镜清斋。1913年经修葺，作为北洋政府外交部宴请外国人的地方，改称"静心斋"。此园的布局有一明显的中轴线，大门、长方形水庭院、镜清斋主体建筑和花园的中心建筑沁泉廊依序坐落在中轴线上，主体突出。花园呈横向在中轴线的东西两边展开。东边有画峰室和抱素书屋两个庭园，西边有枕峦亭和迭翠楼景区，景色丰富。中部沁泉廊前为水景，廊后是小池和以太湖石迭成的假山，山中有洞，山上还建有爬山廊直通迭翠楼。临水的石山景玲珑剔透，与花木和廊桥相互辉映，景色秀丽。尤其是登上枕峦亭和迭翠楼，可看到园外北海和优美的开阔景山景色。静心斋内外景色交融，堪称"园中之园"。

Heart-East Study

It is located on the north bank of Beihai Park and was built in the 23rd year of Emperor Qianlong (1758). It has an obvious central axis. There are Huafeng Room and Baosu Study in the east of the garden, Zhenluan Pavilion and Diecui Pavilion in the west. There is a waterscape in front of Qinquan Gallery in the middle, followed by a small pool and rockery, from which you can climb to Diecui Pavilion. The rockery near the water is exquisitely carved, reflecting each other with flowers and trees and covered bridges. The scenery inside and outside connected with each other, making it an exquisite "garden in the garden".

静心斋水庭
Waterscape

沁泉廊前景观
Qinquan Gallery

西部迭翠楼景色
Diecui Pavilion

实例10　北京颐和园

该园位于北京西北郊，始建于清乾隆十五年（1750年），1765年建成，名为清漪园。1860年清漪园被英法侵略军焚毁，清光绪十二年（1886年）又重建，改名颐和园。1900年颐和园又遭八国联军破坏，1901年修复。颐和园于1961年列为第一批国家重点文物保护单位，并于1998年12月被列入世界文化遗产名录。颐和园占地约290公顷，其特点是：

（1）依山开池，模仿西湖。这里原称瓮山西湖，明时建有好山园。乾隆十四年（1749年），为疏通北京水系，引玉泉山水注入瓮山前的西湖，再辟长河引水入北京城。乾隆十五年（1750年）决定在此建造清漪园，并拓宽西湖水，在前山的中部建大报恩延寿寺，将瓮山改名为万寿山，将西湖改称为昆明湖，这就是依山开池的因由。其总体布局完全是模仿杭州西湖风景格局，湖中建有西堤、支堤和3个岛，象征东海三神山——蓬莱、方丈、瀛洲，此西堤及堤上6桥是仿杭州西湖苏堤和"苏堤六桥"。大片昆明湖水为北面万寿山、西面玉泉山及其后面西山环抱，真好似杭州西湖的缩影。

（2）借景西山，建筑呼应。西边近景为玉泉山，山顶建一宝塔，远景为西山峰峦，景色十分深远。这开阔的园外美景皆借入园中，扩大了此园的空间，这是该园造园的一大特色。

（3）东面为宫殿，东北为居住区。此离宫别苑，采取的仍是前宫后苑的布局。宫廷区又分成朝寝两部分，位于东部布置以勤政殿（光绪时改名为仁寿殿）为中心的建筑群，是上朝处理政务之地。在东北两面扩建了后廷部分，其中慈禧居住之处乐寿堂，布置格外精美，可透过对面廊道墙面上的什锦窗，望到湖光美景。

（4）长廊连接，景观丰富。在前山脚下布置一长廊，将北部自东向西的建筑群连接起来，共有273间、长728m，是中国园林中最长的长廊。它是一条很好的游览路线，可观赏到许多变化的景观。阴雨时避雨淋，烈日时防日晒。它也是观赏的对象，在每间中都可欣赏到彩绘的各地山水画卷。

（5）园中有园，仿无锡寄畅园。全园造景100余处，园中有园，沿湖东岸向北行有十七孔桥、知春亭、夕佳楼等园景，沿西堤北行是六桥景色和一片田园风光。在万寿山前山山腰东部有景福阁园景，可俯视开阔的全园山水景色；前山山腰西部有画中游等园景，同样可横览全园的湖光山色。在东边安排一园景，它在清漪园时称惠山园，在颐和园时名为谐趣园，这是仿无锡寄畅园而建。

Example 10　Summer Palace

It is located in the northwest suburb of Beijing and covers an area of about 290 hectares. It was built in the 15th year of Emperor Qianlong (1750). It was listed as the first batch of Key Cultural Relics Site under the State Protection in 1961 and was listed as World Cultural Heritage in December 1998. It is characterized by the following aspects:

(1) The pool was built along the mountain and imitated the West Lake in Hangzhou. The water of Jade Fountain Hill was introduced into Kunming Lake. The overall layout imitates the landscape pattern of "One Pool and Three Mountains".

(2) Borrowing the scenery and echoing each other. Close-up to the west is Jade Fountain Hill, and a pagoda is built on the top. The distant view is Xishan Mountain with far-reaching scenery.

(3) The palace is located in the east, with the Qinzheng Hall as the center, and the Hall of Happiness and Longevity is the representative building of the back palace in the northeast.

(4) The long corridor connects the buildings in the north from east to west, with a total of 273 rooms and a length of 728 m, where the painted landscapes can also be seen in the corridor.

(5) There are gardens within the garden. There are more than 100 scenic spots in the whole area, including Seventeen Hole Bridge, Spring-feeling Pavilion, and others.

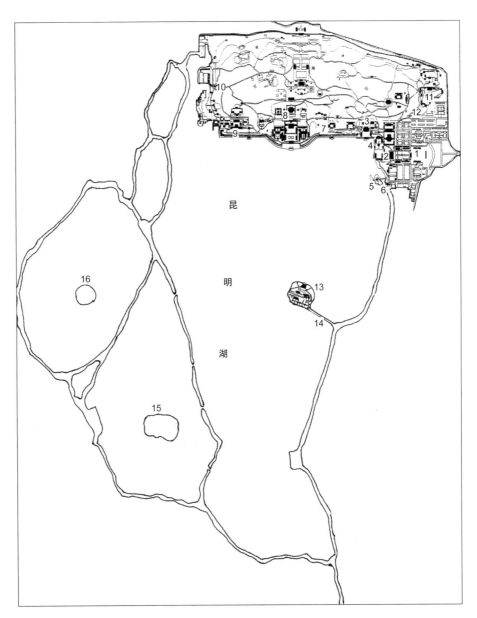

平面 1-东宫门；2-仁寿殿；3-乐寿堂；4-夕佳楼；5-知春亭；6-文昌阁；7-长廊；8-佛香阁；9-听鹂馆（内有小戏台）；10-宿云檐；11-谐趣园；12-赤城霞起；13-南湖岛；14-十七孔桥；15-藻鉴堂；16-治镜阁

Plan
1-East gate; 2-Hall of Benevolence and Longevity; 3-Hall of Happiness and Longevity; 4-Xijia Pavilion; 5-Spring-feeling Pavilion; 6-Wenchang Pavilion; 7-Long corridor; 8-Pavilion of the Fragrance of Buddha; 9-Hall for Listening to Orioles Singing (a small drama stage inside); 10-Suyun Pavilion; 11-Garden of Harmonious Interest; 12-Rosy Clouds Rising; 13-Nanhu Island; 14-Seventeen Hole Bridge; 15-Zaojian Pavilion; 16-Zhijing Pavilion

从佛香阁侧面观湖山景色
Scenery from the view of Pavilion of the Fragrance of Buddha

十七孔桥
Seventeen Hole Bridge

颐和园乐寿堂

它是面临昆明湖的一个四合院落的主体建筑,是慈禧居住的地方。此院落布局,中轴线突出,堂居正中设有宝座、御案、屏风等,堂内西套间为慈禧寝宫,东套间是更衣室。堂外台阶两侧对称布置铜制梅花鹿、仙鹤与大瓶,取其谐音"六合太平"之意。庭院中种植着玉兰、西府海棠和牡丹名贵花木,取"玉堂富贵"之意。院落南部中轴线上陈设一块巨石"青芝岫",使院落更加生辉。院外有慈禧乘船的码头。

Hall of Happiness and Longevity

It is the main building of a quadrangle courtyard facing Kunming Lake, where Empress Dowager Cixi lives. There are thrones, imperial tables, screens in the hall, the west suite is the bedroom, and the east suite is the dressing room. Bronze sika deer, crane and big bottle are symmetrically arranged on both sides of the steps outside the hall, which means peace and tranquility. Magnolia, begonia and peony are planted in the courtyard, meaning abundant wealth, and a huge stone "Qingzhixiu" is displayed in the south of the courtyard. There is a port for Cixi to take a boat outside the courtyard.

乐寿堂前盛开的玉兰树和铜制"六合太平"雕饰物
Blooming magnolia and bronze statues

从院落东南角观庭院全景（左为"青芝岫"巨石）
Panorama from the southeast of the yard (the huge stone on the left)

颐和园谐趣园

位于颐和园万寿山东麓，清乾隆十六年（公元1751年）仿江苏无锡惠山脚下寄畅园的模样建造，名惠山园，其景观以自然山水为主。清嘉庆十六年（1811年）重修，取"以物外之情趣，谐寸田之中和"之意，改名谐趣园。后被英法联军毁坏，光绪时重建，为慈禧观鱼垂钓之处。但此次改建，增加了亭台楼阁建筑，失去了原有造园以自然山水为主的景色，仅园的西北角仍保存着山泉流水的自然景观。

Garden of Harmonious Interest

It is located at the eastern foot of Longevity Hill and was built in imitation of Wuxi Jichang Garden in the 16th year of Emperor Qianlong (1751). It was destroyed by the Anglo-French allied forces and rebuilt in Emperor Guangxu's reign. However, due to this reconstruction, pavilions and halls have been added, and the original landscape dominated by natural landscapes has been lost. Only the northwest corner of the garden still preserves the natural landscape of mountain springs.

东部知鱼桥
Zhiyu Bridge in the east

夏日荷景
Lotus scenery in summer

实例11　北京天坛

天坛位于北京永定门内大街东侧，始建于1420年明永乐迁都北京之时，是明清两代帝王祭天祈谷祈雨的坛庙，占地280多公顷。天坛于1961年列为第一批国家重点文物保护单位，并于1998年12月被列入世界文化遗产名录。其园林特点是：柏林密布，烘托主题，在主要建筑群轴线的外围，即祈年殿、丹陛桥、圜丘的四周密植柏树林木。设计者将殿、桥、丘的地面抬高，人在其上看到的外围是柏树顶部，创造出人与天对话的氛围，以达到祭天的目的。天坛总体布局规则整齐，园林随建筑布局，建筑群严整规则，轴线突出。其象征格局体现"天圆地方"的理念，诸如总平面北面南向两道坛墙都作成圆形，象征天；南面北向的两道坛墙作成方形，象征地。天坛祈年殿、皇穹宇、圜丘都做成圆形，而四周的围墙做成方形，亦与天地呼应。天坛体现了封建宗法礼制思想，即"天"对人间是至高无上的主宰，祭天是神圣的。

Example 11 Temple of Heaven

It is located on the east side of Yongdingmennei Street and was built in 1420. It is an altar temple covering an area of more than 280 hectares, where emperors of Ming and Qing Dynasties worshipped heaven, prayed for grain and rain. In 1961, it was listed as the first batch of Key Cultural Relics Site under the State Protection, and in December 1998, it was listed on the World Cultural Heritage List.

It is characterized by dense cypress trees. Cypress trees are planted on the periphery of the main

祈年殿全景 Panorama of the Hall of prayer for Good Harvests

buildings, and the tops of the trees are seen around, creating an atmosphere of dialogue between man and heaven. The general layout is neat and the concept of "round sky and square earth" is reflected in many places, such as the Hall of Prayer for Good Harvests, the Imperial Vault of Heaven, and the Circular Mound Altar are in round shape and the surrounding walls are square, echoing the heaven and earth.

圜丘至丹陛桥、祈年殿轴线景观（新华社稿）
The axis From Circular Mound Altar to Red Stairway Bridge and Hall of Prayer for Good Harvests (from Xinhua News Agency)

实例12 长城——八达岭

　　八达岭在北京延庆，它是长城的一个关口，建于明弘治十八年（公元1505年）。长城全长6700km，于1961年列为第一批国家重点文物保护单位，现已列为世界文化遗产项目。秦始皇统一六国，将赵、燕、秦曾建起的高大土城墙连接成一道万里长城。到了明代，为加强防御，部分长城改为砖石构造，西起嘉峪关，东到山海关。八达岭一带的长城，依山势而建，平均高6~7m、宽5m，以大块城砖砌筑墙体表面和墙顶地面，外侧作成垛墙，便于防御。下层有空屋，可驻兵和贮存武器。至清代，长城逐渐失去防御作用，日渐荒废。新中国成立后补修了八达岭关城城楼与居庸外镇和城台。八达岭一带长城，每逢秋季有红叶相衬，更显其气势宏伟、优美壮观。

Example 12 Great Wall – Badaling

　　It is located in Yanqing of Beijing and was built in the 18th year of Emperor Hongzhi in Ming Dynasty (1505), with a total length of 6,700 km. It has been listed as World Cultural Heritage and the first batch of Key Cultural Relics Site under the State Protection. The first emperor of Qin Dynasty unified the six kingdoms and connected the tall earth walls into the Great Wall. Badaling, built according to the mountain, has an average height of about 6 m to 7 m and a width of 5 m. The wall surface and the top floor are built with large bricks, and the outer side is made into a crib wall, which is convenient for defense. There are empty houses on the lower floor, where soldiers can be stationed and weapons can be stored. Badaling, with red leaves in autumn, shows its magnificent momentum and beauty.

长城秋色
Autumn scenery of Great Wall

八达岭长城墙身高陡的地段
The steep part of Badaling

实例13　北京四合院和鲁迅故居

北京四合院有明确的中轴线布局。正房北房3间布置在主轴线上，正房两侧各有一间较低的耳房，正房前边左右对称安排东西厢房，并与南墙、垂花门形成中心院落，这是此宅的核心部分。在这中心院落前面有一扁长形的前院，其东南面为大门入口，进门前行左转即可进入此前院，前院南面为倒座房，面对垂花门，此房为接待客人使用。在正房后面有一排北房，称为后罩房，其院落叫后院，这后罩房和耳房一般为女眷的住房。此前院、主院、后院组成典型的三进院落。房屋结构为木梁柱式，墙不承重。屋顶多用硬山清水脊，瓦一般为阴阳瓦，房屋为三面墙，面向院落面开门窗。院落中种有花木，上通天、下接地，人与大自然融合在一起，这是一种很好的居住方式。

鲁迅故居在北京阜成门内西三条胡同21号。1924年5月—1926年8月，鲁迅先生在此居住。此故居为一简单四合院，大门设在东南角，进门左拐，可见正屋北房3间，这是鲁迅的工作室和卧室；南房亦3间，分别是客厅和东西两间厢房。四合院中有鲁迅亲自栽植的丁香树。其北房后面接有一间小屋，称"老虎尾巴"。后院还有一水井，并植有刺梅。今鲁迅故居已重新修缮，按原样布置，并对外开放。

小院风貌
Scenery of the yard

Example 13 Quadrangle Dwellings and Former Residence of Lu Xun

Quadrangle dwellings have a clear central axis. The main building is in the north, with a lower wing on each side. The front left and right sides are called east and west wing rooms, which form a central courtyard with the south wall and the hanging flower gate. Further forward is the front yard, the entrance is in the southeast, and there is a room in the south of the front yard for receiving guests. The structure of the house is wooden beam column, and the wall is not load-bearing.

Lu Xun's former residence is located at No. 21 Xisantiao Hutong in Fuchengmen. He lived here from May 1924 to August 1926. It is a simple courtyard house, with its main gate in the southeast corner, three main rooms, as study and bedroom, three south rooms and two east and west wing rooms. There are clove trees planted by Lu Xun himself in the courtyard. The place has been renovated, laid out as it is, and opened to the public.

实例14　北京西四北六条幼儿园四合院

该院原为一中上层人士的住宅，后改作幼儿园使用。其布局特点同前面叙述的四合院完全一致，只是规模大一些，为四进院落，多了中间一进院落，且主院落的正房为5开间，约16m宽。在这四合院住宅的东西两边，整齐并列着类似的四合院。院外形成6m左右宽的胡同，胡同的另一边同样并列着这种样式的四合院，只是入口大门开在西北边。从下面的胡同照片可以看出，其空间尺度宜人，安静优雅，适宜居住。

Example 14　Quadrangle Dwellings of Beiliutiao Kindergarten in Xisi

It was originally the residence of a middle-class person, and was later used as a kindergarten. Its layout features are exactly the same as those of the quadrangles described above, except that it is a four-courtyard style with a larger scale, and the main building has five rooms. On its east and west sides, similar quadrangles are lined up neatly, and an alley with a width of about 6m is formed outside the courtyard.

中篇：中国建筑科学文化精品图说/北京

幼儿园四合院及其东面四合院落群
The kindergarten and the quadrangle dwellings on the east

西四北六条胡同，左下部为幼儿园入口
Beiliutiao Hutong and the entrance of kindergarten on the bottom left

实例15　北京中国美术馆

位于北京五四大街东端北侧，西面和故宫博物馆、景山、北海公园遥遥相对，建成于1962年，设计师是著名建筑大师戴念慈先生。该馆主要是为了展出我国自五四运动以来在美术方面的各种优秀作品，包括绘画、雕塑、版画、工艺美术品和民间艺术品等，同时也可举办国外来华的美术展览会。全馆总面积16000m^2，其中展出面积7000m^2。展览建筑布置在美术馆中心及其两侧，方便广大观众参观，办公建筑则布置在美术馆两端。东门可供出入，避免与观众相互干扰。该建筑造型舒展、比例尺度适宜、虚实对比有序、彩色明快亮丽，充分体现了现代建筑与传统民族风格的有机融合，并与周围历史文化建筑相协调。

Example 15　National Art Museum of China

It is located on the north side of the east end of Wusi Street, facing the Palace Museum in the west, and was built in 1962. The main purpose of the museum is to display all kinds of outstanding works in fine arts in China since the May Fourth Movement, and also to hold art exhibitions from abroad. The total area of the whole museum is 16,000 m^2, of which the exhibition area is 7,000 m^2. The exhibition part is arranged in the center and its two sides in the architectural layout, which is convenient for the audience to visit.

外景
Exterior

中心圆形大展厅
Circular exhibition hall in the center

两侧凸出展厅
Exhibition halls on both sides

正面中心部分
Front central part

实例16　清华大学建筑设计研究院办公楼

　　这是一座探索与尝试"生态绿色办公楼"的设计,其目的是为该院提供一个健康、节能、高效的工作场所。为此,清华大学设计研究院的设计小组,进行了两年的设计与研究工作,从1997年开始到1999年完成设计方案。该设计方案获得1999年北京市"十佳"建筑设计方案奖。该建筑设计重点考虑了四方面的策略:一是缓冲层策略。在南向设计了一个边庭,过渡季节它是一个开敞空间,即冬季它是一个暖房,夏季就成为一个凉棚,同时朝西设防晒墙。屋顶设两条东西向长条天窗,朝南面设遮阳板系统;二是利用自然能源策略。充分考虑利用太阳能,预留了太阳能光电板系统的设施位置,并通过分析研究,提出在合适条件下要利用地下水作为天然冷源;三是无害化、健康化策略。通过计算机模拟,各层办公室在过渡季节完全可以依靠自然通风维持室内舒适条件,同时,南侧内庭的绿化平台不仅提供一个休息与交流的场所,其本身也是一个健康的自然调节器,可优化室内空气条件;四是整体的节能策略。诸如采用节能照明灯具、联合电位处理、消防泵自动巡检、照明与空调可调节、生活引用水分质供水等方式或措施。

　　该设计小组认为,这座绿色办公室好比一辆"自行车",它使用最少的能源,并对人的身心健康有益。

外景
Exterior

Example 16 Office Building of the Architectural Design and Research Institute of Tsinghua University

It is to explore and try the design of "eco-green office building", which aims to provide a healthy, energy-saving and efficient workplace. It focuses on four strategies: 1) buffer layer strategy, a side court is designed to the south, which is an open space in the transition season, a greenhouse in the winter, and a permafrost in the summer, 2) utilization of natural energy, full usage of solar energy and groundwater, 3) harmlessness utilization of natural ventilation and greening platform to optimize the overall air quality, 4) adoption of energy-saving lighting fixtures, adjustable air conditioners, etc.

屋顶长条天窗
Long strip skylight on the roof

边庭
Side court

实例17　北京国家体育场

该体育场是举行第29届奥林匹克运动会开、闭幕时的会场，也是奥运会田径项目的比赛场，并作为奥运会后大型体育与文化活动的场所。国家体育场可容纳观众10万人，其中永久性观众席位为8万个。该项目设计举办了国际性建筑概念设计方案竞赛，共有13个单位按时提交了设计方案，2003年3月由国际专家评委会对13家的设计方案进行评审，通过无记名投票方式评选出3个优秀方案，最后在此3个优秀方案中推荐出瑞士Herzog & deMeuron建筑设计公司+中国建筑设计研究院联营体设计方案为实施方案。该设计方案展现了体育场空间的独特性，其结构组件相互支撑，形成网格状的构架。看台是一个完整的、没有任何遮挡的碗状型。国家体育场外壳正如小鸟筑巢时用柔软物填充树枝的构造一样，采用ETFE四氟乙烯气垫膜安装在构架外部，使屋顶达到防水要求，并可自然采光、自然通风，其生态、环保、节能效果好，但用钢量大。现已作为北京2022年冬奥会开、闭幕式的会场。

Example 17 National Stadium

It was a venue for the opening and closing ceremonies as well as track and field events of the 29th Olympic Games, and now is a place for large-scale sports and cultural activities. It can accommodate 100,000 spectators. The space of the stadium is unique, and the components of the structure support each other to form a grid-like

鸟巢外观
Exterior

framework. The grandstand is a complete bowl without any shelter. The external shape is like the surface of the branch structure with soft objects filled in by a bird for building a home, commonly known as the Bird's Nest. The ETFE air cushion film is installed outside the framework, so that the roof can meet the waterproof requirements, and it can be used for natural lighting, which is beneficial to the lawn growth in the stadium. It can also make the stadium get natural ventilation and has good ecological, environmental protection and energy saving effects. Now is the place for opening and closing ceremonies of the Beijing 2022 Winter Olympic Games.

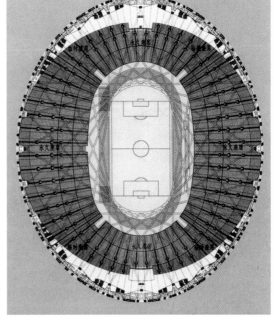

座席平面
Plan of grandstand

三、河　北

承　德

实例18　避暑山庄

（又名承德离宫、热河行宫）

该山庄位于河北省承德市北部、武烈河西岸，占地5.6km²，与西湖面积相仿，始建于康熙四十二年（1703年），1708年初具规模，于乾隆五十五年（1790年）建成。该山庄的特点是：

（1）夏季避暑，政治怀柔。这里山川优美、气候宜人，清代皇帝满族人原居关外，而这里正适合帝后夏季避暑享乐。选择此处靠近蒙古族，也便于藏族来往。为了加强对边疆的管理，采取怀柔政策，常邀请蒙、藏族头目来此相聚，友好相处。

（2）山林怀抱，山水相依。山庄四周，峰峦环绕。山庄西北面为山峦区，占全部面积的4/5；平原位于东南面，占全部面积的1/5。平原中的湖面又约占平原面积的一半，此湖水是由热河泉汇集而成。避暑山庄造园是以山林为大背景，创造了山林景观，它在湖面上创造山水景色，在平原处创造草原景观。

（3）前宫后苑，前朝后寝。宫殿区在南端，苑囿在其后，为"前宫后苑"格局。正宫主殿为"澹泊敬诚"殿，乾隆皇帝在此朝政。后面"烟波致爽"为寝宫，仍按"前宫后寝"的形制布局。

（4）湖光山影，风光旖旎。湖泊区是山庄园林的重点，位于宫殿区北面，湖岸曲折、洲岛相连、楼阁点缀，其景观极其丰富。康熙、乾隆数下江南，将一些江南名胜景观移植于此。游至仿镇江的金山和仿嘉兴的烟雨楼高视点处，这里视野开阔，向外眺望可欣赏群山环抱的湖光山影和南秀北雄的园林景色。

（5）北部平原，草原风光。湖区北岸碧草如茵，驯鹿野兔在穿梭奔跑，真是一片北国草原风光。乾隆在此搭建蒙古包，邀请蒙、藏等少数民族首领野宴、观灯火。

（6）西北山岳，林木高峻。大片山岳区位于山庄西北部。在山岳区西部可观赏到"四面云山"景色，在北部可远眺"南山积雪"景色，在西北部可望见"锤峰落照"景色。

避暑山庄于1961年被列为第一批全国重点文物保护单位。

3. Hebei

Chengde

Example 18 Chengde Mountain Resort

It is located in the north of Chengde in Hebei Province, covering an area of 5.6 km², and was built in the 42nd year of Emperor Kangxi (1703). The characteristics are as follows:

(1) It is a place to avoid summer heat and implement conciliation policy. The climate and scenery

行宫入口
Entrance

热河泉
Rehe Spring

are suitable for summer resort, and since it is close to the border that is convenient for cultivating good relations.

(2) It is surrounded by mountains and rivers and has created many landscapes around the lake.

(3) The layout is "front hall and back garden" as well as "front imperial court and back palace".

(4) The scenery here is beautiful. It has imitated many scenic spots in the south of the Yangtze River.

(5) The north shore of the lake area is covered with green grass, reindeer and rabbits, running back and forth.

(6) A large mountain area is located in the northwest and the top of the mountains can be seen among the clouds.

In 1961, it was listed as the first batch of Key Cultural Relics Site under the State Protection.

《避暑山庄图》（清 冷枚绘）
Picture of Mountain Resort (by Leng Mei in Qing Dynasty)

金山亭
Jinshan Pavilion

从金山亭俯视湖光山色
View from Jinshan Pavilion

四、山　东

曲　阜

实例19　阙里宾舍

建在山东曲阜城中心，北临孔府，西对孔庙，占地2.4hm^2，建筑面积近1.4万m^2，共有客房175间，于1985年建成，设计师是著名建筑大师戴念慈先生和傅秀蓉女士。设计方案经过老一辈建筑专家杨廷宝先生等审阅。杨廷宝先生在看方案前曾说，在这个重点地区修建宾馆，很难处理好，也不易通过。但杨廷宝先生等在审看了该设计方案之后，一致赞赏其方案设计得巧妙，也使得该设计方案很快获得了通过。该建筑以两层为主，采取传统民居四合院的布局，这与孔府的传统屋顶风格类似，其尺度、体量、材料、色彩等皆与孔府、孔庙协调一致，同周围融为一体。此宾馆具有深厚的历史文化特色，在入口处悬挂着"有朋自远方来，不亦乐乎"的匾额，在大厅陈放着战国出土文物"鹿角立鹤"复制品，有欢迎宾客之意。阙里宾舍大厅二层栏板上有12面铜锣，大厅上方环绕着长幅描绘孔子一生的"孔迹图"等。该项目获建设部优秀设计一等奖，并选入《20世纪世界建筑精品集锦》东亚卷中。

入口外景
Scenery at the entrance

4. Shandong

Qufu

Example 19 Queli Hotel in Qufu

It is located in the center of Qufu and covers an area of 2.4 hectares, with a building area of nearly 14,000 m^2 and 175 guest rooms. The building is mainly composed of two floors, adopting the layout of traditional quadrangle dwellings, and its style and color are integrated with the surrounding Confucius Temple and Confucius House. This hotel has profound historical and cultural characteristics, with a plaque hanging at the entrance, "It is always a pleasure to greet a friend from afar", duplicates of cultural relics in the Warring States Period displaying in the hall, and a long "track map" describing Confucius' life activities hanging on the top. It won the first prize of Excellent Design by the Ministry of Construction, and was selected into the East Asian Volume of the *World Architectural Masterpieces Collection in the 20th Century*.

餐厅
Dining room

客房院落
Courtyard of guest house

五、江 苏

1. 南 京

南京城，它是东吴建业城和东晋、南朝建康城。公元229年9月，孙权定都建业。建业都城北依覆舟山、鸡笼山和玄武湖，东凭钟山，西至石头，东晋、南朝皇帝在这些山湖中大造皇家园林，使这一都城为众多的自然山水园林所环绕。长江位于南京城西面，它哺育了南京名城。秦淮河在南京城的南边，它是南京人文的摇篮。南京城内"十里秦淮"一段孕育着这座古城的成长，长江、秦淮河对航运、水利、商业等起着重要的作用。

南京是我国著名的六大古都之一，也是重要的国家历史文化名城。南京已发展了多种工业，特别是化学工业在全国占有重要地位。此外，还有电子工业、高新技术产业等。通过城市工业经济的发展，也带动了城市历史文化的保护。特别是前些时制定出《南京老城保护和更新规划》，它突出保护和发展老城文化，以"十里秦淮"、夫子庙、城南小街巷显现传统文化；以原有空间尺度、优美环境将明故宫片区打造出明代文化、民国文化；明城墙内40km²老城的保护与更新展现了2400多年南京城建的历史文化。

实例20 南京城墙

南京城墙是六大古都中保留最多的一处，约有21km，占原有城墙的2/3，是明洪武年间修筑的，距今已有600余年的历史，并于1988年被列为第三批全国重点文物保护单位。在现存城墙中有3处在全国城墙中是仅有的，其一是城区玄武湖畔的城墙，其大片湖水与高大城墙相依，形成湖墙山景，风光如画、别具一格（见161页上图）；其二是石头城，位于城西北清凉山后，全长3km。城基遗迹为赭红色水成岩，内有河光石，中间突起处好似鬼脸，故有"鬼脸城"之称；其三是城南的中华门，是明洪武初年在南唐都城南门的旧址上重建的，名"聚宝门"。辛亥革命之后取今名，平面近方形，南北长128m、东西宽118.5m，前后四重，每道城墙正中开拱门一座。拱门除设双扇门外，还有上下移动的千斤闸，十分坚固。此三处堪称"南京城墙三绝"，应格外珍惜，并加以整修保护。

5. Jiangsu

5.1 Nanjing

Nanjing is the ancient capital of several dynasties. The Yangtze River and Qinhuai River play an important role in shipping, water conservancy and commerce. As an important national historical and cultural city, Nanjing has also developed a variety of industries, which has promoted the protection of urban history and culture through economic development. "Nanjing Old City Protection and Renewal Plan" highlights the protection and development of the old city culture.

中华门门洞
Doorway of Zhonghua Gate

石头城突出部分
Protrusion of Stone City

中华门城墙
City wall of Zhonghua Gate

Example 20 City Wall

Nanjing city wall is the most preserved place among the six ancient capitals, which is about 21km long and has a history of more than 600 years. It was listed as the third batch of Key Cultural Relics Site under the State Protection in 1988. There are three unique features in the existing city walls. One is the city wall near Xuanwu Lake, where large lakes and tall city walls depend on each other and have picturesque scenery. The second is Stone City. The ruins of the city base are ochre-red water-rock with river light stone inside, and the middle protrusion is like a grimace, so it is called "Grimace City". The third is the Zhonghua Gate in the south of the city, which has quadruple doors.

实例21 玄武湖（公园）

位于南京东北玄武门外，湖周长15km，总面积444hm²，其中陆地面积49hm²。湖水源于钟山（紫金山）北麓，下游是金川河，至下关入长江，另一支入城经秦淮河入长江。宋文帝刘义隆时，因在湖中见到黑龙（鳄鱼），改称玄武湖。1911年，玄武湖辟为公园。玄武湖中有5洲，大片是湿地、绿地，山湖洲岛景色佳丽，不仅是广大市民休息和游览胜地，而且也是城市自然生态平衡的重要之地。玄武湖位于南京城市北端，也起到了"肺"的作用。

Example 21 Xuanwu Lake (Park)

It is outside Xuanwu Gate that has a circumference of 15 km and a total area of 444 hectares. Lake water originates from the northern foot of Bell Mountain, downstream of which is Jinchuan River, and then enters the Yangtze River. In 1911, Xuanwu Lake was turned into a park with large wetlands and green spaces. It is not only a scenic spot for the general public to rest and visit, but also an important place to balance the natural ecology of the city, playing the role of "lung".

Nanjing Bell Mountain Scenic Area is a national key scenic spot.

中篇：中国建筑科学文化精品图说/江苏

山水墙（城墙）景观
Landscape wall (city wall)

山水塔景观
Landscape pagoda

2. 苏　州

苏州，又称姑苏，是著名的"水乡"，也是国家历史文化名城。苏州历史悠久，距今已有2500多年。春秋时它是吴国都城，越王勾践灭吴后，又成为越国都城。秦统一中国后，置吴县为会稽郡治，隋唐时称苏州，宋代改称平江，清代则设苏州府。苏州有悠久的历史，留下了丰富的文化遗产，如著名的虎丘、寒山寺、西园罗汉堂以及代表宋、元、明、清的著名园林沧浪亭、狮子林、拙政园、留园等。苏州园林是我国古代建筑文化的精华。

实例22　"水乡"

从宋平江府图碑中，可以看到当时苏州城内三横四直共7条贯穿全城的干流河道水系。姑苏城内外各有河道，干流河道之间又有密布的分渠，分渠大部分为东西向，以利住宅、商店、作坊形成前街后河。河道进入城墙之处共建有7座水门与闸。城内外也有桥梁300多座。苏州是我国南方典型的水乡城市，其河道曾一度严重污染，后经多次治理，现已好转。

5.2 Suzhou

The Beijing-Hangzhou Grand Canal and Suzhou River meet here, and it is a famous "water town" with rivers interwoven in the city and lakes scattered outside. It has a long history of more than 2,500

一边有路的水巷
A water lane with a road on one side

years, and has left rich cultural heritage, which is the essence of ancient Chinese architecture, horticulture, sculpture, literature and other arts.

Example 22 Water Lane

There are seven mainstream rivers running through the whole city in Suzhou and river-ways inside and outside the city walls. There are also densely distributed branches among the mainstream rivers, most of which are east-west to facilitate the formation of front streets and back rivers for houses, shops and workshops. Seven water gates were built where the river enters the city wall, and more than 300 bridges are inside and outside the city. It is a typical water town city in South China.

有桥沟通两院的小水巷
A bridge links the water lanes on both sides

实例23 拙政园

该园位于苏州市北面，建于明正德年间(1506—1521年)，是苏州四大名园之一。明代吴门四画家之一的文征明参与了造园，他作"拙政园图卅一景"，并为该园作记、题字、植藤。由于文人、画家的参与，将大自然的山水景观提炼到诗画的高度，并转化为园林艺术，使拙政园更富有诗情画意的境地，并成为中国古典园林的一个优秀的典型实例。拙政园空间艺术特点是：

（1）沿对景线构图，主题突出，宾主分明。全园布局为自然式，但仍采用对景线构图手法，主要厅堂亭阁、风景眺望点、自然山水位于重要对景线上，次要建筑位于不重要对景线上。采用对景线手法，不是机械地画几何图形，而是按其自然条件、功能与艺术的要求，灵活地运用这一原则。

（2）因地制宜，顺应自然。这是中国造园的一大特点。拙政园是利用原有水洼地建造的，按地貌取宽阔的水面，临水修建主要建筑，同时注意水面与山石花木相互掩映，并构成富有江南水乡风貌的自然山水景色。

（3）空间序列组合，犹如诗文结构。园林空间序列组合，要作到敞闭起伏、变化有序、层次清晰。其组合安排类似诗文结构的组织，同时也有类似诗词平仄音的韵律。这种空间序列安排，主要通过对比以取得主题突出、整体和谐的效果，并构成了富有诗词韵律的流动空间。

（4）景区转折处，景色动人、层次丰富。它是景区变换的地点，也是欣赏景观的停留点，在这里可观赏到层次丰富的前景、中景、远景，因此，常常成为游人留影的地方。

（5）空间联系连贯完整、相互呼应。园林空间的序列是靠游览路线连贯起各个空间点的。拙政园的游览路线是由园路、廊、桥等组成。此外，还可通过高处鸟瞰进行空间联系。

拙政园于1961年被列为第一批国家重点文物保护单位。

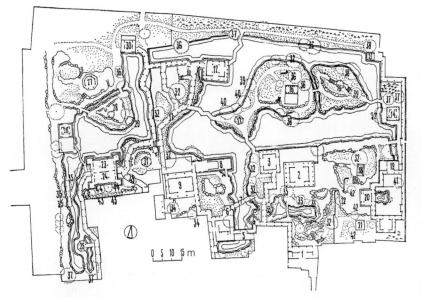

平面
1-腰门；2-远香堂；3-南轩；4-松风亭；5-小沧浪；6-得真亭；7-小飞虹；8-香洲；9-玉兰堂；10-别有洞天；11-柳荫路曲；12-见山楼；13-绿绮亭；14-梧竹幽居；15-北山亭；16-雪香云蔚亭；17-荷风四面亭；18-绣绮亭；19-海棠春坞；20-玲珑馆；21-春秋佳日亭；22-枇杷园；23-三十六鸳鸯馆；24-十八曼陀萝花馆；25-塔影亭；26-留听阁；27-浮翠阁；28-笠亭；29-与谁同坐轩；30-倒影楼；31-宜两亭；32-枫杨；33-广玉兰；34-白玉兰；35-黑松；36-榉树；37-梧桐；38-皂荚；39-乌桕；40-垂柳；41-海棠；42-枇杷；43-山茶；44-白皮松；45-胡桃

Plan
1-Wicket; 2-Drifting Fragrance Hall; 3-South Pavilion; 4-Pine Wind Pavilion; 5-Little Pavilion of Surging Waves; 6-Dezhen Pavilion; 7-Little Flying Rainbow; 8-Xiangzhou; 9-Yulan Hall; 10-Hidden Beautyl; 11-Willow Shadow Road; 12-Mountain in View Tower; 13-Lvyi Pavilion; 14-Bamboo Seclusion Pavilion; 15-Beishan Pavilion; 16-Fragrant Snow and Cloud Pavilion; 17-Pavilion of Lotus Breezes; 18-Xiuqi Pavilion; 19-Begonia Spring; 20-Equisite Hall; 21-Spring and Autumn Pavilion; 22-Loquat Garden; 23-Thirty-six Mandarin Duck Pavilion; 24-Eighteen Mandala Flower Pavilion; 25-Tower Shadow Pavilion; 26-Liuting Pavilion; 27-Fucui Pavilion; 28-Li Pavilion; 29-Pavilion of Sitting with Whom; 30-Refelctin Pavilion; 31-Yiliang Pavilion; 32-Maple and Aspen; 33-Southern magnolia; 34-Magnolia; 35-Black pine; 36-Zelkova; 37-Chinese parasol; 38-Honeylocust; 39-Chinese tallow tree; 40-Willow; 41-Begonia; 42-Loquat; 43-Camellia; 44-White bark pine; 45-Walnut

Example 23 Humble Administrator's Garden

It is located in the north of Suzhou and was built in the Emperor Zhengde's reign in Ming Dynasty (1506-1521). Because literati and painters participated in the construction of the garden, it was more poetic and picturesque, and became an excellent example of Chinese classical gardens. Its artistic features include:

(1) The themes are outstanding by using opposite scenery. Instead of drawing geometric figures mechanically, it flexibly applied the principle of opposite scenery according to the natural conditions, functions and artistic requirements of different places.

(2) Adapt to local conditions and nature. It used the original puddle to build the garden, made the wide water surface according to the landform and built the main buildings near the water.

(3) Spatial sequence combination is just like poetry structure. The garden space changes in an orderly manner with clear levels. There is a rhythm similar to the prosody of poetry, which makes the theme stand out and the whole area is harmonious and unified.

(4) The scenic spots are rich in layers. The turning point of the scenic spot is the place where the scenery changes, and rich front, middle and long-term perspectives can be seen.

(5) Spatial connection is coherent and complete. Tour routes are composed of gardens, corridors and bridges. In addition, it also makes spatial contact through high-altitude sight.

The park was listed as the first batch of Key Cultural Relics Site under the State Protection in 1961.

远香堂、梧竹幽居
Drifting Fragrance Hall and Bamboo Seclusion Pavilion

清晨从雪香云蔚亭望远香堂
Drifting Fragrance Hall from the view of Fragrant Snow and Cloud Pavilion

小飞虹景区
Little Flying Rainbow

西院与谁同坐轩
West yard and Pavilion of Sitting with Whom

实例24 留园

位于苏州阊门外，是苏州大型古典园林之一，其布局与建筑特点是：

（1）主景突出。主体建筑寒碧山房前面对景的是开阔丰富的山水景色。

（2）建筑精美。特别是东北景区的林泉耆硕之馆，门窗、花罩雕刻精美。

（3）奇石珍品。其中冠云峰，是北宋花石纲遗物，是一块难得的奇峰独石。

此园于1961年被列为第一批国家重点文物保护单位。

Example 24 Lingering Garden

It is located outside Changmen and is one of the large classical gardens in Suzhou. Its layout and architectural features are as follows:

(1) The main scenery is outstanding, and the main building Hanbi Mountain Villa is facing open and rich landscapes.

(2) The architecture is exquisite, and the doors, windows and flower covers are exquisitely carved.

(3) There are precious stones, among which the Cloud Capped Peak, a relic of Stone and Flower Tran sortation in Northern Song Dynasty, is a rare unique one.

This park was listed as the first batch of Key Cultural Relics Site under the State Protection in 1961.

冠云峰
Cloud Capped Peak

从寒碧山房旁望山水主景
View from Hanbi Mountain Villa

林泉耆硕之馆室内陈设与落地花罩
Furniture of Linquan Hall of Hermit and Aged

实例25　虎丘

在苏州市阊门外山塘街，距古城3.5km处，旧名海涌山，素有"吴中第一名胜"称誉。相传春秋晚期，吴王夫差葬其父阖闾在这里，葬后3日"有白虎踞其上"，故名虎丘。又一说，丘如蹲虎，以形命名。并传说在"剑池"吴王以三千支宝剑殉葬，秦始皇和孙权曾挖掘，但未发现。过千人石，刻有"别有洞天"圆门内即是"剑池"。池旁徒壁如削，上有横桥，置于绝岩纵壑之间。桥上有双井，名双吊桶，景色险奇。现存完成于宋代初的云岩寺塔，塔高47.5m，全为砖砌，建筑风格别致，它立于40m高的山丘上，气势雄伟。云岩寺塔于1961年被列为国家重点文物保护单位。

Example 25 Tiger Hill

It is located outside Changmen and is known as "the first scenic spot in Wuzhong". According to legend, in the late Spring and Autumn Period, King Fu Chai buried his father King He Lv here, and three days after the burial, "there was a white tiger sitting on it", hence the name Tiger Hill. It is also said that in the "Sword Pool", which is inside the round door carved "Hidden Beauty", King He Lv was buried with 3,000 swords. The walls beside the pool are cut-like and there is a cross bridge. Yunyan Temple Tower in Song Dynasty, with a height of 47.5 m, stands on the hill with magnificent momentum. The tower was listed as a Key Cultural Relics Site under the State Protection in 1961.

"别有洞天"圆门、云岩寺塔（后）
Round door with carved "Hidden Beauty" and Yunyan Temple Tower

"剑池"与横桥
"Sword Pool" and the cross bridge

3. 无 锡

无锡历史悠久。约在公元前1244年，周太王之子泰伯、仲雍，从陕西来到江南，定居梅里（即无锡梅村）并做了首领。公元前223年，秦灭楚，传说秦将王翦带兵过锡山时见锡已被采空，就定名为无锡。清雍正时（公元1723年）把无锡东面分置金匮县，辛亥革命后废金匮县，1927年后直属江苏省。20世纪初期，无锡棉纺、缫丝、染织、机械工业逐步发展，而真正的工业产业蓬勃发展是在新中国成立后。无锡现已成为具有纺织、丝绸、钢铁、机械、轻工、化工、电子等综合性的中等工业城市。随着产业经济的发展，无锡城市建设也有了很大发展。

实例26 寄畅园

该园位于无锡西郊惠山脚下，始建于明正德年间(1506—1521年)，属官僚秦姓私园。寄畅园规模不大，为1hm²，其造园特点有：

（1）小中见大，借景锡山。此园选址在惠山、锡山之间，似惠山的延续，并可将锡山及其山顶龙光塔景色借入园中。

（2）顺应地形，造山凿地。此园地形西面高、东面低、南北向长、东西向短。依此地势，顺西部高处南北向造山，就东部低处南北向凿池。

（3）山水自然，主景开阔。假山位于惠山之麓，仿惠山峰起伏之势，选少量黄石而多用土造山，有如惠山余脉，使假山与天然之山融为一体。假山与水池相映。在知鱼槛中可观赏到开阔的山水主景。水池北面布置有七星桥，增加了水景的层次。

（4）山中有涧泉，水景多出彩。此园的自然景色，吸引了清乾隆皇帝，因此在建造北京清漪园时，他点名仿造此园于清漪园的东北一隅，名为"惠山园"。

此园于1988年被列为第三批国家重点文物保护单位。

5.3 Wuxi

Wuxi has a long history, with records in Zhou, Qin and Qing Dynasties and even the Revolution of 1911, and the name of the city also changed for several times. In the early 20th century, cotton spinning, silk reeling, dyeing and weaving, and machinery industries developed gradually, and it has now become a comprehensive medium-sized industrial city.

Example 26 Jichang Garden

It is located at the foot of Hui Mountain in the western suburbs and was built in Emperor Zhengde's reign in Ming Dynasty (1506-1521). It covers an area of 1 hectare. Its characteristics are as follows:

(1) Seeing the big from the small and taking advantage of the scenery in Xi Mountain.

(2) The construction conformed to the terrain.

(3) The landscape is natural, the main scene is wide, and the rockery and pool are echoing with each other.

(4) There are springs in the mountains with abundant waterscape.

This park was listed as the third batch of Key Cultural Relics Site under the State Protection in 1988.

从南向北望环锦汇漪水景（左为七星桥）
Waterscape when looks from south to north (Seven-star bridge on the left)

从北向南望水景（可借景锡山塔）
Waterscape when looks from north to south (can see the pagoda on Xi Mountain)

实例27　蠡园

位于无锡西南面的蠡湖畔，全园面积近6hm^2，其名因蠡湖而起。蠡湖来自越国大夫范蠡与西施的故事传说，范蠡协助越王勾践灭吴国后，带西施来到太湖，泛舟湖上，留恋于此，后人为纪念他们，称此五里湖为蠡湖。1928年，无锡王禹卿在蠡湖边造园，称为蠡园。园中建有湖心亭，有平桥伸入湖心。湖心亭为长方形，题名为"晴红烟绿"，这是对此处朝夕景色的写照。湖心亭东面还有玲珑小塔，名为凝春塔。1930年，无锡人陈梅芳在蠡园西边建渔庄，故名"赛蠡园"。1952年，无锡市政府在蠡园与渔庄之间修建了300m长廊，将两园合二为一。后不断扩建，又增加新的景观，如1954年在"百花山房"南面的柳堤上建起春、夏、秋、冬四季亭，四季花开，春天有梅花，夏天有夹竹桃，秋天有桂花，冬天有腊梅花。四季清香色艳，配以晶莹湖水，景色更加秀丽。

Example 27　Li Garden

It is located in the southwest of Wuxi and covers an area of nearly 6 hectares. Its name comes from Li Lake, and the latter comes from the legend of Fan Li and Xi Shi in Yue Kingdom. In 1928, Wang Yuqing built a garden near Li Lake, which is called Li Garden. There is a lake pavilion in the park, with a flat bridge extending into the center of the lake, and a small exquisite tower named Ningchun to the east of the lake pavilion. In 1930, Chen Meifang built a fishing villa in the west of Li Garden, and the later municipal government built a 300 m long corridor, which combined the two spots into one and expanded continuously.

合并扩建后景观
Scenery after combination and expansion

湖心亭、凝春塔
Lake Pavilion and Ningchun Tower

4. 扬　　州

实例28　扬州瘦西湖

该湖位于扬州府城外西北部，1765年前盐商们为取悦皇帝在原保障河两岸造二十四景，形成沿河弯曲带状的两岸景区，为清乾隆第四次南巡的游览区。

这一瘦西湖区景色，以纵向景观为多，相互对景和借景。最精彩的对景有中心位置的吹台，吹台上建一方亭，并有两个圆洞门，正好面对五亭桥和白塔这两个重要的景点建筑。

中心景区五亭桥，桥上建有5个亭子，形似莲花，造型既稳重又透剔。白塔为砖结构，其整体造型较北京白塔为瘦，显得更加清秀。

5.4 Yangzhou

Example 28 Slender West Lake

It is located in the northwest of Yangzhou and the landscape of the lake area is mostly vertical and uses a lot of opposite and borrowed sceneries. The most wonderful opposite scenery, such as the square pavilion on the platform in the central part, is built with two round doors, just facing the Five-pavilion Bridge and the White Pagoda. Five-Pavilion Bridge, the central scenic spot, has five pavilions, which are like lotus flowers. The shape is steady and transparent. The White Pagodas is of brick structure, and the overall shape is thinner than that of Beijing White Pagoda, which is delicate and pretty.

白塔、五亭桥夕阳剪影
White Pagoda and Five-pavilion Bridge in sunset

从吹台方亭看白塔、五亭桥晨景
Morning view of the White Pagoda and Five-pavilion Bridge from the square pavilion

六、上　海

上海位于长江出海口处，东临东海，黄浦江、苏州河流经市区汇合流入长江，进入东海。唐代中期由于农、渔、盐业的发展设立了华亭县，南宋末年华亭县升为松江府，府内设有上海镇。后上海镇升为上海县，仍由松江府管辖，至明末上海已是一个手工业生产和商贸中心。1842年鸦片战争后，清政府被迫与英国签订不平等条约，将上海作为5个通商口岸之一，后又在上海划出租界地。20世纪初30年代中期，上海的外国工业、商业与金融业迅速发展，成为远东第一大城市。20世纪下半叶，特别是新中国成立后在上海市区外围发展了一些新工业区，建造起卫星城镇，同时对旧市区进行了调整和治理。进入20世纪90年代之后，上海开发了浦东新区，即在浦东350km² 的范围内逐步发展国际金融贸易区。上海既是中国重要的工业基地和港口，也是一个现代化国际性的金融贸易中心。同时，上海还是国家历史文化名城。

实例29　上海豫园

位于上海市旧城东北部，是上海老城保存下来的较完整的一组中国古典园林建筑群。它始建于明嘉清三十八年（公元1559年）。明末园废，清乾隆二十五年（1760年）重建，鸦片战争后又不断受毁，一部分已变为商场。新中国成立后，豫园陆续整形，但已非原有范围。豫园景观有3处独到之处：其一是仰山堂面对的大假山，它是一件精品之作，系以武康石堆叠、气势雄浑、重峦迭嶂、洞壑幽谷，宛如自然天成。大假山顶上建一六角望江亭，可眺望江景与全园景色；其二是水洞。从大假山东南部山脚下沿溪流向东行，进入东部另一景区，有一白色粉墙横跨溪流中间，下设半圆形券洞，让溪水从洞中流过，其墙上开漏窗，景物倒映水中形成美丽水洞景观；其三是南面园中的玉玲珑石，它峻峭漏透，相传系宋代花石纲遗物，从其造型和规模上来看，是件艺术珍品。豫园于1982年被列为全国重点文物保护单位。

6. Shanghai

It is located at the mouth of the Yangtze River and facing the East China Sea to the east. Huating County was established in the middle of Tang Dynasty, and was promoted to Songjiang Prefecture at the end of Southern Song Dynasty. Shanghai Town was located in the prefecture. During the Qing government, Shanghai became one of the five trading ports. In the later period, Shanghai built satellite cities and developed Pudong District. Now it is an important industrial base and port in China, and also a modern international financial and trade center.

Example 29 Yu Garden

It is located in the northeast of the old city of Shanghai and is a relatively intact group of Chinese classical garden buildings. It was built in the 38th year of Emperor Jiaqing in Ming Dynasty (1559). The garden has three unique features. One is the big rockery facing Yangshan Hall, with a hexagonal Wangjiang Pavilion on the top, overlooking the river view and the whole garden. The second is the water tunnel. The white wall spans the middle of the stream, and there is a semi-circular arch hole, which allows the stream to flow through the hole. The wall opens a leak window and the scenery is reflected in the water. The third is the Jade Exquisite Stone, which is steep and revealing. It is said that it is a relic of the Flower and Stone Transportation in Song Dynasty. This Garden was listed as a Key Cultural Relics Site under the State Protection in 1982.

从仰山堂望大假山
The big rockery from the view of Yangshan Hall

玉玲珑石
Jade Exquisite Stone

水洞
Water Tunnel

实例30 上海里弄

上海里弄式住宅出现在19世纪下半叶后期至20世纪上半叶，是引进英国工业革命后在城市中出现的联排式住宅形式，并同中国传统三合院、四合院的样式混合而成的一种住宅形式。这种形式的住宅，大多是造价一般，适应广大市民的需要。这其中既有较高标准的里弄式住宅，也有更低标准的里弄式住宅，以便满足不同阶层市民的需求。早期的里弄式住宅，每户有一石门框和黑色板门，称为石库门里弄住宅。建筑采用立贴式封火山墙砖木结构，且多为二、三层，有前后小院。进入20世纪之后，里弄规模扩大，但每户单元面积却减少为一开间长条形格局。20世纪20年代后，外观又逐渐西化，并考虑进出汽车的需要。到了20世纪40年代之后，又转向花园里弄，每户有庭院绿地，建筑外观已变为欧式，其层数也在增多。这个实例说明，住宅的样式一定要同当地居民生活方式和经济水平相适应。

Example 30 Lanes and Alleys

The residence in lanes and alleys appeared in the city from the second half of the 19th century to the first half of the 20th century. It is combined with Chinese traditional quadrangle dwellings, and most of them are of average cost and meet the needs of the general public. Residential buildings are made of brick and wood structure with enclosed volcanic walls, mostly with two or three floors, and front and rear small courtyards. After that, the scale gradually expanded, the appearance gradually westernized, and the need of getting in and out of cars was considered. After the 1940s, it turned to the garden lane, and every household had courtyard green space. The architecture and appearance had become European-style, and the number of floors was also increasing.

上海里弄住宅区
Residence in lanes and alleys

实例31　上海体育中心

位于上海市徐汇区，可以举办全市、全国、国际性的体育竞赛和大型文艺演出，以及公众集会活动。上海体育中心内的主要建筑有体育馆、体育场和游泳馆等。体育馆于1975年建成，设计师是著名建筑师汪定曾先生。该建筑平面为圆形，直径114m、高32m，屋盖为钢管网架结构，采用整体吊装就位，技术先进。上海体育中心有双层看台，可容1.8万名观众。游泳馆平面为六角形，有4000个座位。上海体育场于1997年建成，在上海体育馆东面，设计师是著名建筑大师魏敦山先生。上海体育场平面为圆形，直径300m，建筑面积17万m^2，是一座高达70多米的马鞍形建筑，设有标准足球场和400m环形跑道，可容观众8万人。上海体育场马鞍形半透明乳白色的"膜结构"屋顶同白色构架墙面形成强烈对比，形似皇冠，富有时代气息，体现了现代大型体育建筑的特点。这个项目已选入《20世纪世界建筑精品集锦》东亚卷中。

Example 31　Sports Center

It is located in Xuhui District and can hold sports competitions, large-scale cultural performances and public gatherings. The main buildings are gymnasium, stadium and swimming pool. The gymnasium is round in plane and has two grandstands, which can accommodate 18,000 spectators. The swimming pool is hexagonal in plane and has 4,000 seats. The stadium has a circular plane with a building area of 170,000 m^2 and can accommodate 80,000 spectators. The saddle-shaped translucent milky white "membrane structure" roof is in contrast to the white frame wall, and looks like a crown. This project has been selected into the East Asian Volume of the *World Architectural Masterpieces Collection in the 20th Century*.

体育馆外景
Exterior

实例32　上海陆家嘴金贸区、上海东方明珠电视塔等

　　陆家嘴金融贸易区位于上海黄浦江东岸，它是实施上海城市开发战略的中心。此金贸区的总体规划，经国际咨询后由上海城市规划设计研究院等单位完成。下面介绍几个重要项目。于1995年建成的东方明珠电视塔，设计师是华东建筑设计研究院凌本立等。东方明珠电视塔塔高468m，建筑总面积6.5万m^2，塔身构架是3个圆筒体，以大小不同的11个球体作为连接体，形成和谐明快、积极向上的设计氛围，并显示出高科技的形象。东方明珠电视塔三面为黄浦江环绕，隔江面正对上海最繁华的商业街南京路，而且又是上海路、北京路、福州路、建安路、四平路等重要街道的视线终点。另外，还有金茂大厦、上海世界金融中心和620多米的全国最高建筑——上海中心大厦，这些标志性建筑共同构筑了上海浦东新区的立体轮廓线。

Example 32 Lujiazui Finance and Trade Zone and the Oriental Pearl TV Tower

　　The zone is located on the east bank of Huangpu River and has been the center of Shanghai's urban development strategy since 1987. The Oriental Pearl TV Tower, built in 1995, has a tower height of 468 m and a total building area of 65,000 m^2. The frame of the tower is three cylinders, with 11 spheres of different sizes as connectors. The tower is surrounded by Huangpu River on three sides, facing Nanjing Road, the busiest commercial street in Shanghai across the river, and it is also the sight end of other important streets. In addition, there are Jinmao Tower, Shanghai World Financial Center and the tallest building in China - Shanghai Center Tower. These landmark buildings constitute the three-dimensional contour line of Pudong New Area.

夜景（左为上海东方明珠电视塔）
Night scene (Oriental Pearl TV Tower on the left)

实例33　上海博物馆

位于上海市人民广场的中轴线南端，总建筑面积3.8万m^2，建成于20世纪90年代，设计师是著名建筑师邢同和。为了适应这开阔的广场空间，建筑采用横向宽平的体型，既平稳又舒展。建筑下部作成方形，呈两级台阶状，似为基座；顶部作成圆状，有如中华铜镜，似古代青铜器造型与浮雕纹饰。其下方上圆的建筑造型，寓意着"天圆地方"的设计理念，并使其成为一座功能齐全、技术先进、造型现代而又具有东方特色的大型文化建筑。这是一座继旧创新、具有地域文化特点的新建筑。该项目已选入《20世纪世界建筑精品集锦》东亚卷中。

Example 33 Shanghai Museum

It is located at the southern end of the central axis of People's Square with a total construction area of 38,000 m^2. It adopted a horizontal wide and flat shape, which is both smooth and stretched. The lower part of the building is square, with two steps, which is like a pedestal. The top part is round, like a Chinese bronze mirror. The whole part is like an ancient bronze ware. The natural integration means "round sky and square earth". It is a large-scale cultural building with complete functions, advanced technology, modern modeling and oriental characteristics. This project has been selected into the East Asian Volume of the *World Architectural Masterpieces Collection in the 20th Century.*

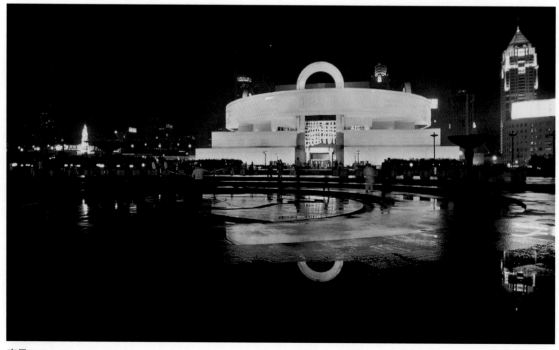

夜景
Night scene

实例34 上海大剧院

位于上海中心地带人民广场的西侧，总建筑面积6.8万m^2，设计单位是法国夏邦杰建筑师事务所和华东建筑设计研究院，建造年代为1998年。其屋顶向天空展开，象征着上海对世界文化艺术的热情追求。平面采用中国传统手法，环绕观众厅和舞台组成"井"字型划分，可进行歌剧、芭蕾舞和交响乐三种演出。上海大剧院充满着活力和梦幻，它运用现代高科技与新材料来营造自身的形象，并已成为象征上海新标志性建筑之一。该项目已选入《20世纪世界建筑精品集锦》东亚卷中。

Example 34 Grand Theatre

It is located on the west side of People's Square with a total construction area of 68,000 m^2 and was built in 1998. The roof unfolds into the sky, which symbolizes Shanghai's passionate pursuit of world culture and art. The plane adopts the traditional Chinese layout technique, which surrounds the audience hall and the stage to form a Chinese character "well"-shaped division. Opera, ballet and symphony can be performed. The theatre is full of vitality and dreams, and uses modern high-tech means and new materials to enrich and create its own image. It has become one of the new landmark buildings symbolizing Shanghai. This project has been selected into the East Asian Volume of the *World Architectural Masterpieces Collection in the 20th Century*.

全景
Panorama

实例35　上海世博会中国馆

此馆设计是华南理工大学著名建筑大师何镜堂教授领衔创作的方案,是从全球华人建筑师征集的344个有效方案中选出的。根据世博会主题,旨在创造自然的地域生态环境和社会的人文生态环境。而在自然地域生态环境创新方面,该建筑有着一套完整的环保与节能措施。第一,自遮阳体型实现夏季最大限度遮阳和冬季最大限度透光的效果。层层出挑的造型设计,使得建筑在太阳高度角较高的夏季遮阳效果明显,而在太阳高度角较低的冬季又能让太阳辐射最大限度进入室内,以此实现节能效果。自遮阳的结构只需采用普通中空玻璃就可以满足节能设计要求,这大大降低外窗的投入。第二,用被动式节能技术为地区馆提供冬季保温和夏季拔风,而地区馆屋顶——"中国馆城市花园"则运用生态农业景观措施实现有效隔热。第三,利用自然通风的架空中庭空间改善热环境。第四,充分利用太阳能,将屋面遮阳板与太阳能电池方阵结合,形成光电遮阳板,使全馆实现照明用电自给。第五,在景观设计层面,加入循环自洁要素。国家馆屋顶可以实现雨水的循环利用,而地区馆大台阶水景和园林设计则引入小规模人工湿地技术,为局部环境提供生态景观。第六,冰蓄冷技术实现用电的移峰填谷。该馆整体方形,主体居中,似中国周王城图的格局。汉以后的唐长安城、元大都城、明清北京都城的设计皆沿用这一形制。中国馆建筑底部及周围配以水体,如都城外围的大地乡村田野,隐喻着中国城乡的自然面貌,它体现着"天人合一"的中国哲学思想和自然宇宙观。

中国馆外观
Exterior

Example 35 China Pavilion of Shanghai World Expo

In the aspect of ecological environment innovation, it has a complete set of environmental protection and energy saving measures. First, self-shading modeling achieves maximum shading in summer and maximum light transmission in winter. Second, the roof of the regional pavilion is effectively insulated by ecological agriculture landscape measures. Third, it improves the thermal environment by using the naturally ventilated overhead atrium space. Fourth, it makes full use of solar energy, and combines the roof sun visor with the solar cell phalanx, so that the whole hall can achieve self-sufficiency in lighting power. Fifth, rainwater can be recycled on the roof of the national pavilion, and small-scale constructed wetland technology is introduced into the regional pavilion to provide ecological landscape for the local environment. Sixth, ice storage technology realizes shifting peak and filling valley of electricity consumption.

The pavilion is square as a whole, and its main body is centered. Several capitals in ancient China are all designed in this shape. The bottom of the building and its surroundings are equipped with water bodies, which symbolizes the natural features of urban and rural areas and embodies the concept of symbiosis with nature.

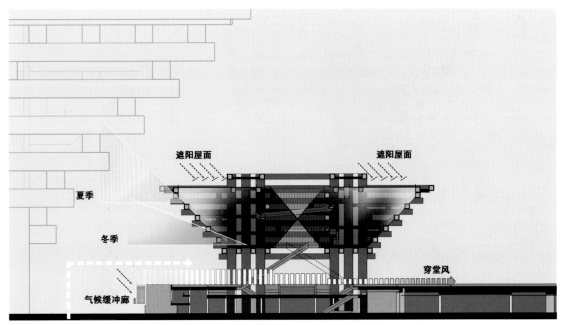

中国馆技术节能
Energy-saving technology

七、浙　江

1. 杭　州

杭州位于钱塘江下游北岸，大运河南端。它是我国六大古都之一，曾为五代时吴越国和南宋时的都城。现为浙江省省会、浙江省政治、经济、文化中心和国家历史文化名城。新中国成立后，发展了多种轻纺工业和重工业，并以丝绸工业为中心，发展绸伞、檀香扇、剪刀等传统手工业。杭州是世界著名的游览城市，现发展了钱塘江新城区，使以西湖为中心的旧城区得到较好保留。

实例36　杭州西湖风景名胜区

杭州西湖风景名胜区位于杭州市的西面，因湖在城西面，故称"西湖"。在古代西湖是和钱塘江相连的一个海湾，后钱塘江沉淀积厚，塞住湾口，乃变成一个礁湖。直到公元600年前后，湖泊的形态才固定下来。公元822年，唐诗人白居易来杭任刺史，他组织"筑堤捍湖，用以灌溉"。公元1089年，宋诗人苏东坡任杭州通判，继续疏浚西湖，挖泥堆堤。17世纪下半叶，清康熙皇帝多次巡游西湖，又浚治西湖，开辟孤山。唐宋时期奠定了西湖风景园林的基础轮廓，后经历代整修添建，特别是1949年中华人民共和国建立后，挖湖造林、修整古迹，使西湖风景园林更加丰富完整，并成为中外闻名的风景游览胜地。其具体特点有：

（1）一座城市大型园林。西湖紧贴城市，"三面云山一面城"，这是西湖园林的地理位置特点，它同样起着城市"肺"的作用。

（2）湖山主景突出。现西湖南北长3.3km，东西宽2.8km，周长15km，面积5.6km²，湖中南北向苏堤、东西向白堤把西湖分割为外湖、里湖、小南湖、岳湖和里西湖五个湖面，通过桥孔沟通五湖。西湖的南、西、北三面为挺秀群山环抱，这一宏观的湖山秀丽景色是西湖的主要景观。

（3）"一池三山"模式。在外湖中鼎立着三潭印月、湖心亭和阮公墩三个小岛，这是延袭汉建章宫太液池中立三山的作法。后北京颐和园也是仿西湖布局，采用"一池三山"模式。

（4）园中园景观。这是中国造园的一大特点，即园中有许多园景，通过游览路线将

西湖平面 Plan of West Lake

其连接起来形成有序的园林空间序列。

（5）林木特色景观。许多景点，绿树成荫，各有特色。如灵隐配植了七叶树林，云栖竹林格外出名，满觉陇营造了桂花林、板栗林，南山、北山、西山配置了成片的枫香、银杏、麻栎、白桦等，西湖环湖广种水杉、间有棕榈等。

（6）四季朝暮景观。春夏秋冬、晴雨朝暮创造了不同意境的景观，这又是一个造园特点。西湖的春天有"苏堤春晓""柳浪闻莺""花港观鱼"景观；夏日有"曲院风荷"营造出"接天莲叶无穷碧，映日荷花别样红"之园林意境；秋季有"平湖秋月"、桂花飘香；冬天有"断桥残雪"、孤山梅花盛开。此外，还有薄暮"雷峰夕照"，黄昏"南屏晚钟"，夜晚"三潭印月"，雨后浮云"双峰插云"。这著名的"西湖十景"以及其他园中园景观展现了西湖四季朝暮的自然景色。

（7）历史文化景观。如五代至宋元的摩崖石刻，东晋时的灵隐古刹，北宋时的六和塔、保俶塔、雷峰塔，南宋的岳王庙，清珍藏《四库全书》的文澜阁，清末研究金石篆刻的西泠印社等。还有历代著名诗人画家留下的许多吟咏西湖的诗篇和画卷，以及清康熙、乾隆皇帝为"西湖十景"的题字立碑等。

（8）小园与大湖沟通。西湖周围的小园景观丰富而多彩，这些小园景色与西湖的大景观结合，构成了独特的江南景观。如西湖西面的郭庄、西湖北面孤山的西泠印社等都是将园景与大西湖景色巧妙地融为一体的典范。

7. Zhejiang

7.1 Hangzhou

Hangzhou is located on the north bank of Qiantang River downstream and at the southern end of Grand Canal. It is one of the six ancient capitals in China, and is now the capital of Zhejiang Province, the political, economic and cultural center and a famous national historical and cultural city. Now a variety of textile industries and heavy industries have been developed, and traditional handicrafts have been developed with the silk industry as the center, which is very popular.

Example 36 Scenic Area of West Lake

It is located in the west of Hangzhou and the foundation outline is from that during the Tang and Song Dynasties, and later it underwent renovation and construction and becomes a famous scenic spot at home and abroad. Specific features are:

(1) A large urban garden. It is close to the city and plays the role of "lung".

(2) The main scene is prominent. West Lake is 3.3 km long from north to south, 2.8 km wide from east to west, 15 km in circumference and 5.6 km^2 in area. It is divided into five lakes and surrounded by mountains on three sides.

(3) "One Pool and Three Mountains". Three small islands, namely Three Pools Mirroring the Moon, Huxin Pavilion and Ruangong Islet, stand in the outer lake.

西湖全景
Panorama

从南向北望西湖全景
Panorama when looks from south to north

(4) Garden in garden. Tour routes connect the scenic spots in the garden to form an orderly garden space sequence.

(5) Forest characteristics. Scenic spots are lined with trees of different features.

(6) Different seasons, weathers and times. Landscape with different artistic conception caught in spring, summer, autumn and winter, sunny days, rainy days, dusk and dawn.

(7) History and culture. There are ancient stone carvings, temples, and pagodas as well as poems and pictures chanting the West Lake.

(8) Small gardens and great lake. The landscape of small gardens around the West Lake is constantly increasing.

西湖与杭州旧城（1975年现状，旧城未建高层建筑，城湖协调）
West Lake and old Hangzhou city (in 1975 with no high buildings)

实例37　杭州小瀛洲三潭印月

小瀛洲位于西湖偏南湖面上，是中国传统造园"一池三山"的写照，其名瀛洲就是传说中神岛的名称。小瀛洲面积为7hm^2，为田字形。此洲是明代万历时期（公元1607年至1611年）疏浚西湖，挖湖泥筑堤埂形成的。小瀛洲在不同季节形成异样的景观，其夏季荷景最为艳丽，正如诗句赞美西湖的荷景："毕竟西湖六月中，风光不与四时同，接天莲叶无穷碧，映日荷花别样红"。在小瀛洲的我心相印亭前，有3座石塔立于湖中，塔高2m，塔身球状中空。每当中秋时节，塔内点亮蜡烛形成了许多小月亮，小月亮又同湖中的明月相映，故得名"三潭印月"。

Example 37　Three Pools Mirroring the Moon in Yingzhou

Yingzhou is located in the south of the West Lake, which is an island with the pattern of "One Pool and Three Mountains", covering an area of 7 hectares. It was formed by digging mud from the lake to build a bank in Ming Dynasty. Different seasons form different landscapes, and the summer lotus scenery is the most gorgeous. There are three stone towers on the island standing in the lake, with a height of 2 m and a spherical hollow body. Every Mid-Autumn Festival, candles are lit in the towers to form many small moons, which set off against the bright moon in the lake, so it is called Three Pools Mirroring the Moon.

立于湖中的3座石塔
Three stone towers in the lake

"三潭印月"碑亭
Monument Pavilion in Three Pools Mirroring the Moon

九曲桥、三角亭前夏日荷景
Summer lotus in Jiuqu Bridge and Triangle Pavilion

实例38　西泠印社

位于西湖孤山的西端，依山傍湖，是中国金石篆刻研究的学术团体，1913年正式成立，吴昌硕任首任印社社长。此处为一台地园，依地势自然布置，各层台地之间以自然曲径联系着。1956年2月，笔者从林木密集的孤山后面通过石洞门进入西泠印社最高台地，在这里见到耸立的华严经塔。华严经塔下是一湾池水，名曰闲泉。其西为吴昌硕纪念馆，东为题襟阁，东南前方是四照阁。走进四照阁，可以俯览开敞的西湖全景，其对比强烈，令人心旷神怡。由四照阁下山，再经曲径通幽的山林景色盘旋而下，便来到以柏堂为中心的底层庭院。

Example 38 Xiling Society of Seal Arts

It is an academic group that studies the seal cutting of gold and stone in China, which was formally established in 1913. The place is a platform garden located at the western end of Gu Hill, which is naturally arranged according to the landform, and the platforms of each layer are connected by natural winding roads. Among them, there is the towering Huayan Sutra Pagoda, with Wu Changshuo Memorial Hall in the west, Tijin Pavilion in the east and Sizhao Pavilion in front. People can get a panoramic view of the West Lake by walking into the pavilion.

从四照阁望西湖景色
West Lake from the view of Sizhao Pavilion

最高台地（左为题襟阁，中为四照阁，右为雕像和华严经塔）
The highest platform (from left to right are Tijin Pavilion, Sizhao Pavilion, the statue and Huayan Sutra Pagoda)

实例39 郭庄

在西湖西岸,环湖西路卧龙桥北,有座庄园为清代宋瑞甫所建,后为郭姓所有,故称郭庄。全园占地近1hm^2,分为"静必居"和"一镜天开"两个景区。"静必居"是宅园部分,位于西南部,为四合院建筑,主厅是"香雪分春"。

这里着重介绍"一镜天开"景区。二层楼的"景苏阁"是这一景区的主体建筑,其后右面是"赏心悦目"假山亭,左边与对面配以亭、榭、廊和山石花木,围合成富有变化的水庭主景。登上景苏阁,向东、北、南望去,就能俯览开阔的西湖敞景。同样,登上赏心悦目亭,亦可领略湖光山色的西湖美景,这是郭庄借景的独到之处。郭庄小园林与西湖大园林融为一体、相得益彰。

Example 39 Guo's Villa

It is located on the west bank of West Lake and was built by Song Ruifu in Qing Dynasty and later owned by someone with the last name of Guo, so it is called Guo's Villa. Covering an area of nearly 1 hectare, it is divided into two scenic spots, namely, "Static Residence" and "One Mirror Open the Sky". The former one is a quadrangle dwellings and the main building of the latter one is Jingsu Pavilion. On the right side of the pavilion, it is the Delightful Pavilion, and on the left, it is surrounded by rocks, flowers and trees, forming a varied landscape. People can see the West Lake when board Jingsu Pavilion and Delightful Pavilion.

主庭园"一镜天开"景区
Scenic spot of One Mirror Open the Sky

从"景苏阁"二楼望西湖
West Lake from the view of Jinsu Pavilion

"赏心悦目"假山亭
Delightful Pavilion

与西湖相连的"景苏阁"后院
Backyard of Jinsu Pavilion connecting with West Lake

实例40 杭州黄龙饭店

位于西湖风景区宝石山北面,离黄龙洞很近,规模为570间客房,可容1000多位旅客,建成于1988年,设计师是著名建筑大师程泰宁。

该饭店的建筑设计是成功的,它同杭州自然环境协调,是座具有地方特色的现代化旅馆。主要特色在:

(1)整体完美。首先同外部大环境和谐。宝石山不高,附近的建筑也不宜高大,设计师很好地掌握着尺度,并化整为零,采用小体量分散式布局,以满足杭州西湖风景区的整体设计要求。

(2)内部富有地方特色。分散的建筑围合成一个内庭园,营造出自然景色,并与外部宝石山结合。在黄龙饭店庭园还可观赏到保俶塔,在黄龙饭店大堂和客房里也可看到江南式内庭园。其建筑与庭园融为一体,水池自然、庭园简洁,叠石完整而有气势;黄龙饭店内部装饰高雅,材料与色彩统一,很有杭州地方特点。

(3)空间层次丰富。无论进入大堂北望,还是从餐厅向南看,或是走出多功能厅向西看或是从健身房东望,东南西北四个方向都有比较丰富的空间层次。

(4)富有新意。黄龙饭店在内向布局、平面构图、空间序列、庭园设计、建筑造型、构件细部等方面都吸取了中国传统文化特点,其设计既有中国传统意味,同时又有创新的地方特色。

Example 40 Huanglong Hotel

It is located in the north of Stone Mountain and was built in 1988 with 570 guest rooms. It is a modern hotel which is in harmony with the natural environment and has local characteristics.

(1) Overall perfection. It adopts a small and distributed layout to adapt to the overall scale of West Lake Scenic Area.

(2) Rich in local characteristics. Scattered buildings surround a Southern China style garden with unified interior decoration, materials and colors.

(3) There are spatial layers. The layout in the four directions of south, east, north and west has rich spatial layers.

(4) Innovative. Tradition has been absorbed in layout, composition, space, modeling and details, but innovations with local characteristics have been made.

从庭园望塔楼
Buildings in the view from the garden

2. 绍 兴

是一座位于宁绍平原西部会稽山麓的古老城镇，古迹很多。大禹在此治水，尚有禹庙在。公元前400多年，越王勾践伐吴前誓师之处尚在，即城南投醪河。公元前210年秦始皇南巡，到山阴登会稽山，并立碑而归，后人称此处为秦望山。东晋著名书法家王羲之为会稽内史时，曾与亲友在兰亭赋诗饮酒，撰书《兰亭集序》。为此，明嘉靖年间还修建了兰亭胜地。南宋时期，绍兴是陪都。到了近代，绍兴是革命先驱周恩来祖居，也是鲁迅、秋瑾的故乡。绍兴已被列为国家历史文化名城。绍兴地区，水网密布，是一个出色的江南水乡。新中国成立后，这里兴建了一批工业企业，改建了市内一些主要街道，也使旧城部分地区失去了原有水乡的特色。现已恢复原有的地域文化特点。2008年，绍兴市获得联合国人居荣誉奖。

实例41 绍兴水乡

绍兴古城不大，南北长5km、东西宽3km，北部湖塘较多，鉴湖水系环绕城市四周。市内河道纵横，府山、蕺山、塔山成鼎足之势。水网依山而过，构成一座集山、水、城市于一体的江南水乡。城市居住街坊处于街道河之间，家家户户有台阶下河，生活方便。绍兴城周边的小镇，亦为水乡之城。绍兴郊区的湖中还建有长长的纤塘桥，这便于拉纤行船或作步行之路。

7.2 Shaoxing

It is located at the foot of Kuaiji Mountain and has many historical sites. Wang Xizhi, a calligrapher of the Eastern Jin Dynasty, once wrote the famous Orchid Pavilion Preface here. It is also the hometown of Lu Xun and Qiu Jin. It is a famous historical and cultural city with dense water network. Now a number of industrial enterprises have been built and some main streets have been rebuilt. In 2008, it won the UN Habitat Honor Award.

Example 41 Watertown

The area here is small, with a length of 5 km from north to south and a width of 3 km from east to west. There are many lakes and ponds in the north. The rivers in the city are vertical and horizontal. Fu Hill, Mang Hill and Ta Hill stand

一边有街水巷
A water lane with water on one side

纤塘桥一角
Part of the Xiantang Bridge

in the urban area, and the water network passes by the hills. The residential areas are located between streets and rivers, and most households have steps to go down the river, making life convenient. There is a long Xiantang Bridge in the lake of a small town of the city, which is convenient for pulling the fiber and sailing or walking.

水巷桥
Water Lane Bridge

实例42 绍兴兰亭

此园位于绍兴市西南14km的兰渚山下。晋永和九年（公元353年）三月初三，书法家王羲之邀友在此聚会，并写了《兰亭集序》。序中写道："此地有崇山峻岭、茂林修竹，又有清流激湍、映带左右，引以为流觞曲水。""流觞曲水"之做法自此相传下来。每逢三月初三日，好友们相聚水边宴饮，水上流放酒杯，顺流而下，若停于谁处谁就取饮，这被认为可被除不祥。兰亭园林的特点有：

(1) 自然造景。进入此园，系依坡凿池建亭，创造一鹅池景。相传王羲之爱鹅，亭中碑文"鹅池"二字系王羲之所书。转过山坡可见兰亭，右转即为"流觞曲水"景。

(2) 历史文化丰富。北面建筑的中心形成流觞亭、御碑亭轴线。御碑亭内有康熙御笔的《兰亭序》和乾隆诗《兰亭即事一律》。其西为兰亭碑亭，碑上"兰亭"两字为康熙所书。其东为王右军祠，系一对称的水庭院落，内有王羲之画像，院落回廊壁上还嵌有唐宋以来10多位书法名家临摹《兰亭集序》之石刻。

Example 42 Orchid Pavilion

It is located at the foot of Lanzhu Mountain, 14 km southwest of Shaoxing. Wang Xizhi, a great calligrapher of Jin Dynasty, wrote the famous *Orchid Pavilion Preface* here. Specific features are:

(1) Natural landscaping. The construction is based on the landform. There is a goose pond and it is said that Wang Xizhi loves geese. When turns over the hillside people can see the Orchid Pavilion, and turns right, that is the scenery of floating wine cup along the winding water.

(2) Rich historical and cultural contents. In the north, there are Floating Wine Cup Pavilion and Royal Monument Pavilion with the inscriptions of Emperor Kangxi and Gianlong inside. In the west, there is a tablet pavilion and in the east there is a memorial temple with a statue of Wang Xizhi and stone carvings copied by more than 10 famous calligraphers since Tang and Song dynasties.

Plan
1-Gate
2-Goose Pond Pavilion
3-Goose Pond
4-Scenery of Floating Wine Cup along the Winding Water
5-Floating Wine Cup Pavilion
6-Orchid Pavilion Monument
7-Royal Monument Pavilion
8-Memorial Temple of Wang Xizhi

平面　1-大门；2-鹅池亭；3-鹅池；4-流觞曲水；5-流觞亭；6-兰亭碑亭；7-御碑亭；8-王右军祠

兰亭
Orchid Pavilion

鹅池
Goose Pond

"流觞曲水"（中间坐者为北京市建筑设计院张镈先生）
Scenery of Floating Wine Cup along the Winding Water (Mr. Zhang Bo sitting in the middle)

实例43　绍兴青藤书屋

在绍兴市前观巷大乘弄,有明代文学家、书画家徐渭(公元1521—1593年)之读书处。入门为一小竹园并有假山,山后墙上嵌有徐渭手书"自在岩"刻石。书屋为一砖木石平房,中隔一墙,分为外南、内北二室,外室为夏凉之屋,南面为方格长窗,下有方格洞直通室外;内室是冬暖之屋,陈列着徐渭的书画作品。前有一小院,青石架上摆放着花木盆景,阳光明亮。外室前有一小天井,种有青藤一棵,徐渭喜爱青藤,故此书屋取名"青藤",并作为自己的号称。这里还设一石砌水池,长2.75m、宽2.64m,中立方石柱,上刻徐渭手书"砥流中柱"。徐渭称"此池通泉,深不可测,水旱不涸,若有神异。"这就是"天池",后来"天池"亦成为徐渭的又一别号。

Example 43 Green Vine Study

It is located in Qianguan Lane and is the reading place of Xu Wei, a writer and painter of Ming Dynasty. The entrance is a small bamboo garden with rockery. The study is a bungalow made of brick, wood and stone, separated by a wall and divided into two rooms. The outer room is a cool house in summer and the inner room is a warm house in winter. There is a small courtyard in front of the outer room with a green vine. Xu Wei loves green vine, so the place is named after it. There is also a stone pool, which is a natural air conditioner. In summer, the cold air in the pool enters the room through the square hole at the lower part of the outer room.

入口
Entrance

外室前小院
Courtyard in front of the outer room

外室（下部透空）
The outer room

内室北小院
North yard of the inner room

实例44　绍兴鲁迅故居

位于绍兴东昌坊周家新台门内，主要建筑现存两楼两底木构民房一幢。鲁迅于1881年9月25日诞生于此，直至1899年外出求学前一直生活在这里。1910—1912年，鲁迅回乡任教绍兴府中学堂并担任绍兴师范学校校长期间，仍居住在此。他的第一篇小说《怀旧》和《会稽郡故书杂集》等都是在这里写成的，鲁迅在家中还经常接待青年学生。此故居后边有一百草园，是鲁迅童年时的乐园。1918年房屋卖给了东面邻居朱姓，1919年冬鲁迅全家迁到北京。新中国成立后，当地政府将鲁迅故居多数家具找回，并按原样陈列，而未拆故居房屋则经过整修恢复了旧貌。

Example 44　Former Residence of Lu Xun

It is located in Dongchangfang. Lu Xun was born here on September 25, 1881, and lived here until he went out to study in 1899. His first novel Nostalgia was written here. There is a Baicao Garden behind, which is the paradise of Lu Xun's childhood. In the later period, the local government recovered most of the original furniture and displayed it as it was. The house of the former residence that had not been demolished was restored to the original appearance after renovation.

外景
Exterior

实例45 绍兴三味书屋

"三味书屋"是清代末年寿镜吾先生家中的木构平房,它临河,且位于新台门东。寿镜吾利用此屋开办一所私塾,少年时代的鲁迅于1892—1897年就在此师从寿镜吾先生求学读书。

屋正中悬挂一块匾额——"三味书屋",系清人梁同书手笔,匾下挂一"松鹿图",中间放一大方桌,左右各放太师椅,屋周围及窗下放有八九张学生课桌。其中一张课桌刻有"早"字,这就是当年鲁迅的书桌。这"早"字是鲁迅亲手所刻,曾有一次因家事而迟到,鲁迅受老师批评,于是鲁迅刻字以提醒自己,从此再不迟到。

此临水书屋,具有绍兴水乡的特色。

Example 45 Sanwei Study

It is a wooden bungalow owned by Mr. Shou Jingwu in the late Qing Dynasty. He used this house to set up a private school where Lu Xun studied from 1892 to 1897. A plaque, "Sanwei Study" is hung in the center of the house, with a painting of pine and dear hanging under. A large square table is in the middle and an old-fashioned armchair is on the left and right side each, and then are the desks and seats for eighty to nine students. One of desks was engraved with the character "early". Lu Xun engraved it by himself, reminding himself not to be late.

临河外景
Exterior along the river

八、安　徽

合　肥

　　1949年前合肥是一座小城市，当时人口才5万，现是安徽省省会，是安徽省政治、经济、文化、交通中心。20世纪50年代末，该市在发展一些工业的同时，重视改善旧城的生活环境和主要干道长江路的设施，成为全国小城市规划与建设的典范，并于1959年在全国城市建设十周年成就展览会上展出其成果。改革开放后，合肥市建立了经济技术开发区，工业发展更为迅速，除改造了原有的钢铁、矿山、机械制造、电机、纺织、印染、化工、建材工业外，又增加了电器、电子等工业。市区人口已由小城市变为大城市。合肥市的规划与建设还有一个最大特征，就是重视环境建设。合肥市已成为全国首批3个园林城市之一。下面着重介绍其园林绿地建设。

实例46　合肥环城绿地及全市绿地系统

　　半个多世纪以来，合肥市十分重视环城绿地和全市绿地系统的规划与建设，具体内容是：

　　（1）围城建造环状绿带。合肥旧城周长8.3km。为保留其护城河，拆除了城墙，西面将城墙土与拓河之土堆成自然山峦，其余造势河岸两侧，形成环城绿化带，既改善了旧城环境卫生，又为居民创造了休息之地。

　　（2）结合古迹发展公园。在旧城环形绿带的东北角，是公元3世纪三国时期"张辽威镇逍遥律"之地。今借题发挥，开辟成30多公顷的综合性公园。旧城东南隅外侧的香花墩为宋包拯早年读书之处，后人修建了包公祠。此处水面宽阔，地形起伏，遂发展成近30hm^{2}的包河公园，并将包公墓重建在此。结合古迹的修建公园，这丰富了环旧城河的绿色林带。

　　（3）西部森林水库绿地。距市中心9km的西郊大蜀山，高280多米，面积550hm^2，在原有林木的基础上开辟森林公园，并在此山北面的董铺水库、大房郢水库发展大片绿地，以保护水库。在

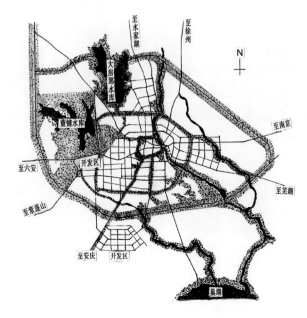

合肥市绿地系统规划
Planning of the green space system in Hefei

大蜀山下发展了果园、桑园、茶园和苗圃，在水库周边发展了经济林，此风景林与经济林的结合将开辟郊区风景区的新局面。

（4）东南巢湖引风林区。在离市中心17km的南郊巢湖，面积达782km²，为中国五大淡水湖之一，它将逐步发展成为具有特色的风景旅游区。另外，拟在市区东南方向逐步发展引风林区，顺应城市主导风向，将新鲜空气引入城市，并有利于城市的自然通风。

（5）二环、三环连接郊外绿地。环绕西郊、东南郊绿地的内边缘与外边缘，逐步建成二环、三环绿化带，并将郊区绿地联系起来。

（6）城市生态绿地系统。前述已建成的一环旧城绿带及其古迹公园，还有正在发展的西郊与东南郊风景林经济林大片绿地以及逐步建设的二三环绿化带，通过引向城市中心的绿带组成了合肥市有机的绿地系统。这一系统非常可贵，它对改善城市生态系统起着重要的作用，它既能改变温室效应、净化空气、美化城市，更为城市居民提供休闲游览场所。城市生态绿地系统建设，这将是21世纪城市的发展方向。

8. Anhui

Hefei

It is the capital of Anhui Province and the center of politics, economy, culture and transportation. In the late 1950s, while developing industry, the city attached importance to improving the living environment of the old city and the construction of main roads, which became a model for the planning and construction of small cities in China. After the reform and opening up, the economic and technological development zone was established, and the industry developed more rapidly. It is one of the first three garden cities in China.

Example 46 Green Space System

Hefei attaches great importance to the planning and construction of green space system, with specific characteristics as follows: (1) The green belt around the city. The moat is retained while the city wall was dismantled. And the green belt was built up around the city on both sides of the river bank. (2) Developing parks with historical sites. In the northeast corner of the old city, a comprehensive park of more than 30 hectares was opened up on the remains of the Three Kingdoms period. (3) Green space of western forest and reservoir. In the western suburbs of Dashu Mountain, a forest park was opened up and developed. A large green space has been developed in the reservoir to the north of this mountain. (4) Wind-induced forest area of Chaohu Lake in southeast China. In Chaohu Lake, one of the five freshwater lakes in China, wind-induced forest areas are gradually developed, and fresh air is introduced into the city along the dominant wind direction of the city. (5) The second and third ring roads connect the suburban green space. (6) Urban ecological green space system. The aforementioned green spaces and related constructions have formed an organic green space system in Hefei.

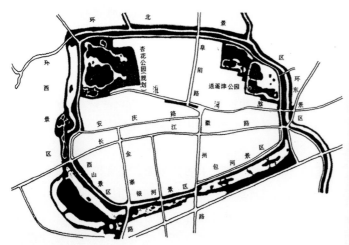
已建成的环旧城公园
Complete park around the old city

包河景区南部丛林
Forest in the south of the Bao River Scenic Area

银河景区
Galaxy scenic spot

经济开发区明珠广场绿地喷泉
Fountain of Pearl Square in Economic Development Zone

实例47 黄山风景名胜区

　　黄山位于安徽省南部,面积为154km²。黄山以"奇松、怪石、云海、温泉"四绝著称,它有着泰山之雄伟、衡岳之烟云、华山之峻峭、匡庐之飞瀑、峨眉之清凉,故明地理学家徐霞客称赞"五岳归来不看山,黄山归来不看岳"。笔者观五岳及四大佛教名山之后,再同黄山相比,确实感到黄山为"天下第一山"。黄山现为国家重点风景名胜区,亦被列为"世界自然遗产"。

　　黄山的奇松、怪石、云海三绝,大多集中在玉屏楼至北海游览路线的两侧。此中心游览区内有黄山的三大高峰,其中,莲花峰最高海拔为1867m;光明顶居第二,海拔为1840m;天都峰列第三,海拔为1810m,且最为险峻,其名取意为天上之都会。除此三大高峰外,还有其他动人的风景,如始信峰的"琴台",笔锋的"梦笔生花",狮子山的"猴子观海"等。

Example 47 Yellow Mountain

　　It is located in the south of Anhui Province with an area of 154 km². It is famous for its "legendary pines, picturesque rocks, the sea of clouds and hot springs". It is a national key scenic spot and is now listed as World Natural Heritage.

　　Yellow Mountain has three peaks, and the highest altitude of Lotus Peak is 1,867 m; Bright Summit is the second, with an altitude of 1,840 m; Tiandu Peak is the third, with an altitude of 1,810 m. There are other moving sceneries, such as "blossom flower on pen in the dream", "Monkey watching the sea", etc.

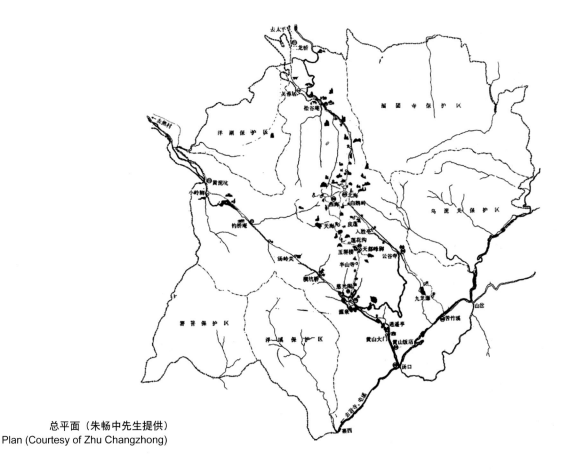

总平面(朱畅中先生提供)
Plan (Courtesy of Zhu Changzhong)

猴子观海
Monkey watching the sea

天都峰
Tiandu Peak

石柱奇松
Strangely-shaped pine on the stone column

九、福建

厦门

厦门位于福建九龙江海口,面对台湾岛,面积为1516km^2。厦门岛是厦门的母城,面积131km^2。1958年我们国家城市设计研究院规划工作组来该市进行城市总体规划设计,当时笔者是工作组成员之一。此时厦门岛旧城区仅10多平方公里,常住人口为10多万人。古时白鹭常栖集这里,故厦门亦称鹭岛。明代初始建厦门城,清代乾隆至嘉庆期间厦门港兴旺,1840年鸦片战争后厦门港被英国划为五口通商之一。1933年设为厦门市,1981年创办厦门经济特区。20世纪90年代笔者多次到厦门,了解到从1982年至1997年,厦门半岛旧城区从12km^{2}扩大到70km^2,城市人口从20多万增到50万人,国内生产总值从10多亿元增到371亿元。厦门市区远期的总体规划已扩大到150km^2,形成了大城市的框架。以港兴市,厦门已成为现代化国际性海港与风景城市。其城市结构以厦门岛为中心,同时辐射四周卫星城镇,形成一城多镇的布局。这种分散式格局,将有利于创造出良好的生活环境。

实例48 厦门筼筜中心区

位于厦门市本岛中心偏西部,背山面水,筼筜湖融于其中。厦门市府大楼南面中心广场于20多年前建起大会堂,这座重要的中心建筑是经过设计竞赛评选出来的。当时有5个设计单位参加,最后选出的是上海建筑设计研究院所做的设计方案,该设计方案简洁开放,前后可以穿通,内部布局也很适合大会堂的使用,并满足民众参观学习。笔者同东南大学建筑系鲍家声教授等参加了这次评选工作,我们都力主开放式建筑,即大会堂公共建筑要便于大众使用,便于内部空间发展变化,同时又要使空间开敞明快,还要有一个较好的绿地环境。厦门市领导和规划部门是有远见的,他们采纳了我们的意见,将市中心大会堂周围规划的建筑群用地都改作绿地,使这里具有良好的生态环境。在筼筜湖对岸,规划部门确定高层建筑要后退,高层建筑之间还要有一定的距离,以减少高层建筑对湖面空间的压抑感。

9. Fujian

Xiamen

The city is located at the mouth of Jiulong River in Fujian, facing Taiwan Island, where egrets often lived in ancient times, so it was called Egret Island. It was one of the five trading ports in Qing Dynasty and became a special economic zone in 1981 and developed rapidly. Xiamen's urban planning is to make it a modern international seaport and scenic city. Its urban structure is centered on Xiamen Island, radiating around satellite towns, forming a layout of one city with many towns like stars twinkling around the moon.

Example 48 Yundang Center Area

It is located in the west of Xiamen Island with mountains and water on its back. The Great Hall was built in the central square in the south of the city building more than 10 years ago. This open building is convenient for the public to use, and it is convenient to change the internal spatial layout due to the development needs. At the same time, there is a better green environment around it. On the other side of the Yundang Lake in the south of the center, the high-rise buildings have retreated, and there is a certain distance between them, which reduces the sense of repression on the lake space. However, as the related construction continues, the rhythm of the three-dimensional outline of the building on the south bank of the lakeside has not yet appeared.

厦门市远期总体规划
Master plan of Xiamen in the long-term

大会堂外景
Exterior of the Great Hall

大会堂前湖景
Lake in front of the Great Hall

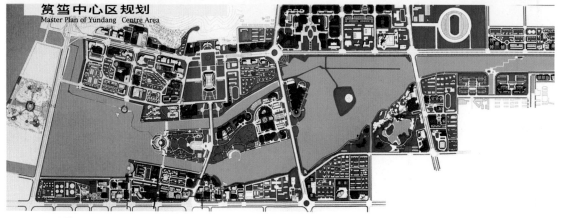

筼筜中心区规划
Master plan of Yundang center area

实例49　厦门康乐新村等居住区

1997年我们参观了厦门新建起的几个居住区，有康乐新村、吕岭花园小区、体育东村等，其特点是：（1）控制适度的建筑容积率，并未过多地追求建筑面积以实现经济利益，建筑以六、七层居多；（2）配套公共服务设施比较齐全，生活方便；（3）住宅建筑设计简洁明快、自然通风，适宜居住并节能；（4）自然环境优美，特别是山青水秀、水粼粼、树葱葱、花艳艳，人居环境极佳，为此，厦门市获得了联合国人居奖。介绍此实例，就是希望各地学习厦门解决居民居住问题的思路和做法。解决城市居民安居乐业问题，这是政府的职责，而借此谋取暴利的政府管理者和开发商等都将会受到国家法律的严惩与制裁。

Example 49 Kangle New Village and Other Residential Areas

Newly built residential areas in Xiamen, such as Kangle New Village, are characterized by: (1) controlling moderate building floor-area ratio and not pursuing too much building area; (2) complete supporting public service facilities and convenient living; (3) the design is simple and bright, naturally ventilated, livable and energy-saving; (4) beautiful natural environment and good living environment. The city won the United Nations Habitat Award, and especially rewarded the city for solving the housing problem for vulnerable residents.

康乐新村中心湖面
The center lake of Kangle New Village

实例50　厦门高崎国际机场候机楼

该建筑设计具有创新性，其最大特点是采用新结构、新材料创造出具有闽南大屋顶曲线的造型。屋顶陡峭的曲线非常突出，让人联想到闽南"倒水"大屋顶的形象特征，其姿势如鸟儿展翅或飞翔状。该楼的结构像闽南木构架穿斗式做法，又如抬梁立柱式屋顶构造，但又完全是新的钢筋混凝土结构，这种既传统而又非传统的设计就是一种创新。它的第二个特点是大空间。在这个长方形大空间内，可根据发展的需要在平面和立体两个层面上自由地分割或加层，以适应新的变化要求；另一个特点是节能减排，其层层的高侧窗自然采光，正脊下的通风口也是自然通畅排风，可以减少能源耗费和有害气体的排放量。这幢候机楼的建筑设计是加拿大B+H国际建筑师事务所，其结构创新设计师是我国结构设计大师孙芳垂先生，这里既有闽南构架的意味，也有新材料、新结构的应用，它经济实用，为节能减排创造了条件，对创新设计也有启示作用。

Example 50　Terminal of Ximen Gaoqi International Airport

The biggest feature of the building is that it uses new structures and materials to create a curved roof with artistic conception in southern Fujian. The steep curve of the roof is very prominent, and it is as light as a bird's wings. The building structure is like a wooden frame in southern Fujian, but it is completely a new reinforced concrete structure. The second feature is the large space, which can be divided or added at

外景
Exterior

two levels of plane and three-dimensional space according to the needs of development, so as to adapt to the new changing requirements. It is also of energy saving and emission reduction functions by high side windows, natural lighting, vents under the ridge, smooth natural exhaust to reduce energy consumption and harmful gas emissions.

大厅 Lobby

实例51 鼓浪屿

位于厦门市西的一个面积为1.84km²的小岛，中间隔有700多米的厦鼓海峡。山多奇石，幽岩洞壑，海浪入洞则声如雷鸣，故取名鼓浪屿。其最高峰为高90m的日光岩。清晨旭日东升，日光正好照射到山岩顶，故取名日光岩。在此山顶，极目远望，厦门市和海上诸岛尽收眼底。1962年在日光岩麓建郑成功纪念馆，以纪念这位明末清初的民族英雄。郑成功曾在此屯兵，并操练水师。在山上还建有莲花庵，庵侧岩石上刻有明人所题"鼓浪洞天"四字，这是厦门的八景之一。近代面对厦门市区一面还建有一八角楼建筑，并构成了这一带优美的立体轮廓线。1958年笔者到厦门开展城市规划设计工作时拍下了这一美景。当时，岛上住户较少，林木葱郁、鸟语花香，在这幽静的山峦环境中，还不时传来悦耳的钢琴声，一些著名的音乐家就出自鼓浪屿，因而这里以"海上花园""音乐岛"等享誉世界。鼓浪屿现为国家重点风景名胜区，2017年7月还被联合国教科文组织列为世界文化遗产。

面海日光岩
Sunlight Rock

面对厦门市区的鼓浪屿一角（中间突出建筑为八角楼，1958年摄）
A Part of Gulangyu Island facing Xiamen (the outstanding eight-corner building in the center, shot in 1958)

Example 51 Gulangyu Island

It is a small island with an area of 1.84 km² located in the west of Xiamen City, which is on the top of Sunlight Rock, the highest peak, with far-reaching view that Xiamen City and the offshore islands are all in sight. In 1962, Zheng Chenggong Memorial Hall was built at the foot of this rock to commemorate this national hero in the late Ming and early Qing Dynasties. The Lotus Temple here is facing Xiamen city and an eight-corner building is built in modern times, which constitute a beautiful three-dimensional contour line in this area. Some famous musicians come from Gulangyu Island, so it is famous in the world in the name of "Sea Garden" and "Music Island". Gulangyu Island is now a national key scenic spot and was listed as a World Cultural Heritage in July 2017.

实例52 南普陀寺

位于厦门市五老山下，遍山苍松翠竹，峭岩深壑，相映成景。南普陀寺的第一个特点是大自然环境幽美。南普陀寺的第二个特点是台地寺。它建于坡地上，各个院落依纵向轴线处于不同高度台地上，从寺右后方的太虚亭远望，可看到景色壮美的厦门港。南普陀寺的第三个特点是建筑具有闽南特色。其建筑曲线、装饰、起翘极其优美，主体建筑天王殿、大雄宝殿等虽为楼阁式、宫室，但屋脊为曲线型，起翘欲飞，体现了闽南建筑的特点。南普陀寺的第四个特点是尚存有历代的珍贵之物。该佛寺始建于唐代，宋重建，元、明历有兴废，清康熙时期靖海将军施琅重建，改称"南普陀"。南普陀寺内藏经阁珍藏有经典宋钟，寺内还保存碑记石刻、乾隆御制碑等历史实物。南普陀寺的第五个特点是发展佛教教育和社会公益事业。现已办起佛教教育学院，分男女二部，男部就设在寺内。同时，它还建立了慈善事业基金会和义诊医疗机构，为社会群众服务。

Example 52 Southern Putuo Temple

It is located at the foot of Wulao Mountain and covered with pines and bamboos. Its first feature is beautiful natural environment. The second is that it was built on the terrace. The courtyards are located on terraces with different heights along the longitudinal axis. The third is that it has the characteristics of southern Fujian, and the roof of the building is curved like a flying bird. The fourth is that it was built in Tang Dynasty and there are still precious relics from past dynasties. The fifth is it develops Buddhist education and social public welfare undertakings, and now a Buddhist education institute has been set up, and a charity foundation and a free clinic medical institution have been established as well to serve the public.

南普陀寺全景（地方提供）
Panorama (Courtesy of the local institution)

十、广　东

1. 广　州

位于珠江三角洲北缘，东江、北江、西江三江交汇处，是华南地区最大的城市，现为广东省省会。广州建城已有2000多年的历史，始自战国至秦汉即为对外贸易港口。随着对外贸易的发展，以及商业、手工业等经济的发展，广州成为了南方的政治、经济和文化中心。广州还是中国历史文化名城。新中国成立后，广州现代工业和城市建设发展迅速，城市工业企业和城市建设布局采取分散的方式，大力发展郊区工业城镇，注意环境保护，同时控制市中心区人口规模。自1957年起，每年春、秋季在广州举办中国出口商品交易会。广州外港建成黄埔新港，进一步促进了广州对外贸易和港口城市的发展。广州的建筑创作在全国一直处于领先地位，其建筑创新体现在新材料、新结构、新技术的应用，同时也重视同园林绿化的结合以提高环境质量，并不断发展岭南地域建筑文化。

实例53　广州矿泉客舍

此客舍靠近广州市区西北边，原为接待外宾的地方，总客房数量为100间，建造年代为1965年，设计师是著名建筑大师莫伯治先生。其建筑设计特点是，重视序列空间的变化和采用庭园式布局。矿泉客舍客房大部分设在二、三层楼上，其下为支柱层。在进入主庭之前，要经过一段前导空间，穿过月洞门，豁然开朗，呈现出一派疏朗高雅的主体景观。这主庭的设计，是将传统与现代做法糅和在一起，既体现建筑的现代感，又表达着"临溪越地，虚阁堪支"的诗情画意意境。同时，还运用竹、石等地方材料和花木素材去塑造庭园建筑的新时代形象和优美的自然环境。莫伯治先生是岭南庭园式新建筑的创始人，其设计的广州矿泉客舍建筑也已被选入《20世纪世界建筑精品集锦》东亚卷中。

10. Guangdong

10.1 Guangzhou

It is located in the northern edge of the Pearl River Delta and is the largest city in South China and the capital of Guangdong Province. The city has a history of more than 2,000 years. It was a foreign trade port from Warring States to Qin and Han Dynasties. With the development of foreign trade, it became the political, economic and cultural center of South China. Guangzhou's architectural innovation is in a leading position in China, which is characterized by adopting new materials, new structures and new technologies, attaching importance to the combination with landscaping.

从支柱层望主庭园和月洞门
Main courtyard and Moon Hole Door

支柱层后庭
The backyard of the pillar floor

Example 53 Guangzhou Mineral Water Guest House

It is near the northwest of the city and was originally a small residence for foreign guests, with a total number of 100 guest rooms. Its design features are attaching importance to the change of sequence space and adopting garden style. Most of the guest rooms are located on the second and third floors. It blends traditional and modern techniques together that not only reflects the modernity of architecture, but also expresses poetic and artistic meaning. It has been selected into the East Asian Volume of the *World Architectural Masterpieces Collection in the 20th Century.*

实例54　广州白天鹅宾馆

位于广州沙面岛南侧，南临珠江白鹅潭，是座拥有1040间客房的国际五星级宾馆。广州白天鹅宾馆1983年建成，设计师是著名建筑大师佘畯南、莫伯治等。其建筑功能、空间和环境协调、统一，特别是门厅、休息厅、咖啡厅、餐厅等公共活动部分临江布置，使旅客便于欣赏江景。中庭作为白天鹅宾馆的景观核心，所有流动空间、餐厅、休息厅、商场等都围绕中庭布置，构成上下盘旋、高旷深邃的园林空间。中庭设计将动与静有机融为一体，加上流动的瀑布与"故乡水"的题字，使得中庭气势雄伟、雅致和谐，并富有岭南庭园特色，常唤起人们的思乡之情，让旅客流连忘返。白天鹅宾馆采用高低层结合的手法，其高层为客房主楼，建筑外墙喷涂白色饰面，颇有天鹅白羽重叠之意。白天鹅宾馆建成使用后，一直坚持对公众开放，体现了社会的公平公正。此建筑也已被选入《20世纪世界建筑精品集锦》东亚卷中。

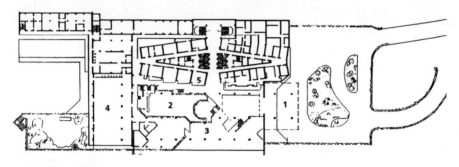

白天鹅宾馆底层平面 1-入口门廊；2-中庭；3-临江休息厅；4-餐厅；5-上部为客户层

Plan of the ground floor 1-Entrance porch; 2-Atrium; 3-Lounge along the rive; 4-Restaurant; 5-The upper part is the customer floor

中餐厅水庭
Water court of the dinner

Example 54 White Swan Hotel

It is located in the south of Shamian and is an international five-star hotel with 1040 rooms. The foyer, lounge, coffee shop, restaurant and other public activities are arranged along the river, and the atrium is designed as a whole multi-layer garden. With the flowing waterfalls and the inscription "Hometown Water", it is full of the characteristics of Ling-nan Garden, which can often arouse people's homesickness. The top floor is for the main guest room and its exterior wall is painted in white, like the feather of white swan. This building has been selected into the East Asian Volume of the *World Architectural Masterpieces Collection in the 20th Century*.

总统套房卧室
Bedroom of the presidential suite

总统套房花园
Garden of the presidential suite

白天鹅宾馆中庭
Atrium

2. 珠海

实例55 珠海规划与建设

珠海市位于珠江口右岸，它是一个新兴的海滨城市，也是中国最早设立的4个经济特区之一。珠海市拱北区是内地与澳门之间的连接口岸，而香洲区渔港则是广东最大渔港之一。1984年应珠海市委的邀请，中国建筑学会组织了一个规划专家工作组对珠海特区的总体规划进行了技术咨询，该工作组成员有郑孝燮、周永原、金瓯卜、陶逸钟、邵华郁、邹德慈和张祖刚等。大家对珠海市提出要有一个合理的经济区域规划，人口结构要合理调整，还要增加信息、科技和服务人口，道路交通要通畅，绿化覆盖率要高，要形成花园城市，要控制建筑层数，并保护好环境等。40多年过去了，珠海确实走出了一条营造美好家园、适宜人居之路。根据总体经济区域规划，珠海市采取"一城多组团（镇）的规划布局"，塑造出一个优秀海滨花园城市的实例。联合国人居中心在颁给珠海市"国际改善居住环境最佳范例奖"的公报中这样写道："珠海经受住了城市人口迅速增长的考验，从一个贫穷落后的渔村，发展成为一个环境和居民生活质量都得到全面提高的城市，成绩斐然。同时，它为中国其他城市的建设树立了榜样，也是全球全面改善人居质量的模范城市。"珠海市还获得"全国园林城市""国家卫生城市""全国环境综合治理优秀城市""全国环境保护模范城市"等的荣誉称号。

珠海市总体规划
Master plan of Zhuhai

10.2 Zhuhai

Example 55 Plan and Construction of Zhuhai

It is located on the right bank of the Pearl River Estuary and is one of the earliest four special economic zones established in China. According to the overall economic regional planning, "the layout of one city with many groups" is adopted to create an example of an excellent seaside garden city. The UN Habitat awarded Zhuhai the "International Best Example Award for Improving Living Environment". It also won the honorary titles of "National Garden City", "National Health City", etc.

实例56 珠海渔女

这是象征珠海的石雕塑，位于珠海香洲区和拱北区之间的滨海公园海水中，有桥与情侣路相连。这一雕塑为写实的作品，它简洁朴素、神态亲切、尺度宜人、环境自然，是个成功的范例。它也有一个动人的神话故事。相传古时有一仙女来到这里，遇到勤劳的青年海鹏，两人相互爱慕。后海鹏听信谗言，让仙女摘下手腕上的玉镯作为信物，仙女无奈脱下此镯，当即晕倒在海鹏怀中，幸好被九洲长老看到，告海鹏去找还魂草。海鹏历尽艰辛，救醒了仙女，从此仙女便成为珠海的渔家女。我国的许多神话故事，实际上就是寓意人类追求美好生活、向往高尚情操的愿望。

Example 56 Statue of Fisher Girl

It is a stone sculpture symbolizing Zhuhai. It is located in the sea water in the north of Binhai Park between Xiangzhou and Gongbei District, and is connected with Lovers Road by a bridge. The statue is of simple shape, friendly manner and pleasant scale, which is a successful example. It also has a moving fairy tale, which expresses a desire to pursue a better life, noble morality and social justice.

实例57 珠海滨海情侣南路

20世纪90年代，沿滨海向南开通，修建起一条7km长的情侣南路，沿路建有宽窄不同的绿化带和绿地，而在绿地后并未建设高层建筑，这一片环境显得格外优美，让人视野开阔、心情舒畅。自然、和谐、美好的建筑空间环境，大大方便了广大市民的生活。情侣路源自上海。上海黄浦江畔有一情侣墙，每逢傍晚或节假日，便有很多情侣到此约会、休闲、漫步，谈情说爱。珠海的滨海情侣南路，同样具有这一功效。

Example 57 Lovers Road

In 1990s, the Lovers Road of 7 km was built along the south coast, with green belts and green spaces of different widths alongside. The name of the road comes from the Lovers Wall in Shanghai. Every evening or holiday, many young people and couples come here to date, relax and stroll.

珠海渔女全景
Panorama

情侣南路晨景
Morning view of Lovers Road

实例58 珠海石景山景区

位于吉大区的东北部，满山是花岗岩的巨蛋石，这是因长期海水浸蚀而形成的光滑外形。山石的组成及外形千姿百态，有的似虎、有的如熊、有的像鱼，这是此处景观的根本特点。这里有水生植物白莲，还有一白莲洞古迹。白莲是此处的一个特色。在珠海被列为特区的初期，于其景区旁修建了石景山宾馆，设计师重视石景山宾馆同山景的配合，采用低层分散式布局，建筑外观亦很简洁，同周围的环境相协调。于另一边又新建了珠海宾馆，设计师是著名建筑大师莫伯治先生，他同样采用低层分散布局的方式。莫伯治大师运用中国传统的结合庭园的空间序列总格局，创造出建筑融合于大自然环境中的空间设计。这两幢新建筑对石景山景区的保护与建设做出了表率作用。

Example 58 Stone Scene Hill Scenic Spot

It is located in the northeast of Jida District. The whole mountain is full of granite egg-shaped stones, which has a smooth shape due to long-term seawater erosion, and white lotus is planted on the water surface. Later, Stone Scene Hill Hotel was built next to the scenic spot, which adopted a low-rise decentralized layout and coordinated with the surrounding environment. After that, Zhuhai Hotel was newly built, which also adopted the low-rise decentralized layout. These two new buildings have achieved harmonious coexistence between architecture and nature for the protection and construction of this scenic spot.

石景山宾馆
Stone Scene Hill Hotel

白莲景
White lotus scenery

石景
Stone scene

珠海宾馆入口
Entrance of Zhuhai Hotel

十一、海　南

三　亚

实例59　三亚天涯海角

在海南岛的最南端，三亚市之西24km处。这里是一条长长的海滩，在海水中和沙滩上散布着成组的巨石。巨石为花岗岩，经海水浸蚀，表面光滑。海滩后面还种有椰林。在海边端部耸立着一组特别高大的巨石，上面刻有"天涯""海角""海阔天空""南天一柱"等字，这是清代雍正十一年（公元1733年）崖州知州程哲等题刻的。此处蔚蓝的天空与海水连在一起，海天一色，又有奇石矗立、海浪翻滚，再加上点点船只和渺小的人影，天涯海角景观看上去显得格外壮丽。笔者在大连、烟台、青岛、宁波、厦门、汕头、珠海等地的海滨似乎从未见到如此壮观的景色，这是由于天涯海角处的空间序列及其高潮点的空间气势所造成的，让人们有走到天涯海角尽头之感。因此，我们在造景或组织天然风景时要注意空间序列的变化和运用对比的手法。

11. Hainan

Sanya

Example 59 Sanya Ends Area

It is located at the southernmost tip of Hainan Island. It is a long beach, with groups of granite boulders scattered in the sea water and on the tall boulders engraved with the words "Skyline" and "End of Earth". Here, the blue sky and sea water are connected togetherbeach, and the surface is smooth after seawater erosion. There is a coconut grove behind the beach, and at the end of the beach stands a group of particularly , the waves roll, and with little boats and small figures, the landscape is exceptionally magnificent. It makes people feel like walking to the horizon and the end of the sea.

天涯海角远望
Distant view

近观天涯石
Boulder of "Skyline"

十二、湖 北

荆 州

实例60　荆州古城

该古城即湖北江陵县城，在长江中游北岸沙市的西面，春秋楚王曾在此建"诸宫"，秦灭楚后成为秦汉封王置府的重镇。传说三国时期刘备借荆州，还有关羽大意失荆州，足见荆州就是兵家必争之地。关羽镇守此地时在诸宫西南筑城，始称其为荆州城。南宋时将土城外包砌砖块，成为砖城。元初时拆除，明初又重建。荆州城周长9km、高近9m，明末时被毁，清顺治三年（公元1646年）又依旧基重建，即为现在保存的实物。荆州古城外形随河流地势呈偏长自由状，东西长、南北短，共设6座城门，上建城楼，东称宾阳、望江，南为曲江，西为九阳，北称朝宗、景龙，现城墙保存完好，而城楼大都已废圮，唯清道光年间（1838年）重修的景龙楼（大北门）尚存，屹立至今。现又将荆州古城东面宾阳城楼修复，巍峨壮观。1995年6月，国际建筑师协会（UIA）理事会在长江武汉至重庆段的游船上召开，中途我们与会成员陆续登岸参观了荆州古城，观赏了完整的荆州古城墙与新修建的宾阳城楼等。这座古城也已列为国家历史文化名城。

12. Hubei

Jingzhou

Example 60　Ancient City of Jingzhou

The ancient city is Jiangling County, which has been a battleground for military strategists since ancient times. The shape is long with the topography of the river. It is long from east to west and short from north to south. There are six city gates with towers built on it. Now the city walls are well preserved, and most of the towers have been abandoned. Only Jinglong Tower rebuilt during the Qing Dynasty (1838) still exists. The Binyang Tower in the east is restored. This ancient city has been listed as a famous national historical and cultural city.

荆州古城城墙
City wall

宾阳城楼
Binyang Tower

江陵碑苑
Jiangling Monument Courtyard

十三、湖 南

长 沙

实例61 长沙岳麓书院

位于长沙市湘江西面的岳麓山下,潭州太守朱洞于北宋开宝九年(公元976年)创建。它是我国保存最完整的一处古代书院,于1988年被列为第三批全国重点文物保护单位。宋真宗祥符八年(1015年)召见岳麓书院山长(院长),赐内府书籍,并亲书"岳麓书院"匾额。南宋时理学家张栻主持书院,著名理学大师朱熹由闽来此讲学,从学者达千人,以致有"潇湘洙泗"之称,是为书院鼎盛时期。到了清代,康熙、乾隆御书"学达性天""道南正派"匾额,赐予书院。光绪二十五年(1903年),书院改为湖南高等学堂。民国十五年(1926年)正式成立湖南大学。

岳麓书院占地2.5hm^2,建筑面积7000m^2,现有讲学、藏书、供祀三部分建筑,大都为清代重建。建筑布局是中国传统的纵向轴线合院式,主轴线上的建筑依次是赫曦台、大门、二门、讲堂、御书楼,其中讲堂为主体建筑,大门与讲堂两侧为各层庭院的半学斋和教学斋。在此中部的左面,是由大成殿组成的四合院。而中部的右后面是一自由布局的园林。主体部分的建筑结构以穿斗为主,没有斗栱,灰墙青瓦、简洁朴素,岳麓书院具有湖南地方建筑特色。

13. Hunan

Changsha

Example 61 Yuelu Academy in Changsha

It is located at the foot of Yuelu Mountain and is the most well-preserved ancient academy in China. The Southern Song Dynasty was the heyday of its development, and Hunan University was formally established here in 1926.

The Academy coverooks and offering sacrifices. The layout is the traditional Chinese longitudinal axis courtyard style, in whichs an area of 2.5 hectares with a building area of over 7,000 m^2. It has three buildings for lecture, collection of b the lecture hall is the main building, and there is a small hall on each side of the gate. On the left side of the middle part, there is a quadrangle with Dacheng Hall in the middle. At the right back of the middle part, there is a freely arranged garden. The building has grey walls and blue tiles, which is simple and has the characteristics of Hunan local architecture.

讲堂
Lecture Hall

从大门望二门
The first gate and second gate in a row

十四、广 西

1. 桂 林

桂林在广西东北部，北起兴安，南至阳朔，在这长100多公里处，平地拔起千姿百态的奇峰，山内多岩洞。

实例62 桂林山水 漓江风光

漓江发源于桂林东北面的兴安猫儿山，流过桂林到梧州并入西江。从桂林到阳朔的40km区间，山峰奇异、山水相依，这是构成桂林山水、漓江风光的两个基本因素。在漓江风光中还有6个其他因素，即翠竹浓林、霞光空濛、轻轻小舟、似锦田野、点点农舍和直下飞瀑，这6个因素中的任一个同前面两个基本因素一起组合成千变万化的桂林山水美景，真可谓"一江伴山千幅画"。伏波山位于市区偏北漓江畔，山峰挺拔，是桂林市区的一个重要景观。桂林是国家历史文化名城，桂林漓江风景名胜区则是国家重点风景名胜区。

14. Guangxi

14.1 Guilin

Guilin is located in the northeast of Guangxi, from Xing'an in the north to Yangshuo in the south. There are different types of mountains and many caves inside.

Example 62 Guilin Landscape and Li River

Li River originates from Mao'er Mountain in the northeast of Guilin, with strange peaks all the way at the sides of the river. In addition, there are six other characteristics in the scenery, namely, dense forest of bamboo, sunglow, light boat, beautiful fields, little farmhouses and straight waterfalls. Fubo Mountain is located on the bank of Li River in the north of the city, and its single mountain is tall and straight, which is an important landscape in Guilin city.

桂林山水景观
Landscape

桂林市区景观
Urban landscape

实例63　桂林芦笛岩

位于桂林市西北郊，距市区约6km。取此名是因为其周围生长芦草，用芦制笛，故叫芦笛岩。岩洞深长，有500多米，曲折上下、盘回变幻，洞顶、洞底、洞壁，耸立或倒挂着千姿百态的石钟乳，有的像幽景听笛、多彩罗帐、早霞狮岭，还有的像桂林山水，真是五彩缤纷，宛若仙境。

Example 63　Reed Flute Cave

It is located in the northwest suburb of Guilin, about 6km away from the urban area. The name is given because reeds grow around and flute is made from reeds. The cave is more than 500 m long, with various stalactites standing or hanging upside down on the top, bottom and wall, which is colorful and just like enjoying a magical fairyland.

石钟乳柱景观
Stalactites

似桂林山水景观
Just like the landscape of Guilin

2. 三江

实例64 马鞍寨

位于桂林以北三江县的林溪河畔。这是一个侗族村落，村寨选在平缓坡地上，地形是前低后高、北高南低，寨前溪水环抱，左右为知名的程阳桥和平岩风雨桥跨水横卧，沟通寨外交通。这里背山面水，山水环抱，坐北朝南，与大自然山水融为一体。具体建筑布局是以鼓楼及其前广场为中心，住宅围绕这个公共活动中心依地形成排顺列。鼓楼及其前广场，是居民聚会、议事和交往的地方。这里的居住房屋是穿斗式木构架，建筑不加粉饰，体现着侗族淳朴的民风。建筑多2~4层。为避潮湿，建筑底层是堆放农具柴草和饲养禽畜之处，上几层则作为卧室和存粮等使用。

程阳桥建于1916年，是三江县内最大最壮丽的一座风雨桥。桥全长76m，有五墩四孔，宽3.7m，桥上建有避风遮雨的廊道，它的功效是多方面的，既用作交通、避风雨，又可以交往与休息。这座桥的构造也很有特色，以小材建大跨梁。每个桥墩宽2.5m、长8.2m，两头做成锐角，以便于水流，并减轻水对墩的压力。桥墩间距17.3m，桥墩上为木梁结构，由于没有大木材，便采取分层、小材连接与悬挑的手法，将木材架于两个桥墩之间，墩上廊亭对下部连续木梁的支撑点起到关键作用，其设计巧妙合理。桥上建高亭5座，这5座高亭的造型优美，似群雁立于溪水之中，展翅欲飞。

14.2 Sanjiang

Example 64 Ma' an Village

This is a village of Dong nationality, surrounded by streams in the front. The specific architectural layout is centered on the Drum Tower and its front square, and the houses are arranged in a row according to the terrain. Without whitewashing, the building embodies the simple folk customs of Dong nationality. The buildings are mostly two to four floors. To avoid moisture, the ground floor is where farm tools, firewood and livestock are piled up, and the upper floors are used as bedrooms and grain storage.

Chengyang Bridge, which communicates with the traffic outside the village, was built in 1916, with a total length of 76 m, and a corridor for shelter from the wind and rain is built on the bridge. Each pier of the bridge is 2.5 m wide and 8.2 m long, and its two ends form an acute angle, which is convenient for water flow and reduces the pressure of water on the pier. There are five high pavilions built on the bridge, which are beautiful in shape, like geese standing in streams, spreading their wings and flying.

三江鼓楼
Drum Tower

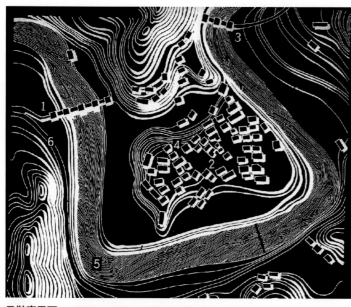

马鞍寨平面　1-程阳桥；2-村寨；3-平岩桥；4-鼓楼；5-河；6-山
Plan　1-Chengyang Bridge; 2-Village; 3-Pingyan Bridge; 4-Drum Tower; 5-River; 6-Mountain

临水民居建筑群
Residential buildings along the river

三江程阳风雨桥
Chengyang Bridge

3. 龙　胜

实例65　龙胜金竹寨

位于桂北龙胜县境内，是一个壮族干栏式民居村寨。全寨建在山腰上，山上翠竹成荫，山下溪水淌流，建筑依等高线灵活布置，背阴向阳，有充足的日照，整体村寨与自然环境浑然一体。解决山腰处用水是个重要问题，该处建设巧妙地从几里外的高山上引来山泉，以竹筒翻山越岭，跨过沟坎，让泉水流进山寨和梯田，既可以饮用，又能够灌田。中心山路，顺地势曲折有序，结合古木参天的风水树组成敞敞空间。建筑空间环境富有变化，同时也便于活动和使用。民居建筑为木构架干栏式，因建于山坡，采取吊脚、附岩、悬挑、筑台、镶嵌等方法建造，形式多样、自然和谐。因寨内外丛竹滴翠，故名金竹寨。

14.3 Longsheng

Example 65 Jinzhu Village in Longsheng

It is a residential village of Zhuang nationality. The whole village is built on the mountainside, shaded by bamboo trees, with streams flowing down the mountain. The buildings are flexibly arranged according to contour lines. From the mountains several miles away, spring water is introduced, which flows into cottages and terraces through bamboo tubes and can be used for drinking and irrigation. The central mountain road combined with ancient trees constitutes an open space, convenient for activities and use. The residential buildings are wooden frame.

沿坡民居建筑群
Residential buildings

中心山路风水树景色
The central road and ancient trees

十五、四　川

1. 成　都

成都位于川西平原腹地，河流密布，总面积12390km²，人口近千万，现是四川省省会，是国家历史文化名城。在新中国第一个五年计划经济建设时期（1953—1957年），全国156项重点工程中就有10项建在成都，尤以东郊工业区为重点。同时，成都市城市规划与建设也有了很大的发展，全市进行了较大规模的绿地建设。2007年成都市获得了"国家森林城市"的称号。

实例66　杜甫草堂

杜甫草堂在四川成都市西郊浣花溪畔，为唐代诗人杜甫在成都的居住地。杜甫草堂自然环境优美，总面积近20hm²，于一条纵向中轴上布置照壁、正门、大廨、诗史堂和后部的柴门、工部祠，诗史堂中陈放杜甫塑像。过柴门是工部祠，祠内正中是杜甫泥塑像。杜甫草堂东侧有一"少陵草堂"碑亭，此草亭象征着杜甫所住的茅屋，其旁翠竹丛丛，有溪水流淌，环境幽静。

15 Sichuan

15.1 Chengdu

Chengdu, a city of dense rivers, is located in the hinterland of the Western Sichuan Plain, with a total area of 12,390 km2 and a population of nearly 10 million. It is now the capital of Sichuan Province. In 2007, it won the title of "National Forest City".

后院工部祠
Ancestral Hall in the backyard

Example 66 Du Fu's Thatched Cottage

It is located in the western suburb of Huanhuaxi and is the residence of Du Fu, a great poet of Tang Dynasty, with a total area of nearly 20 hectares. There is a statue of Du Fu in the ancestral hall, and the monument pavilion on the east side symbolizes the cottage of that year.

"少陵草堂"碑亭
Monument Pavilion

2. 都江堰

实例67　伏龙观

该观位于四川都江堰的离堆之上，北宋时称为伏龙观。传说李冰治水在此宝瓶口下降伏了"孽龙"（江水），故称伏龙观。此离堆原与东北对岸之石连为一体，后建分流之水将其凿开，因而叫离堆。此观的特点：一是周围环境自然幽美。地处两水交叉宝瓶口处，四周山水环绕，景色辽阔自然；二是台地庭院林木遮荫。此观中轴线突出，三层台地庭院逐步升高，其中一层台地为老王殿，二层台地为铁佛殿，三层台地是玉皇楼。登楼可饱览自然山水全景；三是侧面小园为大众歇息处。三层台地东侧布置有小园，每逢进香之日，这些小园就作为大众停留、歇息之处。

15.2 Dujiangyan

Example 67　Fulong Temple

The characteristics of the temple are as follows: (1) Beautiful environment. Located at the intersection Precious-bottle-neck of two waters, the scenery is vast and natural. (2) Terrace courtyard and tree shade. The central axis here is prominent, and the courtyard of three terraces rises gradually. The first terrace is Old King Hall, the second is Iron Buddha Hall, and the highest is Jade Emperor Building, from which a panoramic view of the natural landscape can be seen. (3) The side garden is for the rest of the public and is open on the day of pilgrimages.

伏龙观全景
Panorama

离堆外景（左为宝瓶口）
Exterior (Precious-bottle-neck on the left)

玉皇楼前庭院
Courtyard of Jade Emperor Hall

3. 乐　山

实例68　乐山大佛

位于四川乐山市城东凌云山壁，岷江、青衣江、大渡河三江合流处。大佛依山崖凿成，它是一尊弥勒坐佛，始建于唐开元元年（公元713年），完工于唐贞元十九年（803年），历时90年。大佛头与山平，脚踏江岸，通高71m、头高14.7m、头宽10m、肩宽28m、耳长7m、眼长3m、脚背宽8.5m，它是世界上最大的石刻佛像。佛顶有螺髻1000多个，髻髻相连。大佛服饰上刻有折纹，其发髻、衣纹还起着排水的作用。大佛右侧设险峻陡峭的凌云栈道，栈道沿崖迂回而下，可达江边。乐山大佛气势雄伟、雍容大度，又名凌云大佛。凌云山上有凌云寺，建有天王殿、大雄宝殿、藏书楼等建筑群。凌云寺右有座唐代灵宝砖塔，高38m。凌云寺左建有东坡楼，院内陈放苏东坡塑像，环境幽美、典雅。乐山大佛于1982年被列为第二批全国重点文物保护单位。

15.3 Leshan

Example 68　Leshan Giant Buddha

It is located in the east of Leshan City and at the confluence of Minjiang River, Qingyi River and Dadu River. The Big Buddha is carved of mountains and cliffs. Built in the first year of Emperor Xuanzong in Tang Dynasty (713) and lasting 90 years, it is the largest carved Buddha statue in the world. There are more than 1,000 buns on the top of Buddha, which are connected with each other, and the clothing is engraved with fold lines. This bun and clothing lines also play the role of drainage ditch, preventing water accumulation corrosion. On the right side of the Big Buddha, there is a plank road facing the cliff, which meanders down to the river that people can walk up to the Buddha's foot.

近观乐山大佛，其尺度巨大
Huge close shot

远观乐山大佛全景
Distant view

十六、云南

1. 昆 明

实例69 昆明滇池、西山与昆明规划建设

滇池又名昆明池，在昆明市西南郊，海拔1885m，南北长40km、东西宽平均为8km，总面积297km²，水深平均5.5m。滇池池水是由南面盘龙江等20多条河流注入，向北经河川汇入金沙江。滇池是云贵高原著名的高原湖。滇池风景名胜区是国家重点风景名胜区，昆明是国家历史文化名城。2003年昆明市组织滇池国际规划设计竞赛，共有4家参加，经专家评选，中国城市规划设计研究院获胜，其他参与的3家分别是日本黑川纪章、美国SASAKI和澳大利亚PTW。中规院设计的方案采取的是"平行滇池的组团跳跃式线型发展格局"，将滇池通过河流主干与分支的绿地空间系统分隔成组团，这些组团再组成一整体，从而形成城绿交融的春城特色，并使昆明发展成为优秀的生态城市。

16. Yunnan

16.1 Kunming

Example 69 Dian Lake, West Hill and Construction of Kunming

Dian Lake is located in the southwest suburb of Kunming, with an altitude of 1,885 m, a total area of 297 km². It is a famous plateau lake in Yunnan-Guizhou Plateau. The spatial model adopted by Kunming is "paralleled clustering and leapfrogging linear development pattern of Dian Lake", which separates and connects the natural ecological green space with rivers as the main trunk among the clusters, forming the characteristics of the spring city of Kunming.

昆明市核心区概念规划透视图(中国城市规划设计研究院设计)
Scenography of "the plan of the core area in Kunming"

实例70 昆明地区民居建筑

提起云南昆明民居，大家就会想到"一颗印"住宅，这种类型住宅的平面为方形，很像一颗印，其用地经济，而且很实用。云南民居正房为三间两层，坐北朝南，两侧厢房各两间二层，较低为辅助用房。正房对面为入口门楼，中间围合形成一个小天井。外面是土坯墙粉面，内部是木装修，整体为穿斗式木构架，抗震性能好。当地还称此民居为"三间四耳倒八尺"。这里三间指正房，四耳指各两间厢房，八尺指门楼进深不超过此数。这是一个标准模式。笔者曾走访昆明及其周边地区的多家传统民居建筑，虽然很多处具体情况都有变动，但其基本构成与特点，这包括合院布局、正房突出、二层居多、木构框架、外土内木等，这些都没有变，这就是昆明地区传统民居的特征。

Example 70 Residential Area

When mentioned the dwellings in Kunming, you will think of the "seal-shaped buildings". This type of residence is square, which is very similar to one seal. It is economical for land use and very practical. The main room has three floors, facing south, two rooms on both sides have two floors, and the lower one is the auxiliary room. The opposite of the main room is the entrance gate house, and a small patio is enclosed in the middle. The exterior is the adobe wall and the interior is decorated with wood. The whole residence is a piercing wooden frame of good seismic performance.

昆明民居天井
The patio

实例71　筇竹寺五百罗汉塑像

筇竹寺位于昆明市西北10多公里的玉案山上。寺中大雄宝殿两壁和梵音阁、天台来阁有五百罗汉塑像，这些塑像已成为人们前来筇竹寺观赏的景点。五百罗汉塑像是清光绪九至十六年间（1883—1890年），四川民间雕塑艺人黎广修和他的徒弟共同制作的。每个罗汉像高约1m，泥塑彩绘，既有慈眉善目的菩萨、袒胸露腹的弥陀，也有瞪眼怒气的金刚、稳重沉思的比丘，它们情态各异、生动传神、刻画入微、栩栩如生，同其他罗汉堂塑像有所不同，这里的罗汉似佛非佛、是僧非僧，既有神的内在灵气，又有人的生活气息，它是我国民间雕塑艺术的精品。

Example 71　Five Hundred Arhat Statues in Qiongzhu Temple

Qiongzhu Temple is located on Yu' an Mountain in the northwest of Kunming. There are 500 Arhat statues in the temple, which have become the focus for people to watch. The statues were made by Li Guangxiu, a Sichuan folk sculptor, and his disciples during Emperor Guangxu' s reign in Qing Dynasty. Each statue is about 1 m high, painted with clay sculpture, and lifelike, with both the inner aura of God and the characteristic of human beings.

罗汉塑像
Arhat statue

实例72 昆明大观楼

位于昆明市西北部滇池的北岸，南临浩瀚的湖面，并对着西山最高峰太华山。开阔的湖光山色使此楼成为最佳的观景点，因而得名为"大观楼"。最早这里是滇池水面，后水位下降形成小岛，清康熙二十一年（公元1682年）湖北名僧来此讲经，创建了观音寺。康熙二十九年（1690年）巡抚王继文相中了这里的大自然美景，于是筑堤修阁，大观楼便是此处亭台楼阁中的主体建筑。大观楼于咸丰年间遭兵火，现存建筑为同治五年（1866年）重新修建的。大观楼位于岛的最南端，登楼远眺，滇池西山景色尽收眼底，因而此楼成为文人墨客聚会之处，赋诗吟咏、赏景交谈。清乾隆年间文人孙髯翁写下"古今第一常联"，共180字，字句简练、对仗工整、情景交融，上联描述了大观楼四周美好的景物，下联则感叹云南历史的演变与发展。此长联也使大观楼名闻遐迩。

Example 72 Grand View Tower

It is located on the north bank of Dian Lake, facing the vast lake in the south and Taihua Mountain, the highest peak of Xishan Mountain. The broad view of the lake and mountains makes this building the best scenic spot. Where poets and writers meet, they chant poems and enjoy the scenery and talk. During the reign of Emperor Qianlong in Qing Dynasty, Sun Ranweng, a scholar, wrote "the first long couplet in ancient and modern times", with a total of 180 words. The first couplet described the beautiful scenery around the tower, and the second couplet lamented the historical evolution of Yunnan, which made this place more famous.

昆明大观楼外景
Exterior

实例73　西山龙门

从西山山腹林木丛中的华亭寺往上登山，沿人工开凿的别有洞天绝壁石洞隧道，过旧石室，进普陀胜景，再过慈云洞石室到达天阁石室。在这3个石室间都需要沿洞中的蜿蜒路拾级而上，最后登上顶峰龙门，这里的垂直悬崖高出滇池水面300多米。对此处的感受，正如石壁上所书："仰笑宛离天五尺，凭临恰在水中央"，置身其中，真是一个绝妙的境地。旧石室是明嘉靖年间开凿的，后两个石室是清乾隆四十六年（公元1781年）至咸丰三年（1853年）分两期工程间断完成的，其陡峭而艰巨的雕塑艺术工程在国内实属罕见，这些出色的石窟建筑艺术具有非凡的历史文化价值。

Example 73　Dragon Gate of West Hill

People can climb up from Huating Temple among the forest trees in the mountainside of West Hill, pass through the old stone room, enter Putuo scenic spot and then pass through Ciyun Cave Stone Room to Tiange Stone Room. Between these vertical cliff here is more than 300 m above the water surface of Dian Lake, and it feels like being in outer space here. The arduous and steep sculpture art projects are rare in China, and these outstanding grottoes have historical ee stone rooms, you need to climb the winding stone steps in the cave, and finally you can climb the peak Dragon Gate of the cliff. The veand cultural value.

从龙门望滇池
Dian Lake from the view of Dragon Gate

靠近龙门石洞隧道
The stone tunnel near the Dragon Gate

2. 路 南

实例74　石林

在云南路南彝族自治县内，面积有2.7万 hm^2，是典型的岩溶地貌。这里石峰平地拔起，犹如一片片森林，甚为壮观，因而得名"石林"。石峰、石柱的高度20~40m，其组合造型似各种姿态的动植物。著名的游览点剑峰池，池水中立有一石峰，宛如一把利剑直刺青天。最著名的石峰是阿诗玛石峰。当地传说，阿诗玛是撒尼族美丽的姑娘，为反抗富人热布巴拉抢她为妻，被淹死在此地，后来她就变成了这座阿诗玛石峰。路南石林风景名胜区是国家重点风景名胜区。

16.2 Lunan

Example 74　Stone Forest

It is located in Yi Autonomous County of Lunan with an area of 27,000 hectares. It is a typical karst landform, where stone peaks are pulled up like a forest, which is very spectacular. There is a stone peak in the middle of the famous tourist spot Sword Peak, which is like a sharp sword stabbing the sky. The most famous stone peak is Ashima Stone Peak, which is the embodiment of local maid Ashima.

剑峰
Sword Peak

中篇：中国建筑科学文化精品图说／云南

石林前活动广场
Square before the stone forest

阿诗玛石峰
Ashima Stone Peak

259

3. 大 理

实例75　南北城门楼

大理古城位于大理市下关北13km处。明洪武十五年（公元1382年），明将沐英打下大理后，在羊苴咩城[唐大历十四年（779年）至元至元十一年（1274年）为南诏大理国都，1274年都城由大理迁至昆明]范围内缩小规模重建大理城。城为方形，四面各长1500m，城墙高6m，设四座城门楼。南北城门楼由南北中心街道贯通。南北向街共有5条。东西城门楼由东西向街道错开连接，以示"财水"不外流。东西向街共有8条。古城整体布局规整，又有溪水环绕，加上淡雅大方的白族民居，创造出优美自然的生活环境。1982年国务院批准大理古城为首批24个历史文化名城之一，1983年起修复已毁之处，并重建了南北城门楼等。大理风景名胜区是国家重点风景名胜区。

16.3 Dali

Example 75　North-South Gate Tower

The ancient city of Dali is square, with a length of 1,500 m on all sides and a 6 m high city wall. There are four gates on each direction, with five north-south central streets running through the north-south gates and eight east-west streets. The overall layout is regular and surrounded by streams and the elegant houses of Bai nationality, creating a beautiful and natural living environment. In 1982, the State Council approved the ancient city of Dali as one of the first 24 famous historical and cultural cities. Since 1983, the damaged areas have been repaired and the north-south gates have been rebuilt.

大理南北城门楼
North-South Gate Tower

实例76 大理崇圣寺三塔

位于大理城北面1.5km的苍山中和峰之下，面对洱海，背负苍山，三塔耸立在这山光海色的幽美环境中。

大塔名为千寻塔，据多方面考证，建于南诏丰祐年间（公元823—859年），相传唐朝派人指导，按西安小雁塔形制营造，高69.13m，共16级，方形密檐式空心砖塔，内设木楼梯，沿梯可到顶层。在大塔西面两侧对称建有一对小塔，高42.4m，平面八角形，为10级密檐砖塔。三塔形成鼎足之势，塔身以白灰泥抹面，其前有一水潭，笔直的三塔立于苍山洱海之间，此景观显得格外净洁秀丽。规模宏大的崇圣寺因地震与兵火已毁，唯三塔犹存。1978—1980年，由国家拨款修整，恢复了三塔原貌。三塔于1961年被列为第一批国家重点文物保护单位。

Example 76 Three Pagodas in Chongsheng Temple

The largest tower, named Qianxun Tower, is said to be built according to the shape of the Little Wild Goose Pagoda in Xi'an, with a height of 69.13 m and a total of 16 layers. It is a square dense eaves hollow brick tower with wooest of the big tower, which are 10-layer dense-eaves brick towers. The tower is covered with white plaster, and there is a poolden stairs inside. A pair of small towers with a height of 42.4 m and octagonal plane is symmetrically built on both sides to the w in front of it. Three towers stand between Cang Mountain and Erhai Lake, and there are reflections near the pool. The large-scale Chongsheng Temple has been destroyed by the earthquake and the fire, but only the three towers still exist.

三塔外景
Exterior of the three pagodas

实例77　大理白族民居建筑

在大理苍山洱海北部的白喜洲、周村等白族聚居的民居建筑，它根据这里温暖湿润的气候以及无霜期有310天，还有多木、石材料和本民族的生活习惯等共同创造的，具有自己的特色。

民居建筑为合院格局，与中原地带相仿，或许受到中原的一些影响，具体布局有独特之处，多为"三坊一照壁"，即三面有房，主房对面是一照壁，以此四面围合成一个中心院落；"四合五天井"，即四面有房，两面房相交处都有一小耳房和小天井，加上中心的大院落。其构造合理，为梁柱木框架柔性结构，适应该地区多地震的自然条件。人们根据房屋的进深和木材的大小，采用穿斗式或抬梁式木构架。山墙采用卵石、青石、土坯墙粉白，以利防火。建筑就地取材。此地背负苍山，有木材和丰富的石材，这些石材用于墙基、墙角、墙壁、檐口、过梁以及屋顶瓦沟等。民居建筑装修精美，特别是正房客厅的三合六扇格子门，工艺非常精细，上节采用多层镂空透雕，底层为图案形花纹。门楼突出，装饰性的门楼是白族民居建筑的一个重要特征，其突出程度有如各地殿阁的造型，飞檐斗栱、木雕、彩画、大理石屏、石刻、泥塑和凸花青砖等共同组成主体形制，而且精巧和谐。其花木自然，白族民居建筑重视环境的自然优美，天井台砌有花坛，并植有山茶、缅桂、石榴树等，还摆有盆花。

Example 77 Residential Building of Bai Nationality

The buildings are in a courtyard pattern, the most common one is the style of houses on three sides and a screen wall opposite the main house, and then forms a central courtyard; or there are rooms on all sides, and a small side-room and a small patio at the intersection, plus a large courtyard in the center. The structure is reasonable, and it is a beam-column and wood frame structure, which can meet the situation of many earthquakes in this area. It is an important feature to use local materials. The decoration of wood carving, color painting, marble screen, stone carving, clay sculpture and blue brick with raised flowers form the main pattern, which is exquisite and harmonious. At the same time, it also pays attention to the natural beauty of the environment. There are flower beds, camellia, pomegranate trees and potted flowers on the patio platform.

扇门雕饰
Door carving

街面山墙
Building wall along the street

入口门楼
Entrance

院落天井
Patio in the courtyard

实例78　白族村头小广场

我们走访了洱海西岸的喜洲镇、周城村和东岸的两个渔村，发现在其村头或入口处都有大青树和一个小广场，有的广场大一些，而且还有个戏台，这就是全村公共活动的场所。大青树就是高大的榕树，树干粗壮、盘根错节、树顶如伞，在白族人民心中它是生命和吉祥的象征。此小广场上常有摊贩销售商品，大青树还能起到遮阳纳凉的作用，此树已成为白族村的重要标志。从下面所附周村村头小广场照片就可以看到，这些壮美的大青树已将小半个广场覆盖着，其后面还设有戏台，供村民观看演出使用。右页的一张照片是小普陀东面的渔岛村入口，此村为一小岛，以一长桥连通，入口处设门，门后有一小广场和一棵大树。还有一张照片是此渔村附近的沿岸小村，其村头不大的空地上，耸立着一棵大青树，这就是该村的标志物。

周村村头大青树
Banyan tree in Zhou Village

周村村头小广场戏台
The stage in the little square of Zhou Village

Example 78 Little Square in the Village of Bai Nationality

Almost all the small villages along the Erhai Lake have a big banyan tree and a little square, which is the place for public activities of the whole village. Banyan tree is a symbol of life and good fortune in the hearts of Bai nationality, and it can also provide shade and cool in the summer. There are often street vendors selling goods in the little square, and some have a stage behind for villagers to watch performances.

渔岛村入口
Entrance of a fishing island

沿岸小村入口
Entrance of a village along the lake

4. 丽江

实例79　丽江古城

位于云南西北部的玉龙雪山脚下。玉龙泉水经丽江古城西北面玉龙桥，分西、中、东三支河流进入古城区，此三支主河水与城区内的泉水汇合在一起，再分成网状的支渠流入街巷，或穿墙过院，形成"家家泉水，户户花木"的水乡风貌。这三支主河流顺地形坡度由西北向东南方向流去。主街依河伸展，并结合东西横向的支渠组成街巷网络。在中心地区，临西河旁开辟一个东西向长的四方广场，这就是四方街，它是古城进行公共活动的繁华场所，也是古城建筑群的核心部分。丽江市为了保护古城的历史文化，将其他市政府行政机构和经济、商贸等中心都安排在古城外面的新区，确保了丽江古城得到整体的保护。丽江是国家历史文化名城，并已列入"世界文化遗产"，2007年还获得了联合国教科文组织的遗产保护奖。

16.4 Lijiang

Example 79 Ancient City of Lijiang

It is located at the foot of Jade Dragon Snow Mountain. The Jade Dragon Spring passes through Jade Dragon Bridge, and is divided into three rivers, which enter the ancient city area, join with the spring water, and then flow into the streets and lanes, forming a water town style of "every household with spring water, flowers and trees". In the central area, along the river, an east-west long Square Street is a bustling place for public activities in the whole city, and is also the core part of the city's buildings. The ancient city has been listed as World Cultural Heritage and won the UNESCO Heritage Protection Award in 2007.

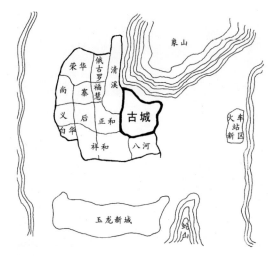

丽江古城与新区发展示意
Development plan of the new district of Lijiang Ancient City

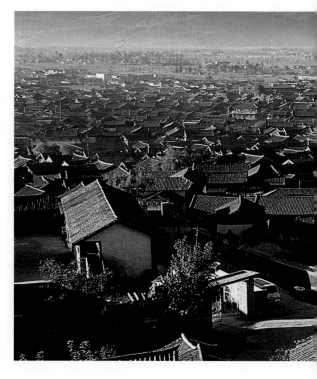

丽江古城中心区
The central area of Lijiang Ancient City

丽江古城鸟瞰
Aerial view of Lijiang Ancient City

实例80　丽江纳西族民居

丽江纳西族民居的平面布局虽然类似白族民居，亦近于中原地带的合院式，但结合当地的自然条件和本族人民的生活习惯与爱好，有着自己的特点，主要表现如下：

一是合院自由。平面布局为内向合院式，但只有正房两层三间不变，其他厢房都随地形自由变化、活泼自然。二是有水相伴。有的水渠沿院落边部而过，也有的水渠穿墙进院，院落中有水池一方，形成庭园。三是抱厦宽阔。正房前出厦（即前廊），比其他处宽大，有的进深3m多，纳西族习惯在这里待客、用餐、聊天饮茶、做家务活等。四是装修精细。建筑构造同样是木构架，有穿斗式或抬梁式，其抱厦内的梁枋、花罩和六扇屏门都有透雕、浮雕装饰。五是山墙精美。山墙多作成悬山式，有深长的悬挑，配以博风板与悬鱼，轻巧遮阳。二层处开窗，采光通风。雕饰木窗配以粉墙，既有对比，又有变化，其尺度和谐。有的木窗加一挑檐，可以美化与保护墙面。

水渠沿墙边小巷
Alley along the canal

院落外景观
View outside the courtyard

院落内景
View inside the courtyard

Example 80 Residential Area of Naxi Nationality

The plane layout is an inward courtyard style. Except for the two floors and three rooms of the main room according to the traditional rules, all other rooms change freely with the terrain. Houses are accompanied by water. Some canals pass along the side of the courtyard, while others pass through walls to enter the courtyard. There is a pool in the courtyard to form a garden. The front porch of the first room is wider than other places, so it is used to entertaining and dining here. On the second floor, windows are opened, with lighting and ventilation. Carved wooden windows are matched with walls, with contrast and harmonious scale.

小巷
Alley

5. 西双版纳

实例81　傣族村寨民居

西双版纳位于云南省南部，属热带北部边缘气候，炎热潮湿，盛产竹木，且有水资源。西双版纳风景名胜区是国家重点风景名胜区。傣族村寨位于山水之间，完全融在大自然环境之中。傣族人民历来重视对自然山水的保护，他们划有"垄林"区，在此区内严禁砍伐、采集、开垦和狩猎。村寨总体规划十分规整，多数成棋盘网格状，内部有十字形主要道路，四面设寨门，中间设寨心，每家每户占一小方格用地。佛寺坐落于入口处附近，是村寨建筑群的统领者。村寨一般建在沟溪旁或江河边，傣族人民崇尚水，在村寨内建有水井，并建亭保护。寨内每户民居的二层有一大的晒台，上面置水缸储水（现许多村寨已接通自来水），供这户人家洗澡或灌洗使用。由于西双版纳地区雨多且急，因而其屋顶大、坡度陡，当地称之为"孔明帽"。在大屋顶檐下开窗，做成活动木板式，可开可关，使用灵活。此处干栏式为二层，楼上住人，分堂屋与卧室两部分；楼下为牲畜圈和堆放杂物使用，多不设围栏。上楼后有宽敞的外廊，以利通风采光。民居建筑为竹木构造，整体为柱网框架结构，因当地生产竹材，故以竹代木。其主体构件为竹，或竹木混用，所以又称此地民居为"竹楼"。

16.5 Xishuangbanna

Example 81　Villages of Dai Nationality

Xishuangbanna is located in the south of Yunnan and is hot and humid and abounds in

傣族民居村寨外景
Outside view of the village

bamboo and wood. The overall planning of the residence is very regular, and most form a chessboard grid. Villages are generally built with waterside. A well is in the front of the village with a pavilion for protection. The buildings have large roof and steep slope, which are adapt to rainy climate. The building has two floors, with people living on the upper floor, livestock and sundries stacked on the lower floor. It is made of bamboo and wood, and the whole structure is a column grid frame structure.

竹楼院落环境
Courtyard

十七、新 疆

1. 乌鲁木齐

实例82　新疆人民会堂

位于乌鲁木齐市新旧区连接处，1985年建成。该会堂工程设计首先考虑同周围建筑的协调关系，因其对面为不对称形体的昆仑宾馆，所以将会堂主体与副体做成不对称的组合，同时也反映了新疆伊斯兰传统建筑不严格对称的特征。该会堂吸取传统的方圆组合形式，除主楼方形、副楼圆形的形体外，主楼四角各设一圆形塔，以增加体形的变化、对比和稳重感。此角塔内为管道，顶以不锈钢做成小穹顶，富有现代感。柱间与外墙饰以连续拱形构件，檐部采用深金黄色琉璃瓦，与挺立的白石柱形成对比。进入大堂，迎面镶嵌着大理石"天山之春"巨幅壁画，这些内容使人们感受到新疆地域建筑文化特色。

17. Xinjiang

17.1 Urumqi

Example 82 People's Hall

It is located at the junction of old and new districts in Urumqi and was built in 1985. Firstly, the design considers the harmonious relationship with the surrounding buildings, and makes the main body and the auxiliary body of the hall into an asymmetrical combination, which also reflects the characteristic that the traditional Islamic architecture is not strictly symmetrical. The hall also draws on the traditional combination of square and round. Except that the main building is square and the auxiliary building is round, there are four round towers with pipes inside and a small stainless steel dome on the top at four corners of the main building. The columns and exterior walls are decorated with continuous arched parts, and the eaves are made of deep golden glazed tiles, which is in contrast to the standing white stone pillars.

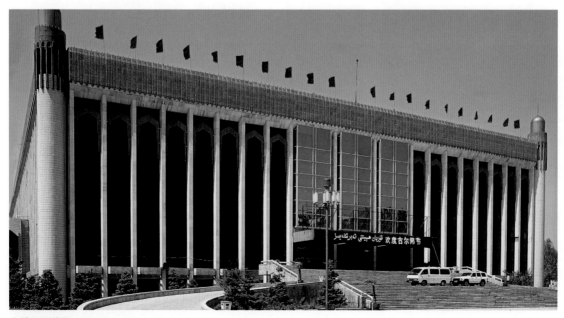

主楼立面外景　Exterior of the main body

实例83　新疆迎宾馆

位于林木茂盛、环境优美的乌鲁木齐市近郊，主要用于接待各国来访的政治领导人，建于1985年。入口朝南，底层是大堂、接待厅、一般客房和厨房等，二层是贵宾房、总统套房、宴会厅和礼拜厅等。西南设一花园，同周围大自然环境融为一体。该馆设计的一大特点是将功能、结构、地域传统建筑风格结合在一起，将传统尖拱做多样变化使用，如入口悬厅底部做成交叉尖拱的曲梁，门窗外形做成尖拱或半尖拱状，特别是侧面的水塔，具有花瓣式的形象，使构筑物转化为表现维吾尔族传统文化美的形象之物，它还起着控制这一建筑群环境空间的作用。此建筑系新疆建筑设计研究院设计。

Example 83
Guest House

It is located in the suburbs and is mainly used to receive visiting leaders from various countries. The entrance faces south, with lobby, reception hall, general guest room and kitchen on the ground floor and VIP room, presidential suite, banquet hall and worship hall on the second floor. There is a garden in the southwest, which is integrated with the surrounding natural environment. Another characteristic of the design is that it combines the function, structure and regional traditional architectural style, and uses the traditional arch in many places.

入口外景
Exterior of the entrance

凉水塔
Water tower

2. 高昌

实例84　高昌故城

位于新疆吐鲁番城东40多千米的阿斯塔那村东面。两汉魏晋的戊己校尉屯兵驻此，前凉时为高昌郡郡治，鞠氏高昌时为王国国都，唐时为西州州治，后做过回鹘高昌王都，元末毁于战火，明初始废，前后共历经1500多年，在历史上曾是吐鲁番地区的政治、经济、文化中心，1961年被列为国家重点文物保护单位。该城规模与布局在前凉到鞠氏王朝时期形成，有内城、宫城和外城，类似唐长安城，总面积约200hm^2，近方形，现城垣残存，以夯土筑。城内建筑遗址遍布，在内城北部见一高台，其上有15m高的夯筑建筑，据分析这是佛寺遗址。在外城西南部又见一寺院遗址，殿堂、高塔、佛龛等残迹清晰，其面积约有1hm^2。城内建筑系用夯土、土坯建造，为土木构造，门窗上部做成拱形，体现着吐鲁番地区的建筑特征。

17.2 Gaochang

Example 84　Ancient City of Gaochang

It is located in the east of Astana Village, more than 40 km east of Turpan. It has been the political, economic and cultural center of Turpan in history for more than 1,500 years. The original inner city, imperial city and outer city are similar to Chang'an in Tang Dynasty, with a total area of about 200 hectares, and there are many architectural sites in the city. In the northern part of the inner city, there is a high platform with a 15 m high rammed building, which should be a Buddhist temple site. The buildings in the city are made of rammed earth and adobe, and the upper parts of doors and windows are arched, which embodies the architectural features of Turpan area.

寺院遗址
Ruins of temple

3. 交 河

实例85　交河故城

在新疆吐鲁番城西10km的亚尔孜沟中，位于两条河汇合环绕的"叶状岛"地带，故名交河。6世纪初，鞠氏高昌时在这里建交河郡城。现存遗址是唐代及其以后的建筑，约在元末逐渐废弃。城沿土崖呈长方形，南北长1000m，东西宽处约300m。南面入口处有一纵贯城内南北的中心街道，宽3m，以此中心街道为界，划分出3个区：西北区多寺庙遗址；东北区为居住区，可看到院落的遗迹和巷口同街道连接的传统布局；东南区为行政管理区，有一较大建筑，似是官衙。从南门入口向北走，正对南北大道为一佛教寺院，它是全城最大规模的建筑，管理人员告诉我们在此处曾发掘出唐代莲花瓦当。此处的遗址和出土文物，对研究新疆的城市建设和历史文化很有参考价值。1961年被列为国家重点文物保护单位。

17.3 Jiaohe

Example 85　Ancient City of Jiaohe

It is located 10 km west of Turpan and the existing ruins here are buildings of Tang Dynasty and later. The city is rectangular, with a length of 1,000 m from north to south and a width of about 300 m from east to west. There is a central street running through north to south, with a width of 3 m. With this central street as the boundary, the city is divided into three areas: temple sites in the northwest, residential areas in the northeast and administrative areas in the southeast. Going north from the entrance of the south gate, there is a Buddhist temple, which is the largest building in the city.

南北中心街道
Central Street from north to south

4. 喀什

实例86 喀什民居

喀什地处帕米尔高原脚下，海拔1335m，西通中亚、西亚，南至印度，是古代丝绸之路的必经之地，也是中西文化交流的重要城市。喀什用地较少，地形起伏，人口多，因而民居密度大、街巷窄，常利用街巷上部空间建房，形成许多过街楼；另二层悬挑楼亦有不少，这是住宅区街巷的特征。其民居院落特点是：多为长方块状，房屋为L形，余下一角为院落；二层楼房居多，有廊、露天楼梯。这个小院落或叫小天井，可作为户外活动的场所。维吾尔族居民崇尚自然环境，当地有句谚语："宁可无书，不可无树"，故此小院都种有花木和盆花。朝向小院的拱形柱廊、门窗、檐口的木装修很精细，有的还配上石膏花饰，色彩丰富，造型富丽。但位于街巷的外部却十分简洁，为砖砌或土墙，很少开窗，可避免强烈日照和冬季寒风，其私密性也好。

17.4 Kashgar

Example 86 Residential Buildings

It is located at the foot of Pamir Plateau with an altitude of 1,335 m and is the only place where the ancient Silk Road must pass. Residential buildings are dense and the streets are narrow, so the upper space of the streets is often used to build houses, forming many across street buildings. Residential courtyards are characterized by rectangular blocks, L-shaped houses and courtyards planted with flowers and trees in the other corner. Most are 2-storey buildings with corridors and open stairs. Decoration is very fine and colorful with gypsum ornaments. However, the exterior of the street is very simple, which is brick or earth wall, and seldom opens windows, which can avoid strong sunshine and cold wind in winter.

喀什民居小巷外景
Exterior of a residential alley in Kashgar

民居院落一层外廊木装修
Wooden decoration of the first floor veranda of residential courtyard

二层外廊木装修
Wood decoration of the second floor veranda

民居院落盆花绿化
Pot flower in residential courtyard

民居院落露天楼梯
Open stairs of residential courtyard

实例87　艾提尕尔清真寺

位于喀什市中心艾提尕尔广场，始建于公元15世纪，至19世纪上半叶形成现在的规模，占地1hm²。这座规模宏大的清真寺留给我的印象是：

（1）入口外观，肃穆宏伟。寺门高大，方中有拱、拱中有方，花纹线条精细，在米黄色的砖面上勾画出洁白的轮廓线，醒目耀眼。在寺门两侧不对称地耸立着两座10多米高的塔楼，并以较低的墙面连成一个有起伏变化的整体，显得肃穆庄严、宏伟壮观。

（2）院内环境，自然清幽。从两边甬道进入院内，迎面苍松翠柏，路边排列着整齐的参天白杨，并配有水池。水池池水清澈如镜，创造出一个净洁、清雅、幽静的环境氛围。

（3）礼拜殿堂，排柱精美。礼拜殿堂分内外两部分，全长160m，进深16m，同时可容6000~7000名穆斯林做礼拜。此殿的排柱最为突出，成网格状排列着140根绿色雕花木柱。在柱间白色密肋顶棚里，有节奏地布置着绘有花卉图案的天花藻井，使林立的绿柱富有韵律感，并创造出列柱精美的空间环境，这很适合穆斯林在这里安静地进行礼拜活动。寺前广场是人们休闲、观赏和交易的地方，但每逢穆斯林节日，上万信伊斯兰教的维吾尔人聚集在寺内外进行礼拜。礼拜仪式完毕后，寺门顶上响起唢呐声和纳格拉鼓的节奏声，身着节日盛装的人群涌向广场，随声起舞，平日里难得一见的维吾尔族"萨满舞"，此时也在广场上出现，并尽情欢乐。

Example 87 Id Kah Mosque

It is located in the downtown square and was built in the 15th century with an area of 1 hectare. The main features are as follows:

(1) The entrance is solemn and magnificent. The temple gate is tall, with a white outline on the beige brick surface. Two towers with a height of more than 10 m stand asymmetrically on both sides, and the lower walls are connected into a whole.

(2) The environment inside is natural and quiet. Entering the courtyard from the tunnels on both sides, the trees are green and the pool water is clear.

(3) There are 140 green carved wooden columns inside, among which smallpox caissons with flower patterns are arranged rhythmically. It can accommodate 6,000 to 7,000 Muslims for worship. After the ceremony, people dressed in festive costumes flocked to the square and danced to the music.

清真寺门外景
Exterior at the entrance

礼拜殿堂排柱廊
Column corridor inside

礼拜殿堂内景
Interior of the mosque

十八、甘　肃

敦　煌

实例88　敦煌莫高窟

在敦煌城东南25km处，它具有深、神、美、玄、秘、谜的特点，集建筑、雕塑、绘画艺术为一体，是世界佛教艺术的宝库、人类文化的宝藏。1961年被列为第一批国家重点文物保护单位。1996年9月，中国建筑学会《建筑学报》编委会委员在对敦煌市规划与建设进行评议之后，委员们到此莫高窟参观。敦煌莫高窟给我的感受是，这颗文化明珠虽经沧桑风雨，但千年不衰、丰姿犹在。现存历代壁画、塑像的洞窟有492个，其中壁画45000多平方米，彩塑2415身，这些画面如按2m高排列，就可组成25km长的巨型画卷，它是我国现存内容最为丰富、规模最大的石窟艺术宝库。洞窟最大者40m高、30m见方，最小者高不过尺。壁画内容为佛像、佛教史迹和经变、神话以及装饰图案等。彩塑均为泥制，有单身像、群像。群像以主佛居中，两侧侍立天王、菩萨等，少则3位，多则11位，最大者33m高，最小者10cm高，其神态各异，人物性格突出。

18. Gansu

Dunhuang

Example 88　Mogao Grottoes

At 25 km southeast of Dunhuang, it is a treasure house of Buddhist art in the world, integrating architecture, sculpture and painting. In 1961, it was listed as the first batch of Key Cultural Relics Site under the State Protection. There are 492 caves, 45,000 m² murals and 2,415 colored sculptures. The largest cave is 40 m in height, while the smallest cave is less than 1 foot. The frescoes include Buddha statues, Buddhist historical sites, fairy tales and decorative patterns. Painted sculptures with different expressions are made of clay, of single figures and group figures, the largest of which is 33 m high and the smallest is 10 cm.

第112窟反弹琵琶伎乐（中唐）
Playing-pipa-behind posture in No. 112 cave (Middle Tang Dynasty)

主入口外景
Exterior at the main entrance

十九、香　港

位于广东珠江口外，它面对南海，包括香港岛、九龙半岛和新界，面积为1104km², 人口700多万。现已发展成为世界贸易的重要港口和重要金融中心。1840年英国发动鸦片战争，1842年英国又强迫清政府签订了《南京条约》，并占据了香港岛，1898年英国再次强迫清政府签订《展拓香港界址专条》等条约，占据了香港整个地区。香港经历了99年的割据后于1997年7月1日回归祖国，中国恢复对其行使主权。

实例89　香港城市交通

香港城市交通组织得比较好，对广大香港居民来说是方便的。城市地下铁路贯穿本岛东西，有两条过海隧道同九龙地区组成环状地铁线路，自1998年香港新机场启用时，此环形地下铁路线已伸出专线通至新机场。地面汽车交通与道路的组织亦很有序，在湾仔港湾处有过海隧道连通本港与九龙地区；于隧道出入口处开辟了大片的立交桥场地，避免了平交，便于分路行使；在主干道交叉口处设立交桥，保证了单行线路行使连贯，只是路线长一些；在公共聚集活动处设有较大的汽车停车场。人行道路逐渐扩大组成系统，1986年7月我第一次到香港，在中环和尖沙咀中心区见到的人行道路只是小范围的连贯，其中间架有一些跨街桥廊。1993年、1995年我曾多次到香港考察，并明显地看到这些中心区的人行道路系统在扩大；当我21世纪初再到香港，并居住在本岛后山香港大学时，已见有爬山廊道通至后山，人行道路系统仍旧在发展，方便了居民。香港在城市交通组织与发展的这些做法，对于我国的大城市很有参考价值。

湾仔运动场东至海底隧道间立交场地
Overpass between Wanchai sports field and cross-harbor tunnel

19. Hong Kong

It is located outside the Pearl River Estuary in Guangdong, facing the South China Sea, including Hong Kong Island, Kowloon Peninsula and the New Territories. It covers an area of 1104 km^2 and has a population of over 7 million. Now it has developed into an important port of world trade and an important financial center of the world. After 99 years of separatism, Hong Kong returned to the motherland on July 1, 1997.

Example 89 City transportation

The place has convenient transportation. The subway runs through the east and west of Hong Kong Island. There are two cross-harbor tunnels and together with the subways in Kowloon area to form a circular subway line. In terms of ground transportation, there is a cross-harbor tunnel connecting Hong Kong Island and Kowloon, with a large overpass at the entrance and exit of the tunnel, and a large car parking lot at the public gathering place. The sidewalks gradually expand to form a system, which is convenient for residents.

中环立体人行廊道
Three - dimensional pedestrian corridor in central HK

实例90　汇丰银行大厦

位于香港中环皇后大道，于1985年建成。该大厦地上48层、地下4层，高178.8m，设计者为英国诺曼·福斯特事务所。汇丰银行大厦设计方案是在7个著名建筑事务所的竞赛中获胜的，其设计构思采用预制装配的新结构与新布局，创造最大的使用面积和可灵活改变的建筑空间。诺曼·福斯特一改传统做法，将中央电梯和服务中心布置在两侧，形成8组"通天钢组合柱"结构。在这8组主柱上牢扣着5层三角形垂悬桁架，分为5个区的各层楼板由这些桁架悬吊着，这5个区的悬吊层数由8个递减至4个，垂悬的桁架本身占两层。这样创新的布局，获得将近10万m^2建筑面积，交通、服务中心在两侧，又获得最大使用面积7万多平方米，占总面积的71%。其中间南北通透，南望山景，北观海景，创造了最好的建筑使用空间。各层大空间内墙均为活动隔断，可根据需要灵活改变。另外，在垂直方向亦可根据高度需要调整悬挂的楼板高度，这是世界建筑史第一次做到能够灵活改变楼层高度的新建筑。此外，该建筑还将底层打通，使南北两个方向的人流穿行而过，以方便交通和居民使用。在其高大中庭的穹顶上镶有一排巨大的镜片，由电脑控制阳光反射，将阳光折射到中庭并照亮地面广场。该设计获英国皇家建筑师学会（RIBA）金质奖，并选入《20世纪世界建筑精品集锦》东亚卷中。虽然这个大厦造价高，但其创新的设计思想与设计手法值得借鉴。

Example 90 HSBC Plaza

It is located in Queen's Road, Central Hong Kong and was completed in 1985, with 48 floors above ground and 4 floors below ground. The design adopted prefabricated new structure and new layout to create the largest use area and flexible building space. The central elevator and service center are arranged on both sides with mountain view in the south and sea view in the north. The inner walls of large spaces on each floor are movable partitions, which can be flexibly changed as required. In addition, in the vertical direction, the suspended floor height can be adjusted according to the height requirements. In addition, the ground floor is opened, and people from both north and south directions can pass through it. The dome of the atrium is inlaid with a row of huge lenses, and the computer controls the sunlight reflection, which refracts the sunlight to the atrium and shines on the ground square, so that there is suitable brightness here. This design was selected into the East Asian Volume of the *World Architectural Masterpieces Collection in the 20th Century*.

中篇：中国建筑科学文化精品图说/香港

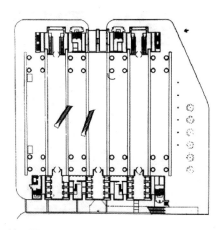

地面层平面
Plan of ground floor

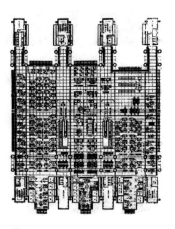

楼层平面
Plan of floors

中庭一角
Part of the atrium

通透办公空间
Transparent office

地面广场
Ground square

285

外景
Exterior

实例91　中国银行大厦

位于香港中环花园道，建成于1989年，共70层、高368m，设计者是世界著名建筑师贝聿铭先生。该建筑的设计要求是：在中环商业金融区要起到统领这一地区的作用。贝聿铭先生采用竹子造型，隐喻节节高。其具体设计方案是分段变形的塔楼、内部无柱的筒形结构。该建筑25层以下为正方形，以对角线分为4个三角形立方体，在26层去掉一个三角形立方体，至38层再去掉一个三角形立方体，到51层还去掉一个三角形立方体，这时只留下了一个三角形立方体直至顶层。该建筑加上顶部的一对桅杆，达到368m的高度，建成时成为香港最高的建筑。它富有创造性的独特构造造型，完全达到了创新规划的要求。中国银行大厦控制着中环CBD区的空间环境，并将这些周围知名建筑组合成为一个整体。此外，在楼层的底部和外部创造性地运用中国园林的手法营造出多样、宜人的花木景观，让人流连忘返。设计是十分谨慎的，它还满足了香港风荷载大的要求以及严格预算为1.3亿美元的控制要求。该建筑已被选入《20世纪世界建筑精品集锦》东亚卷中。

Example 91 Bank of China Building

It is located in Garden Road, Central Hong Kong and was built in 1989 with 70 floors and a height of 368 m. It is a successively high bamboo style. The concrete scheme is a segmented deformed tower of a cylindrical structure with no column inside. It is square below the 25th floor and divided into four triangular cubes diagonally. One triangular cube is removed from the 26th floor, another triangular cube is removed from the 31st floor, and another triangular cube is removed from the 51st floor. At this time, only one triangular cube is left up to the top floor, and a pair of masts connected to the top reaches a height of 368 m, which is the tallest building in Hong Kong when it is completed. At the bottom of the floor and the outside environment, Chinese gardens are creatively used to create a variety of pleasant landscape environments. This building has been selected into the East Asian Volume of the *World Architectural Masterpieces Collection in the 20th Century*.

侧面水木景观
Side view

从底层大厅内看室外花木
Outside trees from the view at the lobby

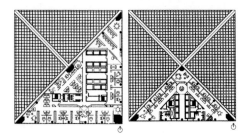

3、26、38、51层平面（Pei Architects事务所提供）
Plan of Floor 3, 26, 38, 51 (Courtesy of Pei Architects)

外景
Exterior

二十、澳　门

澳门紧邻广东珠海市,其地理范围包括澳门半岛、氹仔和路环岛,面积共23.8km^2(现已填海扩大)。明代时这里是中国的一个渔村,由于渔民信奉妈祖,在岛的西南角建有妈祖阁。16世纪葡萄牙人在这里登陆,询问地名,当地人以为是问寺庙名,故答妈祖阁,即英文中"Macau",此名称迄今一直沿用。清光绪十三年(1887年)十二月清政府与葡萄牙签订《中葡通商和好条约》,将澳门割让给葡萄牙。1987年3月中葡两国签订了《中葡联合声明》,确认1999年12月20日回归祖国。

实例92　大三巴牌坊

大三巴牌坊是澳门最主要的标志性建筑,为1835年圣保罗教堂遭遇大火留下的前壁遗迹,由于残存的教堂前壁像中国的牌坊,因而称之为"大三巴牌坊"。1990—1996年,圣保罗教堂遗址进行了考古发掘和遗迹休整工作,并在原教堂主祭坛部分建立了一个天主教艺术博物馆,以加强对此建筑遗迹的保护,并丰富该建筑文物遗迹内涵。此项目已选入《20世纪世界建筑精品集锦》东亚卷中。

20. Macao

Macao is adjacent to Zhuhai, including Macao Peninsula, Taipa and Coloane Island. Macao became a Portuguese colony in the 13th year of Emperor Guangxu in Qing Dynasty and returned to the motherland on December 20, 1999.

Example 92 Ruins of St. Paul

It is the most important landmark building in Macao. It is the remains of St Paul's Church left by a fire in 1835. Later, a Catholic Art Museum was established to strengthen the protection of the ruins. This project has been selected into the East Asian Volume of the *World Architectural Masterpieces Collection in the 20th Century*.

大三巴牌坊前景观
View of the Ruins of St. Paul

实例93　澳门博物馆

位于大三巴牌坊东侧山冈，依圣保罗炮台（大炮台）而建，1998年建成，建筑面积2800m²。建筑共3层，并充分利用地形，有部分在炮台山内，使博物馆与大炮台结合为一个整体。大炮台始建于1616年，当年是为了击退海上来犯的荷兰人而建。该馆设有自动扶梯，它可以到达博物馆入口大厅。这里还有400年前砌筑的挡土墙，设计者马锦途建筑师将此墙给予保留。展厅的内容包括中国古老的东方文化、澳门地区文化的起源至17世纪中叶的历史、16世纪以来与欧洲文化的融合、澳门民间传统艺术和现代澳门的特征等。从三层展厅出口出来，便到了大炮台顶，在此处可看到澳门全景。该设计的最大特点是与环境的结合十分协调。

Example 93　Museum of Macao

It was built on the basis of St. Paul Fortress with a building area of 2,800 m². It makes full use of the terrain and integrates with the fortress as a whole. The museum is equipped with an escalator to enter the entrance hall and there is a retaining wall built 400 years ago. From the exit of the third-floor exhibition hall, people can get to the top of the big fort and see the panoramic view of Macao.

入口大厅前老挡土墙
Retaining wall at the entrance hall

博物馆与炮台顶
Top of the museum and fortress

实例94　市政厅广场

澳门市政厅，始建于1583年，它由一群先进市民组织并管理着澳门的日常生活，这个组织类似于绿色民主体系。在此后的400多年中，它是澳门发展的重要场所，也是澳门的权力中心。澳门市政厅前广场亦是个重要的空间。从1983年起，不断地对市政厅广场及其周围的建筑进行修复、保护和改善，它现已成为重要的旅游地段和广大市民进行各种休闲和文娱活动的地方。澳门市政厅广场自北向南逐步放宽，形成喇叭口状，在南面尽端处中心建起主体建筑——市政厅，厅北面广场布置有绿化的圆形水池喷泉，东西两侧是整齐的券廊式的3层楼房，地面是采用葡萄牙方块石按水纹式横向铺砌，将通长的广场连接为一个整体，这是澳门市政厅广场的特色。澳门市政厅的造型与装饰简洁、庄严，布局为庭园式，内有一绿化庭院，自然环境优美。

Example 94　City Hall Plaza

The plaza gradually becomes wider from north to south, forming a trumpet-mouth shape. The main building city hall is built in the center of the south end. There is a circular pool fountain in the north of the hall. The east and west sides are 3-storey buildings with neat arch gallery style. The ground is paved horizontally with Portuguese square stones like water lines, connecting the whole square as a whole. The main building city hall is simple and solemn in shape and decoration, with a green courtyard inside.

市政厅建筑外观
Exterior of the city hall

广场景观
View of the plaza

市政厅内庭院
Courtyard of the city hall

实例95　澳门岛沿岸城市轮廓

　　澳门本岛是从其西南隅的西望洋山发展起来的，现其沿岸城市轮廓线是由海湾、山丘、历史建筑、现代建筑和林木等构成，体现着自然景色与人工景观的协调以及中西文化的融合。从海面向东北方向望去，画面中心是西望洋山，山顶耸立着西望洋圣母堂及主教私邸，其东北面还建有中国银行澳门分行新楼。该新楼高163m，是一座37层八角形塔楼，也是澳门最新的标志性建筑。在此塔楼北边则是另一座澳门标志性建筑——澳门葡京酒店和葡京赌场。这一展开画面以蓝天为背景，其城市轮廓有韵律、有节奏地变化着。从西望洋山上俯视山脚景色，深色的融和纪念高塔同样呈现出一幅优美轮廓的画面。

Example 95　Outline of the Coastal City

　　The outline of the coastal city is a fluctuating landscape line composed of bays, hills, historical and modern buildings and trees. Looking from the sea to the northeast, the center of the picture is Penha Hill, with Notre Dame Cathedral and Bishop's private residence standing on the top. There is a new building of Bank of China Macau Branch in the northeast, and Portuguese Hotel and Portuguese Casino in the north. This unfolding picture takes the blue sky as the background, and the city outline changes rhythmically. Looking down at the scenery at the foot of Penha Hill from the west, there stands a dark tower of Ronghe Gate on the sea beside the mountain, which also forms a picture with beautiful outline.

从西南隅西望洋山至东北面沿岸城市轮廓
City outline from Penha Hill to the northeast coast

从西望洋山俯视融和纪念高塔
Ronghe Gate Tower from the view of Penha Hill

东北面沿岸建筑群（中间为中国银行澳门分行新楼）
Buildings along the northeast coast (new building of Bank of China Macau Branch in the middle)

二十一、台　湾

1993年，经国务院台湾事务办公室批准，由建设部组团，以中国科学院、中国工程院学部委员、中国建筑学会副理事长、清华大学教授吴良镛为名誉团长，中国建筑学会秘书长张祖刚为团长，清华大学建筑学院院长赵炳时、天津大学建筑系主任胡德君为副团长的"中国大陆建筑师赴台学术交流考察团"一行23位建筑专家，应台湾5个学术团体负责人林长勋、喻肇青、黄祖权、林俊兴、吴夏雄联名邀请参加了台北市、台中市、高雄市举行的"1993年两岸建筑学术交流会"。台湾方面出席交流会的有知名的建筑师和老建筑师近200位，此次学术交流与考察取得了圆满的成功。这是第四次两岸建筑学术交流会，前3次分别是在1988年的香港、1989年的曼谷、1990年的北京(清华大学)召开的。这次学术交流会经过两年多的努力与筹备，并且第一次在台湾召开，更有其特殊的意义。

本次建筑学术交流的面比较广，主要集中探讨了城市与建筑的中国特色问题。我们在台北、在台中、在高雄市共进行了5次学术会议，大陆与台湾学者发表了26份学术报告。研讨会集中讨论如何从中国两岸实际出发去探索具有中国特色的城市规划与建筑设计创作道路，如何保护城市人文环境，如何提高城市环境质量和建筑设计水平，如何改善建筑教育条件暨培养高品质建筑专业人才，如何增强两岸学术交流并开展规划设计的合作实践等。讨论会气氛热烈、坦诚，反映良好。

这次建筑学术交流会实地参观考察的建筑工程项目较多。大陆学者先后参观了台北、台中、高雄、彰化、鹿港等城镇，访问了台湾大学城市与建筑研究所、台湾新竹清华大学及工业园区、台中东海大学。在台北参观了国父纪念堂、中正文化中心、故宫博物院、台湾图书馆、剑潭青年中心、世贸大厦、震旦大厦、宏图大厦、圆山饭店、凯悦饭店、环亚饭店、李祖原建筑事务所、宗迈建筑事务所等。在台中参观了台湾省建筑师公会、台湾省自然科学博物馆、台湾省美术馆、手工艺研究所、环球巨星大厦、全国饭店、鹿港妈祖庙、龙山寺、彰化文化中心、大佛寺及日月潭风景区。在高雄参观了宝成大厦、长谷大厦、圆山饭店、中正文化中心、市政厅、许崇尧建筑事务所、蔡博安建筑事务所以及高尔夫球场等处，并受到热烈欢迎和接待。通过身临其境的参观访问和学习由各单位赠送的书刊资料，使得我们大陆学者能够较全面和具体地了解台湾城市与建筑的发展状况，其经验与教训对我们很有启发。

21. Taiwan

With the approval of the Taiwan Affairs Office of the State Council, 23 architectural experts from the "Chinese Mainland Architects Academic Exchange Delegation to Taiwan" participated in the "Cross-Strait Architectural Academic Exchange Meeting" from July 18 to 25, 1993 at the joint invitation of Taiwan academic groups with more than two years' efforts and preparatious, it was the first time that the meeting wad held in Taiwan after the previous oues in Hong Kong (1998), Bangkok (1989) and Beijing (1990), which has its special significance. Nearly foo famous architects from Taiwan atlended the meeting and it was a complete success.

This academic exchange of architecture covers a wide range, but focuses on the Chinese characteristics of cities and architecture. It explored the creative path of urban planning and architectural design with Chinese characteristics, like how to protect urban humanistic environment, how to improve urban environmental quality and architectural design level, how to improve architectural education conditions and train high-quality architectural professionals, how to enhance cross-strait academic exchanges and carry out cooperative practice of planning and design, etc. Another characteristic of this architectural academic exchange meeting is that there are many construction projects visited and investigated on the spot. Through the visits and books presented by various units, mainland scholars can have a more comprehensive and concrete understanding of the development of Taiwan's cities and buildings.

Scholars from both sides of the Strait held a press conference on July 18 and 25 for introduciy the general situation of the academic exchange and answeviuy questions from the media. It was mainly about the impressions of mainland scholars' visit in Taiwan, the prospects and ways of future exchanges and cooperation, and the achievements and status of female architects in the mainland .

1993年两岸建筑学术交流会代表在台北合影
A group photo of the two coast architectural expert delegates at Taipei in 1993

1. 台　北

实例96　台北国父纪念馆

坐落于台湾省台北市，设计建造时间是1968—1972年，设计人是著名建筑师王大闳。孙中山先生是中国民主革命的先驱者，逝世后民众称先生为"国父"。该纪念馆采用变异的中国传统黄色琉璃瓦大屋顶形式，其南面主入口的屋面在中央断开，变为弯曲起翘的大雨罩，这是其造型的主要特征。整体外观仍采用中国传统的基座、柱子、屋顶三段划分的模式，平面为方形，四周为柱廊，庄严肃穆，很有气势。列柱与梁枋均为钢筋混凝土构造，其节点细部都已变异简化，室内装修、色彩、木格窗等皆有中国传统建筑的风格。这个项目已选入《20世纪世界建筑精品集锦》东亚卷中。

21.1 Taibei

Example 96　Sun Yat-sen Memorial Hall

It adopts the variant yellow glazed tile roof form, which is disconnected in the center of the roof at the south main entrance, and becames a bending heavy rain cover. The overall appearance still adopts the traditional Chinese mode of dividing the base, pillar and roof into three sections. The plane is square and surrounded by colonnade, which is solemn and imposing. However, columns and beams are made of reinforced concrete, and the details of their joints have been changed and simplified. This project has been selected into the East Asian Volume of the *World Architectural Masterpieces Collection in the 20th Century*.

外景
Exterior

中篇：中国建筑科学文化精品图说/台湾

实例97　台北故宫博物院

　　台北故宫博物院坐落在台北一清幽之处，该建筑采用中国传统宫殿式，由高平台、高台阶、高城墙式墙面托着琉璃瓦大屋顶宫殿式建筑群，加上入口的石牌坊和对称的轴线布局，其外观很有庄严雄伟之势。其建筑内部珍藏着许多珍贵的文物，这些文物都是原北京故宫的珍宝，后被蒋介石多次转移，最后从南京运到台北的，包括中国历代字画、陶瓷、金属制品等国宝；1993年7月，我们中国建筑代表团成员参观访问时见到了这些国宝。另外，该博物馆还经常举办学术讲座和学术研讨会等活动。

Example 97　Taibei Palace Museum

First of all, it is located in a quiet place in Taibei's back mountain. Secondly, the building adopts Chinese traditional palace style, with high platform, high steps and high walls supporting glazed tile roof, and with the entrance stone archway and symmetrical layout of axis, its appearance is solemn and majestic. Third, there are many precious cultural relics here, which are treasures that originally existed in the Forbidden City in Beijing. In addition, lectures and academic seminars are often held here, which is beneficial to improve the understanding of Chinese culture.

外景
Exterior

实例98　圆山大饭店

位于台北中山北路，建成于1973年，有客房630间，设计人是杨卓成、修泽兰先生。该建筑为中国宫殿式多层建筑，即在长方形的多层建筑上加一个重檐的宫殿式大屋顶。整座建筑由台基、柱身和大屋顶三段构成，其中柱身为多层红柱廊，檐下斗栱为青绿色，屋顶为黄色琉璃瓦，这些都是以清式营造则例为蓝本设计建造的，所以它是一座将中国传统宫殿形式套在现代建筑上的经典作品。高雄市也有一座圆山大饭店，其建筑造型同台北的大体一样，都是为了体现中华民族特色，但其建筑环境比台北的要优美许多，其近处山水相依、花木茂盛，给人以清新的感受。

台北圆山大饭店外景
Exterior of Grand Hotel in Taipei

台北圆山大饭店餐厅
Dining hall

Example 98 Grand Hotel

It is located in Zhongshan North Road and was built in 1973 with 630 guest rooms. Its shape is a Chinese palace-style multi-layer building, which consists of a platform foundation, a column body and a large roof with double eaves. The column is a multi-layer red colonnade, with turquoise arch under the eaves and yellow glazed tiles on the roof. These are all designed and built on the basis of the palace of Qing Dynasty, so it is a typical work which sets the traditional Chinese palace form on the modern architecture. The Grand Hotel in Kaohsiung has the same architectural style as Taipei.

高雄圆山大饭店外景
Exterior of Grand Hotel in Kaohsiung

从高雄圆山大饭店客房阳台看园林景观
Landscape from the view of the balcony of Grand Hotel in Kaohsiung

实例99　台北剑潭青年活动中心

位于圆山大饭店附近，建于1989年，设计人是台湾建筑师朱祖明先生。该建筑是一座继承传统精神并进行大胆创新的一个好作品。为什么这样评价呢？它没有在大屋顶上做文章，而是在建筑空间环境方面继承中国的传统布局手法，采用合院式，在建筑的内部与外部都重视与自然天地的结合，并在内外配以花木，使建筑室内、室外都融合在大自然的绿色之中，创造出适合在此活动的生态环境。建筑充分利用自然采光、通风，既舒适又节能。建筑墙面选用地方红砖材料。其构造、装饰做法与外形都来自于当地的传统，但又有所革新，并取得了简洁朴素的地域文化特色。它的创作理念与设计手法值得参考。

内院
Courtyard

外景
Exterior

Example 99 Taibei Jiantan Youth Activity Center

It is located near the Grand Hotel and was built in 1989. In the aspect of architectural space environment, it inherits the traditional layout technique of China, adopts courtyard style with flowers and trees, and makes the interior and exterior of the building blend into the green of nature. The building makes full use of natural lighting and ventilation, which is both comfortable and energy saving. Local materials of red brick are selected for building walls. Its structure, decoration and appearance are all derived from local traditions, but they have been innovated and achieved the effect of simplicity and regional cultural characteristics.

2. 台　中

实例100　东海大学校园规划和路思义教堂

该校园位于台中市，设计建造时间是1954—1963年，设计者是世界著名建筑师贝聿铭和陈其宽、张肇康先生。此校园的最大特点是环境好，其格局为中国院落组合式，主要建筑——路思义教堂也很有创新特色。设计者采用一条人行林荫道作为轴线，并将各学院自成院落的组群连接为一个整体。每个学院的院落是由建筑（入口、教室、办公室、走廊）和草坪与古木组成。建筑简洁、高雅，富有传统特色，建筑使用材料有传统的红砖、木材、白灰、灰瓦和现代钢筋混凝土等。路思义教堂被布置在主要轴线的侧旁，这里属于校园的中心区，整体结构是用4片薄壳混凝土墙做成呈锥状的弧形体造型。教堂顶尖留有可开启的带状沟天窗，创造出了一个具有哥特式教堂神韵的现代建筑空间。墙外则铺黄色琉璃砖，它又带有中国传统意味，可谓是一个创新的作品。这个项目已选入《20世纪世界建筑精品集锦》东亚卷中。

21.2 Taizhong

Example 100 Taizhong Donghai University Campus and Luce Memorial Chapel

The campus is located in Taizhong. Its main building, Luce Memorial Chapel, is innovative. A pedestrian avenue is adopted to connect all the departments with their own courtyards as a whole. The building materials used is traditional red brick, wood, lime, grey tile and modern reinforced concrete. The whole structure the Chapel is a conical arc shape with four thin-shell concrete walls. An openable skylight strip groove is at the top, creating a modern space environment with Gothic church charm. The walls are covered with yellow glazed bricks with Chinese traditional meaning. This project has been selected into the East Asian Volume of the *World Architectural Masterpieces Collection in the 20th Century*.

学院院落
Courtyard of a department

路思义教堂
Luce Memorial Chapel

3. 鹿港

实例101　龙山寺

鹿港位于台湾岛西岸中部，清代中期曾是岛内第二大城镇，后因港口淤塞，由盛变衰，现为台湾的一个历史文化"观光"地。其传统建筑龙山寺是台湾省内现存最大、最老的龙山寺寺庙，现位于鹿港三民路龙山街旁。相传龙山寺建于明代永历七年（公元1653年），由福建泉州肇善禅师所建，后因扩建需要，于乾隆五十一年（1786年）迁至现址。在台湾岛内凡是以龙山寺为名的，都是由泉州"安海龙山寺"分灵割香而来，亦成为泉州移民聚居之地。鹿港龙山寺除主祭观音菩萨外，还兼具民间信仰，祭为民保佑安全、富裕等诸神灵，同时，它亦有社会救济的功能。每年农历2月19日是观世音菩萨生日，这一天龙山寺都要举行法事，香客云集，人们祈求息灾降福。该寺总布局为轴线三进院落式，建筑为木构琉璃瓦大屋顶泉州式样，但具体有其独特之处：山门窄而高，表现出佛寺凝重的气息。一进院落的主要建筑前殿（三川殿）的两翼做成45°八字墙，从两侧龙门、虎门向二、三进院落的层层八角门望去，极富穿透感；二进院落，于三川殿后连建一戏亭，戏亭彩绘与藻井肃穆华丽，院落两侧立有两棵茂盛的大榕树，正中坐落着最重要的主体建筑正殿，其前加了一个拜亭，供人多祭拜时免受日晒雨淋之苦。主殿高大宽阔，共有40根柱，供奉着主神观音菩萨；三进院落主要建筑后殿，保持未上彩的原木结构，显得简洁质朴。其院落幽深宁静，具有禅修特色。

21.3 Lugang

Example 101 Longshan Temple

Lugang is located in the middle of the west bank of Taiwan, and its traditional building, Longshan

二进院落（左为戏亭）
The second courtyard (a stage pavilion on the left)

Temple, is the largest and oldest existing Longshan Temple in Taiwan Province. Besides religious use, it also has the function of social relief organization. The general layout of the temple is three-courtyard style, and the building is a large roof with wooden structure and glazed tiles. The main hall is tall and wide, with 40 columns dedicated to the Lord God Guanyin, while the back hall keeps the uncolored log structure, which is simple and unadorned.

拜亭与正殿
Worship pavilion and main hall

戏亭藻井
Caisson

下篇：
外国建筑科学文化精品图说

Part Three:
Illustration of Foreign Architectural Science and Culture

一、埃　及

实例102　开罗吉萨金字塔群

约公元前3000年，上、下埃及王国统一。约公元前2560年，建起最为宏伟的第四王朝吉萨金字塔群。约公元前1500年，建起德尔·爱尔·巴哈瑞神庙（Temple of Deir-EL-Bakhari）。德尔·爱尔·巴哈瑞神庙为台阶大露台式，在其周边植树形成圣林神苑，以祭祀阿蒙神。笔者于1985年1月到德尔·爱尔·巴哈瑞神庙现场参观时仅存建筑，树木皆无。公元前14世纪，即国王阿米诺菲斯（Amenophis）三世时期，首都建在底比斯（Thebes），当时大兴土木，神庙兴旺。下面介绍吉萨金字塔群和底比斯卡纳克阿蒙太阳神庙。

吉萨金字塔群，位于埃及开罗市西南的吉萨地区（Giza），建在尼罗河西岸一块广阔无边的30m高台上，塔形高大、简洁、沉稳，极富表现力。它是"世界七大奇迹"之一，也是联合国教科文组织选入世界第一批文化遗产的项目。

该塔群是埃及第四王朝（约公元前26世纪）存放奴隶制法老（国王）木乃伊的陵墓，共有3座金字塔，均为正方锥体，其中最大的Cheops金字塔高146.6m、底边长230.35m；Chephren金字塔高143.3m、底边长215.25m；Micerinus金字塔高66.4m、底边长108.04m。这3座金字塔是用浅黄色石灰石块砌成的，所用石块巨大，有的长5~6m、重3t以上，两个大金字塔各要使用200多万块巨石。墓室在塔的中心位置。

在3座塔前方有一座巨大的狮身人面像，是就地岩石凿成，高约20m、长约46m，它与对角线相接的金字塔群形成了一个富有变化的整体。关于狮身人面像——斯芬克斯（sphinx）有一个神话故事，即在上埃及底比斯城外出现了一个长着美女头部、狮子身子的怪物斯芬克斯，她是巨人堤丰和蛇怪所生的一个女儿，她对来往底比斯的居民提出了谜语，如果有人猜不出就会被她吃掉，于是全城恐慌。该城张贴告示，谁能除掉此怪物，就可获得王位。原先逃离底比斯的俄狄浦斯勇敢地站在斯芬克斯面前，请求解答谜语，而狡猾的怪物却给

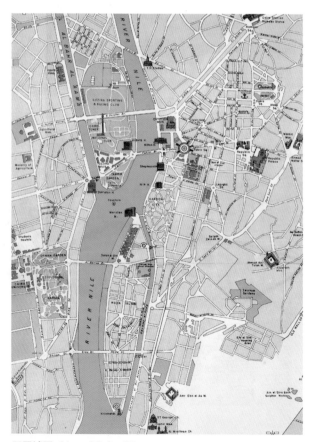

开罗城图　Map of Cairo City

俄狄浦斯出了一个难猜的谜语："早晨四条腿走路，中午两条脚走路，晚上三条腿走路，这是什么生物？"俄狄浦斯随即答出"这是人，人在幼年时恰似人生的早晨，用两条腿两只手在地下爬行；到了壮年就像人生的中午，用两条腿走路；到了老年就如人生的黄昏时刻，需拄拐杖，犹如三条腿走路。"俄狄浦斯准确地破解了谜语，使斯芬克斯不得不坠崖死去。此后，俄狄浦斯就成了底比斯国王。

这组金字塔群的建设历时30年，有10万奴役参与建造完成，它是奴隶制社会皇权统治压迫下的历史产物。同时，它也体现了古埃及卓越的起重与建造技术，以及高水平的建筑艺术。

1. Egypt

Example 102 Cairo Giza Pyramid Group

The Kingdom of Egypt was unified in about 3000BC, and the most magnificent Giza Pyramid Group was built in about 2560BC. During the period of King Amenophis in the 14th century BC, the capital was in Thebes, with great construction and flourishing temples.

It is located in Giza, southwest of Cairo, a platform about 30 m high on the edge of the desert at the west bank of the Nile River. The tower is tall, concise and steady. It is one of the Seven Wonders of the World and the only group of buildings that have been preserved up to now, and it is the first batch of cultural heritage projects selected by UNESCO in the world.

This tower group is the mausoleum where the mummy of Pharaoh (king) was stored in the fourth dynasty of Egypt. There are three pyramids, all of which are square. The largest Cheops pyramid has a height of 146.6 m and a bottom edge length of 230.35 m, Chephren pyramid has a height of 143.3 m and a bottom edge length of 215.25 m, and Micerinus pyramid has a height of 66.4 m and a bottom edge length of 108.04 m. These three pyramids are made of light yellow limestone blocks, and the stones used are huge, some of which are 5 m to 6 m long and weigh more than 3 tons. In front of the three towers, there is a huge Sphinx, which is cut in situ by rock, with a height of about 20 m and a length of about 46 m.

开罗吉萨金字塔群,展示了"大漠落日,壮丽塔影"的意境
Giza Pyramids Group, showing the landscape of "a setting sun in desert and the splendid tower shadow"

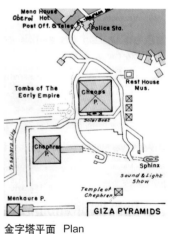

金字塔平面　Plan

细部。透过光影和人物的对比,体现巨大的石块和壮阔的金字塔
Details, showing the huge stone and grand pyramid by the constrast of light, shadow and figures

下篇：外国建筑科学文化精品图说/埃及

斯芬克斯前身　Former part of Sphinx

夜景（1985年1月下旬国际建筑师协会第15届大会闭幕后在此举办音乐晚会）
Night scene (when a music evening was held here after the 15th conference of UIA in late January 1985)

外景　Exterior

实例103　卢克索卡纳克阿蒙太阳神庙

想了解更多的埃及古文化，还要沿尼罗河到上埃及去考察。我们专程到卢克索（Luxor）参观考察，它是古埃及中王朝和新王朝的都城底比斯（Thebes）遗址所在地。现卢克索在尼罗河西岸市区的东西两面保存着卢克索阿蒙和卢克索卡纳克（Karnak）阿蒙(Amon)太阳神庙。阿蒙神于公元前1991年传至底比斯。卡纳克阿蒙太阳神庙建成于公元前14世纪，它是埃及最为壮观的神庙。该神庙规模大、布局对称，有明确的中轴线，将入口、前院、列柱大厅、后院串联起来，空间层次分明，创造出神圣的氛围。前院立有拉姆西斯二世（Ramses Ⅱ）雕像，后院立有方尖碑。中间列柱大厅是神庙的核心建筑，由16列共134根高大密集的石柱组成，列柱上刻有象形文字、太阳神故事和帝王功绩。中间两列12根圆柱，高20.4m、直径3.57m，上面的大梁长9.21m、重达65t。该神庙是继开罗金字塔后建成的又一雄伟建筑，它再次体现了古埃及高超的起重与建造技术。

Example 103　Luxor Amon Sun Temple

Built in the 14th century BC, it is the most spectacular temple in Egypt. It is of large-scale, symmetrical layout and clear central axis, which connecting the entrance, front yard, column hall and back yard in series, creating a sacred atmosphere. There is a statue of Ramses Ⅱ in the front yard and an obelisk in the back yard. The center column hall is the core building of the temple, which consists of 16 rows of 134 tall and dense stone pillars, and the pillars are engraved with hieroglyphs, stories of sun gods and imperial achievements. There are two lines of 12 pillars in the middle, with a height of 20.4 m and a diameter of 3.57 m. The girders above are 9.21 m long and weigh 65 tons.

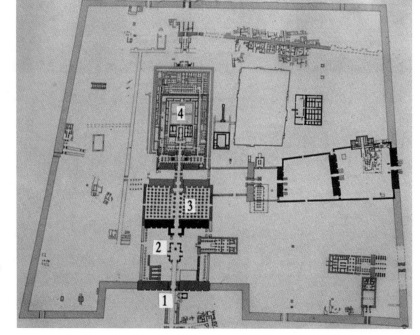

总平面
1-入口；2-前院；3-列柱大厅；4-后院

Site plan
1-Entrance; 2-Front yard; 3-Column hall; 4-Backyard

前院拉美西斯二世雕像，腿部为王后像
Statue of Ramses II and the statue at, the leg part is the Queen

仰视中间列柱顶部
Look up at the top of the middle column

入口门楼前，左边为神庙中轴线　Entrance and the axis on the left

列柱大厅前　Entrance of the columne hall

实例104　埃及亚历山大城

亚历山大城是埃及的最大海港，全国的90%进出口物资经此港吞吐。这里气候温和，是地中海沿岸的避暑胜地。亚历山大城也是历史名城。公元前332年，希腊马其顿国王亚历山大一世统治此地时建成亚历山大城，以其名作为该城名。亚历山大城西部港口建有被称为世界七大奇迹的"灯塔"。此"灯塔"于公元1326年被大地震所毁，公元1477年在其倒塌处建一城堡，纳赛尔时期又将城堡改为埃及航海博物馆。这个城市沿海岸呈带形布局，东西长30km，南北窄，南北最窄处仅有2km，其北面是浩瀚的大海,在城西部还保存着一些古代出土雕塑、建筑构件等。

Example 104　Alexandria

It is Egypt's largest seaport, through which 90% of the country's imports and exports are handled. It has a mild climate and is a summer resort along the Mediterranean coast. The city has a strip-shaped layout along the coast, with the vast sea in the north and some ancient unearthed sculptures, building components and other cultural relics in the west of the city.

西部埃及航海博物馆　Maritime Museum

东部滨海新建筑　New buildings along the east coast

二、伊拉克

伊拉克有悠久的历史，两河流域（美索不达米亚）作为世界古代文明的发祥地之一，公元前4700年就出现了城邦国家。美索不达米亚的北部古称亚述，南部为巴比伦尼亚，而巴比伦尼亚北部叫阿卡德，南部为苏美尔。公元前2000年美索不达米亚先后建立阿卡德王国、乌尔帝国、古巴比伦王国、亚述帝国和新巴比伦王国。公元前550年为波斯帝国所灭，公元7世纪时又被阿拉伯帝国吞并，16世纪时受奥斯曼土耳其帝国统治。

实例105 巴比伦城

它是世界著名的古城遗址，位于伊拉克首都巴格达南90km的幼发拉底河岸处。约公元前1830年古巴比伦王国成立，在此建都，其间汉谟拉比国王（约公元前1810—前1750年）领导制定出具有历史意义的世界上第一部法典——《汉谟拉比法典》，以法治国，正文282条，刻于石碑上，该原件现存巴黎卢浮宫博物馆。公元前604—公元前562年，即国王尼布甲尼撒时期，巴比伦城规模宏伟，建筑壮丽，有3道城墙环绕。现仿建的伊什塔尔门局部，外贴琉璃砖，并以砖组成公牛及其他四条腿走兽的图案，其色彩典雅亮丽，据悉两河流域是琉璃砖的发明地。于巴比伦城北面宫内，尼布甲尼撒为其妻子谢米拉密得建造了被誉为世界七大奇迹之一的"空中花园"。此园为多层露天平台，在平台四周种植花木，整体外形恰似悬空，故称"空中花园"（Hanging Garden）。现仅存"空中花园"遗址，但它对于我们今天的悬空绿地建设仍有启发。巴比伦城中还建有高91m的7层塔庙。现在北面故宫遗址中还立有一座雄狮巨石雕塑，这就是著名的巴比伦雄狮，它也是巴比伦的象征。公元前539年后，巴比伦城被波斯人占领，公元前4世纪末巴比伦城逐渐衰落，至公元2世纪时变成了废墟。

1985年我们来到这座古城遗址，看到了正在加紧修复遗址的情景，还参观了其博物馆。博物馆内展出了巴比伦古城复原的模型，以及刻有《汉穆拉比法典》的石碑复制品，甚为壮观。

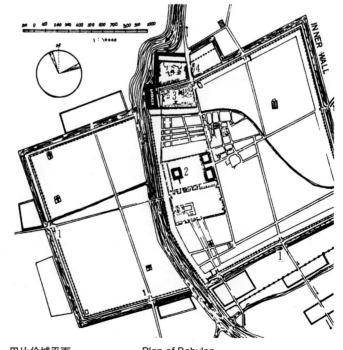

巴比伦城平面
1-城门；2-塔；3-南宫；
4-伊什塔尔门

Plan of Babylon
1-City Gate; 2-Tower; 3-South palace;
4-Ishtar Gate

现立于城北部故宫遗址内的巴比伦雄狮
Babylon lion standing on the ruins of the palace in the north of the city

仿建的伊什塔尔门前部
Front part of the Ishtar Gate

2. Iraq

The new Babylon Kingdom became independent in 625BC. King Nebuchadnezzar II developed the city of Babylon and built the "Hanging Garden", which became one of the seven wonders of the ancient world.

Example 105 The City of Babylon

It is a world-famous ancient city site, located at the right bank of Euphrates River and 90 km south of Baghdad. During King Nebuchadnezzar's period, Babylon was magnificent in scale and architecture, surrounded by three city walls covered with glazed bricks, which made up of bulls and other four-legged animals with elegant and beautiful colors. In the palace in the north of the city, Nebuchadnezzar built a "hanging garden" for his wife. This garden is a multi-layer open-air platform. Flowers and trees are planted around the open air, and the overall shape is just like hanging. There is also a 7-storey pagoda temple with a height of 91 m. Now there is a giant stone sculpture of a lion stepping down on one man in the north of the palace ruins, and the lion is the symbol of Babylon.

博物馆内展出的古城模型，左边为塔庙
The model of the ancient city exhibited in the museum, and the left part is the Pagoda Temple

已修复的遗址局部　Part of the restored site

三、希 腊

希腊于公元前5世纪兴盛起来，著名的雅典卫城就是在这个世纪建成的。雅典卫城祭奉雅典守护神雅典娜，具有神园的风格。此时，住宅庭院也得到发展。因2000年前庞贝城有此类住宅的实物，故在后文庞贝城实例中介绍。希腊雅典人喜欢群众活动，因此，建筑与聚会广场、体育比赛场所等相结合，人们可以在这里聚会、比赛、交换意见、辩论是非。

实例106　雅典卫城

雅典是希腊的首都。雅典作为著名古都，卫城是其重要标志。雅典卫城位于城中心一个高出地面约70~80m的陡峭高台上，重建完成于公元前430年前后，是为纪念波希战争中希腊取得反侵略的胜利而建的，它已成为希腊的宗教和文化中心，其主体是雅典娜胜利女神庙。雅典娜是希腊神话中智慧与战争女神，她与海神波塞冬争夺雅典保护权并取胜，成了雅典保护神。雅典卫城东西长约280m，南北最宽处130m，从西端上下山，主要建筑偏于西、北、南三面，建筑布局因地制宜、自由活泼。

雅典卫城主体建筑是帕提农神庙，这是守护神雅典娜之庙，公元前438年竣工，它凸出在卫城的最高处，采用围廊式，周围列柱为46根多立克式柱，每根柱高10.43m、底径1.905m，比例匀称、刚健有力。此建筑布局符合人的比例尺度，并具有象征性的艺术感染力，其细部的精美等说明了它是一座古典建筑的优秀实例。

伊瑞克提翁庙是卫城的主要建筑之一，传说伊瑞克提翁是雅典的始祖。此庙廊柱系爱奥尼式，建于公元前421—公元前406年。为了接引从帕提农神庙西北角走过来的仪典队伍或上山过雅典娜神像前行的人群，在其南面角上布置一个女郎柱廊，面阔3间、进深两间，同时还立着6个2.1m高的女郎雕像柱子。秀丽的女郎充分表现了爱奥尼柱式的性格，该庙与其南边多立克柱式的帕提农神庙形成了鲜明的对比。在雅典卫城高台南端角下，还建有露天剧场。1993年9月25日晚，世界著名音乐演奏家雅尼（Yanni）在此举办了雅尼音乐会，获得了成功，这再一次证明了在古建筑群中开展音乐、戏剧演出活动，可以取得特殊的效果。

3. Greece

Greece flourished in the 5th century BC, with the rise of democratic spirit, and large-scale constructions were carried out everywhere. The residential courtyard was characterized by the atrium-style courtyard in the surrounding colonnade.

Example 106 Athens Acropolis

The Acropolis was built to commemorate Greece's victory in anti-aggression in the Persian War. It is about 280 m long from east to west and 130 m wide from north to south. Its architectural layout depends on the terrain. The main building is the Parthenon Temple, which is located at the highest place and is of the corridor style. The Erechtheion Temple is for welcoming the team or crowd receiving ceremonies, which is in sharp contrast with Parthenon. Under the corner at the southern end of the Acropolis Tower, there is an open-air theater.

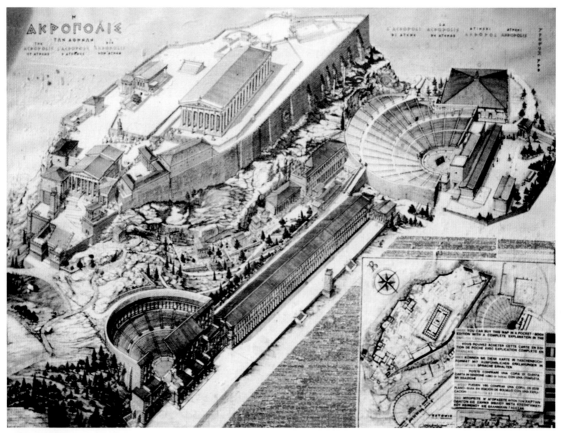

雅典卫城鸟瞰 Aerial view

建在高地上的帕提农神庙　Parthenon Temple

伊瑞克提翁庙秀丽的女郎柱廊　Lady colonnade of the Erechtheion Temple

雅典卫城露天剧场　Open-air theater

四、意大利

1. 罗 马

罗马帝国于公元前200年左右在地中海沿岸占据主导地位。罗马城为古罗马帝国首都，19世纪末至今为意大利首都。它是欧洲古老的城市，人口近300万，面积200多平方公里。相传希腊攻破特洛伊城后，特洛伊王子逃到意大利，王位传至里亚·西尔维娅时，被叔父所篡，新国王害死西尔维娅，并将其二子装筐投入台伯河，后遇救，被一母狼哺育养大，其中一子罗莫洛在台伯河畔建城，罗马即从罗莫洛演变而来的。今罗马城徽是母狼育婴图案，亦是由此故事引来。罗马城于公元前30—公元476年为古罗马帝国首都，公元756—1870年为教皇国的首都，1870年意大利王国统一后成为意大利首都。第二次世界大战之后，随着城市的发展，在老罗马城南7km处另建一座新城，于是老罗马城就成了一座巨型历史古城。

实例107 罗马中心广场

罗马中心广场（Forum）位于罗马城巴拉丁山和卡比多利山山脚下，是在公元前后几个世纪形成的，是罗马奴隶制下共和与帝制时期的政治、经济、文化活动的中心。罗马中心广场布置有政府机构、元老院、作为法庭与会议厅使用的巴西利卡（Basilica）、庙宇、商业市场、金融钱庄等。罗马中心广场初期布局散乱，至共和末期修建了凯撒广场，有明确的轴线和有序的空间组合，之后添建的奥古斯都广场、图拉真广场，更加体现了帝制的威严。它不仅轴线对称，而且作纵深多层次的空间布局。这组中心广场建筑群遗迹，现完整地保护着并对外开放，以供游人凭吊。

4. Italy

4.1 Rome

It is an ancient city in Europe with a population of nearly 3 million and an area of over 200 km^2. It was the capital of the Roman Empire, the Papal State and Italy from the past to present. After the Second World War, a new city was built 7 km south of the old Rome, and the old Rome became a huge historical museum city.

Example 107 Raman Forum

It was the center of political, economic and cultural activities during the republican and imperial periods under Roman slavery, including government agencies, the senate, and temples and so on. There is a clear axis and an orderly spatial combination. Later, Augustus Square and Trajan Square, which are added, further embody the majesty of the imperial system. They are not only symmetrical in axis, but also have a deep and multi-level spatial layout.

下篇：外国建筑科学文化精品图说/意大利

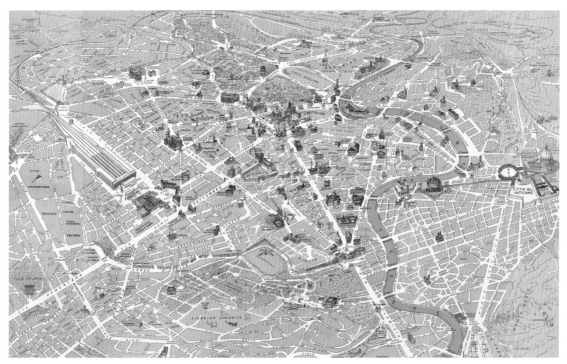

罗马城图　Plan of Rome

图拉真集贸市场
Trajan market fair

赛维鲁斯凯旋门
Severus Triumphal Arch

实例108 大斗兽场

大斗兽场是为人所知最著名的大型建筑之一，是专为观看角斗而造，公元72年始建，公元80年完工。建筑为椭圆形，长轴188m、短轴156m，立面高48.5m。分为4层，下3层各80间券柱式透空墙，第四层为实墙，其结构稳定，功能使用合理，适合人流均匀聚散。叠柱式的水平分割加强了它的整体感，使其外观宏伟、有力、完整，也是一座结构、功能、形式统一和谐的建筑，代表了当时罗马券柱式建筑的最高成就。至今，世界各地的体育建筑基本上还保持着这一形制。

Example 108　Colosseum

It was specially built for watching gladiators, which is oval, with 188 m long axis, 156 m short axis and 48.5 m high elevation. There are 80 arch-type hollow walls in the lower three floors, and the fourth floor is a solid wall. It is stable in structure and reasonable in function, and is suitable for people to gather and disperse evenly. The horizontal division of stacked columns enhances its overall sense. It represents the highest achievement of Roman arch architecture at that time.

下篇：外国建筑科学文化精品图说/意大利

大斗兽场内景
Interior

大斗兽场外景　Exterior

实例109 罗马图拉真纪功柱

位于罗马图拉真广场后面的一个29m×16m的院内，建于公元109—113年间，为罗马多立克式独立柱。柱高29.55m，底径3.70m，连基座总高35.27m。柱身材料为白色大理石，内部空心。柱子设185磴石级盘旋而上，人可登上柱顶。柱身为螺旋形200m长的浮雕带，刻着图拉真远征达奇亚的功绩，柱头上矗立着图拉真像。由于院子小而柱身高，人们对帝王的仰视产生了崇拜之感。此后，这种独立单根纪念柱形式开始在欧洲流行。

Example 109 Column of Trajan

Built in 109 to 113, it is a Roman Doric independent column with a height of 29.55 m, a bottom diameter of 3.7 m, and a total height of 35.27 m. The column body is made of white marble, and its interior is hollow. It is equipped with 185 stone steps to spiral up, so that people can climb the top of the column. The column is a spiral embossment belt with a length of 200 m, engraved with the achievements of Trajan's expedition to Dacia, and the statue of Trajan stands on the stigma.

图拉真纪功柱 Column of Trajan

实例110 君士坦丁凯旋门

凯旋门是为炫耀战争胜利的纪念建筑，立面为方形，多为三开间券柱式，中间券洞高大、两侧矮小。其基座和女儿墙都较高，墙面刻有浮雕。女儿墙上部分刻着铭文，铭文字与浮雕都记录着皇帝战争的军事功绩。君士坦丁凯旋门于公元312年建于大斗兽场旁，其修复从1804年开始，至今看到的就是这个模样。它将其他罗马古代的纪念装饰品移至此处，中间券部分是君士坦丁的功绩，而其他处则是图拉真、奥瑞里俄斯等的功绩。前述罗马中心广场遗址里还保存着公元204年建造的赛维鲁斯凯旋门（Arch of Septimius Severus）和公元81年修建的替度斯凯旋门（Arch of Titus），但这两座门都没有像君士坦丁凯旋门那样修复得很完整，而只保持原状。

Example 110 Arch of Constantine

The Arch of Triumph is a memorial building to show off the victory of the war. The facade is square, and most of them are three-bay arch columns. The middle arch is tall, and both sides are small and short. The pedestal and parapet are high. The wall is engraved with embossments, and the parapet is partially engraved with inscriptions. The text and embossments are all records of the military achievements of the emperor's war. The Arch of Constantine was built near the Colosseum in 312, and its restoration was in the 18th century. During the restoration, other ancient Roman memorial decorations were moved here.

君士坦丁凯旋门 Arch of Constantine

实例111 罗马圣彼得大教堂

它是世界上最大的天主教堂，建于公元1506—1626年，历时120年。经过创新与保守势力的斗争，虽然该创新方案未得到全部实现，但仍保留着集中式穹顶的创新特点，体现了文艺复兴时期的创造活力和建筑发展的新成就。

此教堂通过设计竞赛，1505年选中了伯拉孟特集中式形制方案，1506年按照这一方案动工。1514年伯拉孟特去世，新教皇任命拉斐尔修改设计方案为拉丁十字形，后又委托米开朗琪罗主持这项工程。米开朗琪罗凭借他的声望，摒弃了拉丁十字形制，恢复了伯拉孟特集中式方案，并突出中心穹顶空间，以体现大教堂的雄伟壮丽，加大了支承穹顶的4个墩子，简化了四角布局。穹顶直径41.9m，接近万神庙，但高度高出万神庙3倍。大教堂外部顶尖高138m，是罗马全城的最高点。1564年米开朗琪罗逝世后，由维尼奥拉在四角上各加一个小穹顶，以利采光通风。1606年在保守势力的压力下，拆除了已动工的米开朗琪罗设计的正立面，加上了三跨大厅，这破坏了集中式特点和内部空间以及外部造型的完整性，以致在教堂前看不到完整的穹顶。

教堂前面的广场是由一个梯形加一个椭圆形组成。椭圆形广场中心竖立着一座方尖碑，其两侧各设一喷泉。广场四周的围廊柱高19m，很有气势。广场地面向教堂逐渐升高，使这组大教堂建筑显得格外壮观，它不愧是意大利文艺复兴时期的一个代表作品。

Example 111 St. Peter's Cathedral

It is the largest Cathedral in the world. It retains the innovative characteristics of centralized dome and embodies the creative vitality and new achievements of architectural development in Renaissance.

The design abandoned the Latin cross, highlighted the central dome space, increased the four piers supporting the dome, and simplified the layout of the four corners. The dome is 41.9 m in diameter and 138 m high at its top, which is the highest point in Rome. The square in front of the church is composed of a trapezoid and an ellipse. An obelisk is erected in the center of the oval square, with a fountain on each side. The colonnade column around the square is 19 m high, which is very imposing.

进入广场望教堂正面　Front of the Cathedral

从顶部望室外广场与城市景观
The square and city landscape from the view at the top

高大的室内穹顶空间　Huge dome space

实例112 罗马特雷维喷泉

特雷维喷泉于1762年建成在Corso大街东侧,因三条路交会在此处,故依此意取为特雷维。罗马城内多喷泉和古水道,全市有3000多个喷泉,特雷维喷泉是最大的一个喷泉,所以又称大喷泉。喷泉中央立着海神像,两侧壁龛内立有象征安乐和富饶的女神,喷泉水花飞溅,池水碧绿清澈,有人将其叫作贞洁水。这是一组建筑、雕塑、喷泉融为一体的优秀作品。

Example 112 Trevi Fountain

It was built on the east side of Corso Street in 1762. Among the more than 3,000 fountains in the city, it is the largest fountain, so it is also called the Big Fountain. There is a statue of Poseidon standing in the center of the fountain, and there are goddesses symbolizing happiness and richness in niches on both sides. The fountain is splashed with water, and the pool water is green and clear. Some people call it Virgin Water. This is a group of excellent works integrating architecture, sculpture and fountain.

外貌全景　Exterior

实例113 罗马西班牙广场

　　这是一个17世纪罗马巴洛克式城市设计的好作品，它将台地上的"三一教堂"和台地下的城市街道连接一起的大台阶广场，是人们在此休闲活动的好场所。游人们在这里席地而坐、欢笑交谈，极富悠闲的生活气息。广场由一轴线贯穿，从上至下为"三一教堂"、方尖碑、布满鲜花的大台阶、名为老船的喷泉水池、城市支路。据说，此处的喷泉水池是伯尼尼设计的，轴线正对着的支路现为意大利时装街。

Example 113 Piazza Spagna

　　It is a large square connecting the Trinity Church on the platform and the city streets under the platform. People here have leisure activities, sit on the floor, laugh and talk, and enjoy a leisurely life. From top to bottom, the square runs through an axis, including "Trinity Church", obelisk, big steps full of flowers, fountain pool named Old Boat, and city branch road. The branch road opposite the axis is now Italian Fashion Street.

从"三一教堂"前向下俯视广场
The square from the view of Trinity Church

从喷泉池前往上望"三一教堂"
Trinity Church from the view of the fountain pool

实例114　埃马努埃尔二世纪念碑

于1885年开始至1911年在卡比多山的北坡建成埃马努埃尔二世纪念碑，它是为纪念意大利独立统一而营建的。其规模宏大，面宽135.1m、高70.2m，主体是在高台基上布置列柱廊。柱高11m，全部采用白色大理石贴面。中央大台阶部分为无名战士墓，寓意祖国的祭坛。在它们的上方高基座上立着意大利开国国王埃曼努埃尔二世（Vittorio Emanuele Ⅱ）骑马的青铜镀金像，青铜镀金像的高宽各12m；而在前面台阶两侧则立有两组青铜镀金群雕，左面的寓意思想（Thought），右面的代表行动（Action）。这个建筑是世界上最宏伟的纪念碑之一，因其建筑高大壮观，在罗马许多地方都能看到它，但其建筑风格与周围环境不够协调。

Example 114　Monument to Victor Emanuele Ⅱ

It was built to commemorate the independence and unity of Italy. The main body is a colonnade with a column height of 11 m, all of which are covered with white marble. The central large step is the Tomb of Unknown Soldier, symbolizing the altar of the motherland. On the high pedestal above them stands the bronze gilded statue of the founding king of Italy, Vittorio Emanuele Ⅱ. There are two groups of bronze gilded group sculptures on both sides of the front steps, with the left side implying Thought and the right side representing Action.

外景全貌　Exterior

2. 佛罗伦萨

佛罗伦萨在意大利语中意思是鲜花之城，它是古罗马时期的军事要塞。在1434—1494年美狄奇家族统治时期，佛罗伦萨成为欧洲最大的工商业与金融中心，资本主义萌芽也在这里最早出现，因而新文化、新思想的文艺复兴曙光在此升起，于是成为文艺复兴的发源地，素有"西方雅典"之称。

实例115　佛罗伦萨市中心西尼奥列广场

佛罗伦萨市中心西尼奥列广场，其主体建筑是建于13世纪的碉堡式旧宫，即在雉堞式檐口上矗立着高95m的塔楼，极具中世纪庄严坚固的城堡特点。这座旧宫现为市政大厅，其侧翼走廊连同整个广场已成为一座露天雕塑博物馆。广场中心位置立有海神喷泉，栩栩如生。从广场行至阿诺河，河上有一座著名的桥廊式"旧桥"。

4.2 Florence

Florence means the city of flowers in Italian. It was a military fortress in ancient Rome and later became the birthplace of Renaissance, and was called "Athens in the West".

Example 115　Signoria Square

Its main building is a bunker-style old palace built in the 13th century, with a tower of 95 m standing on the cornice. This old palace is now a municipal hall. The wing corridor and the whole square have become an open-air sculpture museum. Poseidon Fountain stands in the center of the square. From the square to the Arno River, there is a famous old bridge of corridor style.

佛罗伦萨城中心区，灰色为步行街
①西尼奥列广场；②"旧桥"
Central area of Florence, where the gray part is pedestrian street ①Signoria Square ②Old Bridge

由阿尔诺河旁通往旧宫的步行街
Pedestrian street leading from Arno River to the old palace

旧宫西尼奥列广场及其雕像群　Signoria Square and its statues

立于中心广场的海神雕像
Statue of Poseidon standing in the central square

下篇：外国建筑科学文化精品图说／意大利

著名的桥廊式"旧桥"
Famous "old bridge" of corridor style

3. 威尼斯

实例116　威尼斯圣马可广场

威尼斯是世界上有名的水乡城市，圣马可广场则是威尼斯最大的广场，也是世界上最著名的广场之一。圣马可广场获得了"欧洲最漂亮的客厅"美誉，它也基本上是在文艺复兴时期扩建、修改完成的。16世纪的珊索维诺为圣马可广场做了较多的规划设计工作。

圣马可广场由大、小两个广场组合而成。大广场为梯形，长175m，东面宽90m，布置着主体建筑圣马可大教堂；西边宽56m，经改扩建，西、南、北三面为连成一体的新旧市政大厦。小广场同样为梯形，东面是紧挨圣马可大教堂的总督府，西边是连接新市政大厦的图书馆，南边海口处有一对从君士坦丁堡搬来的石柱，海中建有圣乔治教堂和尖塔的小岛，并成为对景。回首北望可见圣马可大教堂和高100m的钟塔，这两座对比强烈的主体建筑起着既分隔又连接大小广场的作用。

4.3 Venice

Example 116 Piazza San Marco

Venice is a famous water town in the world, and Piazza San Marco has the reputation of "the most beautiful living room in Europe". It consists of a large square and a small square. The large square is trapezoidal, and the main building is Basilica di San Marco and the municipal building. The small square is also trapezoidal, with the governor's house in the east, the library in the west and a pair of stone pillars in the south.

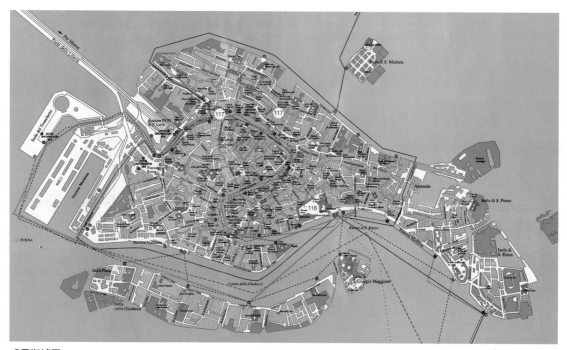

威尼斯城图
City map of Venice

广场东面的圣马可大教堂和总督府
Basilica di San Marco and governor's house in the east

小广场北向南望
Small square

广场东面总督府　The governor's house in the east

圣马可大教堂前的鸽群　Pigeons in front of Basilica di San Marco

大广场西向东望 Large square

实例117 威尼斯水乡风貌

威尼斯本岛中心贯穿着一条S形的大运河，可乘船从西北方向沿大运河到达圣马可中心广场。以这条大运河为主干，分流出中小河道。全市交通由不宽的道路网和水道网交织组成，其中道路网供步行，水道网走船只。岛内无汽车交通，非常安全与环保。沿大运河伸进的中小河道，分为3种类型：一是河道一旁有路；二是河道两旁有路；三是河道旁无路，而建筑立于河道两边。在河道中行驶着两头翘起的小艇，名为"冈朵拉"，小艇轻捷穿梭，构成了具有威尼斯独特风格的画面。

Example 117 Watertown Landscape

There is an S-shaped grand canal running through the city center, which is the main trunk, branching out into small and medium-sized rivers. The traffic of the whole city is composed of road network and waterway network for walking and boats. There is no automobile traffic on the island, which is very safe and environmentally friendly. In the river, a small boat with two upturned ends, named "Gondola", forms a picture with unique Venetian style.

两旁有路的西北部大运河夜景
Night view of the grand canal with roads on both sides

两边无路的水乡风貌
Landscape with no road on both sides

4. 庞 贝

实例118 古城公共建筑和住宅区

庞贝古城位于意大利南部靠近那不勒斯海湾，离维苏威火山南麓约2km，原建于公元前8世纪，于公元79年因维苏威火山爆发被埋毁，18世纪开始被挖掘出，经过200多年的不断发掘，至1960年完成。庞贝城呈一树叶状，建在一个椭圆形台地上，约63hm^2，有3km长的城墙环绕。城墙里布置有纵横两条大街，形成"井"字状。各地区由小街巷有序地划分成长方形或方形地块。城西南长方形广场北端为丘比特神庙（Temple of Jupiter），广场中部西面为阿波罗神庙（Temple of Apollo），广场南端西面是巴西利卡（Basilica）。巴西利卡是城中最古老的重要建筑，起初是做市场使用，后改为法庭。广场长轴南端为市政厅，这一中心广场周围集中了政府、法庭和庙宇建筑，是全城政治、宗教、经济的活动中心。在南北向主要街道Via Distabia与东西向主要街道交叉口处设有公共浴室和健身运动活动中心，再沿Via Distabia大街南行至顶端，便是以大剧场为主的娱乐活动中心。在城东南角还设有可容5000名观众的圆形露天竞技场。从这些公共建筑可以看出，这里开展公共活动已很广泛。

庞贝住宅区十分规整，街巷的路面是用巨石铺成的。在街巷的路口可以看到公共水槽，这是从城外引山泉水通过砖石砌成的引水槽流到公共水槽和富人住宅中。住宅为列柱围廊式（Peristyle）庭院，庭院中有水庭和绿地花园。这种中庭式住宅源于希腊。

4.4 Pompeii

Example 118 Public Buildings and Residence

Near the Gulf of Naples, it was built in the 8th century. It was buried by the eruption of Vesuvius on August 24th, 79BC, and was excavated from the 18th century to 1960. The city is leaf-shaped, built on an oval platform surrounded by a 3 km long city wall. There are two streets in the city wall, and each area is divided into rectangular or square plots by small streets in an orderly manner. There are temples, city hall, public bathroom, grand theater, amphitheater, etc.

The residential area here is very regular, the streets and lanes are paved with boulders, and there are public sinks at the intersections of the streets and lanes.

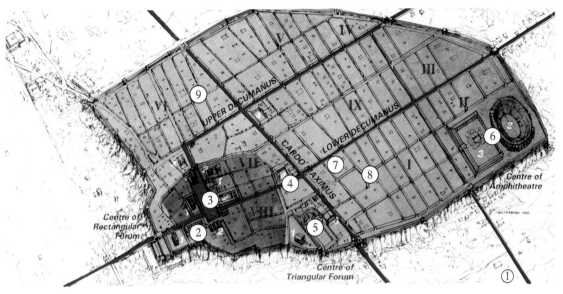

1-庞贝城　Plan of Pompeii

2-中心广场南端西面的巴西利卡　Basilica at the west side of the south end

3-从中心广场远眺维苏威火山　Center square and Vesuvius from afar

4-公共浴室外的健身运动场地　Sport field outside of the public bathroom

5-大剧场　Grand theatre

6-圆形露天竞技场　Amphitheater

7-巨石路面的住宅区街巷
Street in the residential area

8-街巷路口的公共水槽
Public sink

9-苇蒂住宅（列柱围廊式庭院）
Vettii courtyard

5. 巴格内亚

实例119　兰特别墅园

该园位于罗马西北面的巴格内亚村，初建于公元14世纪，16世纪建成。这个台地园的特点是：

（1）台地完整。花园位于自然的山坡，创造了4层台地。最低一层呈方形，由花坛、水池、雕塑、喷泉组成，十分壮观。第二层平台呈扁长方形，种植有梧桐树群。通过奇妙的圆形喷泉水池两边的台阶上到第三层平台，中间为长方形水池，两侧对称地布置种有树木的草坪。第四层台地是最上面的一层，宽度变窄，为下面的1/3宽。其纵向分为两部分，在低部分相当大的斜面上做成连续瀑布，贯穿中轴线；高的部分是平台，中心放一海豚喷泉，其后以半圆形洞穴结束。这4层台地，在空间、大小、形状、种植、喷泉、水池等方面都有节奏地变化着，并用中轴线将4层台地连成一个和谐的整体。

（2）水系新巧。各层平台的喷泉流水达到了极好的装饰效果。

（3）高架渠送水。全园用水是从别墅后山引来流水，即由一条22.5cm宽的小型高架渠输送。

（4）围有大片树林。在此园左侧坡地上栽植树木，形成大片树林，这就是公园Park的来历。这一树林的功效是多方面的，既可以改善气候，保持土水，也可衬托出花园主体。

4.5 Bagnaia

Example 119　Villa Lante

Located in Bagnaia, the villa was completed in the 16th century. Its characteristics are:

(1)The platform is complete and has four layers. The lowest layer is square and consists of flower beds and pool fountain; the second is flat and rectangular, and there are buttonwood groups; the middle of the third is a rectangular pool with trees and lawns symmetrically planted on both sides; the width of the fourth is narrowed, and it is divided into two parts longitudinally.

(2)The water system is ingenious, and the fountains on each platform run water, achieving excellent decorative effect.

(3)Water is supplied by the elevated canal, and the whole garden is supplied by flowing water from the mountain behind the villa.

(4) On the left side, a slope is planted with trees to form a large forest, which is called Park. This is the origin of name Park in today's use.

从三层平台中轴线上俯视一、二层台地园
First and second layers from the view of the third layer

兰特别墅园平面
Plan

四层平台前的连续瀑布
Extended waterfall in the fourth layer

五、法 国

1. 巴 黎

法国首都巴黎跨塞纳河两岸，旧城区分20个区，面积105km²，市郊周围有7个省。区和省为大巴黎，总面积1.23万km²，市区人口200多万，为法国第一大城市。巴黎是历史文化名城，有2000多年历史，最早是在塞纳河中西岱岛上建设居民点。岛上居民是巴黎西族，故城市取名巴黎。

公元486年，日耳曼法兰克人夺取塞纳河流域，至6世纪，法兰克王国定都巴黎。公元14世纪，巴黎才成为法国政治、经济和文化中心。1789年7月14日，资产阶级革命摧毁路易封建王朝。1871年3月28日工人阶级武装起义，建立了人类历史上第一个无产阶级政权——巴黎公社。巴黎有很多世界著名的历史遗迹和建筑与艺术品，也是一个重要的世界文化中心。我有幸5次访问这座城市，对它比较了解，感悟也较多，其中最大的感受是，巴黎历史文化名城整体保护做得好。

实例120　巴黎圣母院

巴黎圣母院是世界闻名的天主教堂，建于公元1163—1345年，坐落在巴黎市中心塞纳河中的西岱岛上，曾遭战火破坏，后来重建，1864年对外开放。它是一座典型的哥特式教堂，平面宽47m、长127m，内部大厅高32.5m，可容9000人进行宗教活动。哥特式教堂的结构特点是，采用骨架券作为拱顶的承重构件，使拱顶荷载集中到十字拱的四角，并使飞券落在侧廊外侧横向墙垛上，因此侧廊拱顶不承担中厅拱顶的侧推力，中厅可以开很大的侧高窗，尖券和尖拱的侧推力比较小，结构大为减轻，材料大量节省，同时形成了自己的轻巧风格。巴黎圣母院的闻名，不仅是因为它的建筑精美壮丽，而且法国作家雨果还写了著名小说《巴黎圣母院》，如今这里已成为法国政治活动的中心。2019年巴黎圣母院被烧坏，现正在修复中。

5. France

5.1 Paris

Across the Seine River, the old city with an area of 105 km² is divided into 20 districts. Paris is a famous historical and cultural city with a history of more than 2,000 years. On March 28, 1871, the armed uprising of the working class established the Paris Commune, the first proletarian regime in human history.

Example 120　Notre Dame de Paris

Located on Cite Island, it is a typical Gothic church with a plane width of 47 m and a length of 127 m, and its inner hall is 32.5 m high, which can accommodate 9,000 people for religious activities. The skeleton arch is used as the bearing member of the vault, so that the load of the vault is concentrated at the four corners of the cross arch, and the middle hall has a large side high window, forming a light style. Unfortunately, it burned out in 2019 and is now being repaired.

下篇：外国建筑科学文化精品图说／法国

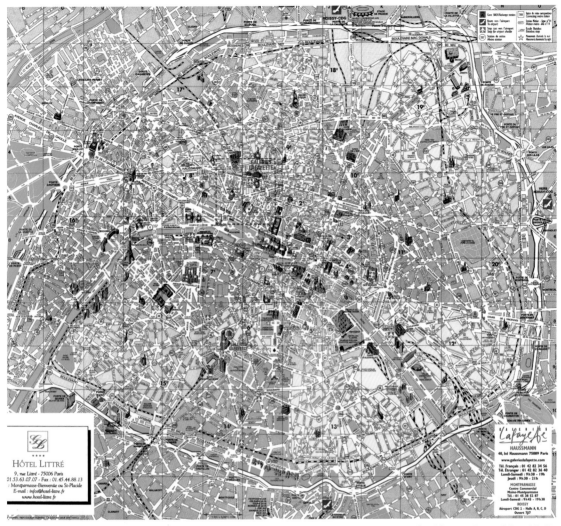

巴黎旧城平面　Plan of the old city

巴黎旧城中心区　Paris old city central area

351

晨曦中的巴黎圣母院。清晨刚升起的太阳照亮了路灯,幽静的环境让人想起这里曾发生的故事
Notre Dame de Paris in sunrise. I walked here in the early morning to see the road lights glowing in the sunrise and the quiet environment brought me to the stories that happened here before

实例121　巴黎沃克斯·勒·维康特园

该园是路易十三、路易十四时期财政大臣孚盖的别墅园，位于巴黎市郊，由著名造园家勒·诺特设计，始建于1656年。由于孚盖当时专权，建此园要显示自己的权威，设计人满足了这一要求，采用了严格的中轴线规划，有意识地将这条中轴线做得突出，不分散视线。花园中的花坛、水池、装饰喷泉十分简洁，并有横向运河相衬，因而使这条显著的中轴线控制着人心，让人感到主人的威严。孚盖的思想追求，也使他丧了命。1661年该园建成后，孚盖于8月17日也请王公贵族前来观园赴宴，路易十四也到别墅园观赏。路易十四看后感到孚盖有篡权的可能，于是借题在三周后的9月5日将孚盖下狱问罪，并判无期徒刑，后孚盖死于1680年。

Example 121 Vaux Le Vicomte Garden

This design met the requirements of the owner Nicolas Fouquet, who was the Finance Minister of Louis XIII and XIV. He was dictatorial at that time and built this garden to show his authority. Strict central axis planning is adopted, and the flower beds, pools and decorative fountains in the garden are very simple, which are matched by the horizontal canal, so that this obvious central axis makes people feel the majesty of their owners. However, Louis XIV was invited to visit the Garden for dinner after the completion in 1661. He felt that Fouquet might usurp the power, so Fouquet was jailed for questioning and sentenced to life imprisonment.

主体建筑
Main building

花园中轴线景观　Central axis of the garden

后部运河处的轴线景观，远处高地有大力神雕像立于轴线末端上
The axis at the canal and the statue of Heracles far away

细雨蒙蒙的花园中部景观
Middle part of the garden in fine rain

实例122　巴黎凡尔赛宫苑

凡尔赛宫苑坐落在巴黎西南部18km处的凡尔赛镇上，镇上3条放射路聚焦在此宫前。这里原是一个小村落，路易十三为打猎时休息，修建了城堡。1661年路易十四开始建宫，1689年完成，历时28年，建筑面积11万m^2。建筑为三层古典主义形式，以东西为轴线，向南北对称展开，主体长达700多米，中间是王宫，两翼是政府办公处、剧场、教堂等，外部简洁壮观，内部富丽堂皇，采用大理石镶砌，以雕刻、挂毯和巨幅油画装饰。宫殿如此雍容华贵，加上大规模的园林，使其成为世界闻名的法国宫苑，联合国教科文组织已将它列为世界文化遗产。当时，路易十四让勒·诺特等规划设计凡尔赛宫苑，并提出：要搞出世界上未曾见过的宫苑，要超过西班牙埃斯库里阿尔宫。为了达到路易十四这一要求，体现君主的绝对的权威，勒·诺特等采取了如下措施。

（1）大规模。大胆地将护城河、堡垒合并，且向远处延伸。占地800hm^2，其中中心区100hm^2，做成了宏大的宫苑。

（2）突出纵向中轴线。3条放射路，焦点集中在凡尔赛宫前广场的中心。穿过宫殿的中心，轴线向西偏北伸延。在这条3km长的纵向中轴线上布置有拉托那喷泉、长条形绿色皇家大道、阿波罗神水池喷泉和十字形大运河。站在凡尔赛宫前台的地园上，并沿着这条中轴线望去，景观深远、气派严整、雄伟壮观，展现了君王的威风。

（3）采用超尺度的十字形大运河。勒·诺特对此设计从未感到怀疑，认为巨大的运河像伸出双臂的巨人，可以给人以无比宏伟开阔的印象。

（4）均衡对称的布局。在纵向中轴线两侧和次要横向轴两侧均衡对称地布置许多小型花园，既有变化，但又是统一的。大运河中轴线左右两侧，布置有一对放射路，它通向十字形运河两臂的末端，其中左端是个动物园，右端是大特瑞安农（Trianon）和小特瑞安农园。这样处理，既突出中轴线，又丰富了大小园景。

（5）以水贯通全园。在纵向中轴线上布置壮观水景，这是凡尔赛宫的一大特点，也是它名闻遐迩的一个主要原因。

（6）遍布塑像。中心的两大喷水池以拉托那和阿波罗神塑像为核心，其雕塑生动细致、神态自如，起到了画龙点睛的作用。在皇家大道两侧各布置一排栩栩如生的塑像，以作陪衬之用。

（7）建筑与花园相结合。除将建筑的长边及其凹凸的外形同花园紧密联系外，有的建筑还将花园景色引入室内。如著名的镜廊，其全长72m，一面是17扇朝向花园的巨大拱形窗门，另一面镶嵌与拱形窗对称的由400多块镜片组成的17面镜子，镜子中反映了花园景色。

（8）路易十六时期，小特瑞安农园改建为田园景观，使小村庄成为此园的中心，其附近还有农场、牛奶场、谷仓等，形成了田园景色村落，并于1784年建成。

Example 122 Château de Versailles

The construction has a building area of 110,000 m². It takes the east and west as the axis, and spreads symmetrically to the north and south. The palace is in the middle, and the government offices, theaters, churches, etc. are on the two wings. The exterior is simple and spectacular, while the interior is magnificent. The main features include:

(1) A large scale. The moat and fortress were boldly merged and extended to a distance, covering an area of 800 hectares.

(2) Highlight the longitudinal central axis. The three radiation paths focus on the center of the square in front, and passing through the center of the palace.

(3) Super-scale cross-shaped Grand Canal is adopted, which is like a giant with outstretched arms.

(4) Balanced and symmetrical layout. Many small gardens are evenly and symmetrically arranged on both sides of the longitudinal central axis and the secondary transverse axis.

(5) Water runs through the whole garden. Continuous spectacular waterscape is arranged on the longitudinal central axis.

(6) Statue everywhere. Among the two fountains in the center, the statues of Latona and Apollo are the core, and a row of statues on both sides of Royal Avenue is vivid.

(7) Combination of architecture and garden. In addition to connecting the long side of the building and its concave-convex shape with the garden, some also introduce the garden scenery into the room.

(8) The Petit Trianon Garden was transformed into an idyllic landscape, with farms, dairy farms and barns nearby.

凡尔赛宫位置　Location of Château de Versailles

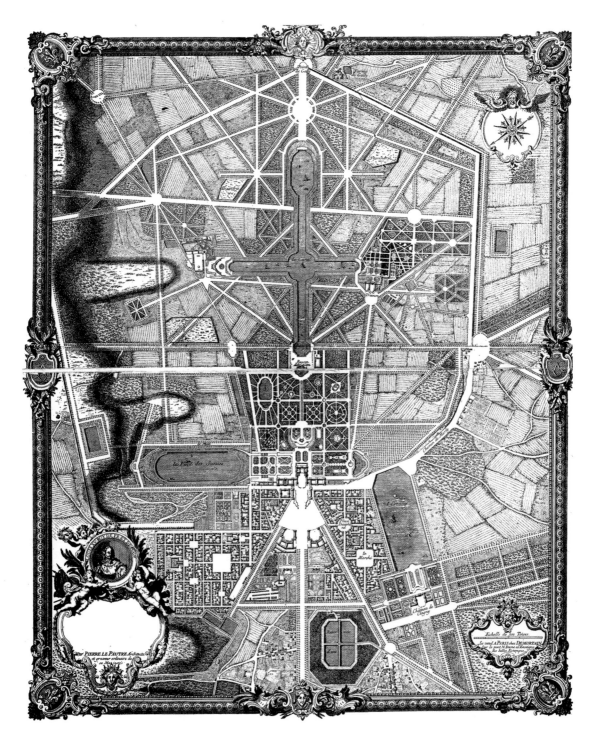

凡尔赛宫苑平面（Marie Luise Gothein, 1928年）
Plan

凡尔赛宫前广场（当地提供）
Front square (Courtesy of the local institution)

中轴线景观
Central axis

下篇：外国建筑科学文化精品图说/法国

从园林望宫殿中心
Center of the palace

拉托那雕像水池（从西向东看）
Pool of Latona Statue

阿波罗雕像水池
Pool of Apollo Statue

下篇：外国建筑科学文化精品图说/法国

从东向西望皇家大道
Royal Avenue

东部大运河水面
Water of the Grand Canal

小特瑞安农
居室及其前面田园景观（左）
Petit Trianon Garden and the farm

小特瑞安农牛奶场景色（右）
Milk house

实例123　巴黎卢浮宫

卢浮宫始建于公元13世纪，当时是为了防御而建造的一座城堡。到了14世纪，查理五世将它改建为王宫，但又相当长时间没有作王宫使用。直到16世纪、17世纪昂利二世、昂里四世、路易十三和路易十四都对此处不断改建、扩建，形成"方形庭院"，并收藏着各种艺术品。到路易十四去世时，卢浮宫已成为各种绘画和雕塑作品展示的场所。到18世纪末，它又作为博物馆对外开放。如今，卢浮宫已成为巴黎市东西向中轴线的起点。20世纪80年代，卢浮宫因扩大规模需要而改建，后来法国总统密特朗选用了美籍华裔著名建筑师贝聿铭先生的金字塔式扩建设计方案。卢浮宫内馆藏的维纳斯雕像和《蒙娜丽莎》画像等经典作品吸引着全世界的观众前来观赏。

Example 123　Louvre Palace

At first, it was a castle built for defense in the 13th century, and then it was continuously rebuilt, becoming a place where paintings and sculptures were often exhibited, and then opened to the public as a museum at the end of the 18th century. In 1980s, French President Mitterrand chose the pyramid expansion design of famous Chinese-American architect I. M. Pei for the need of expanding the scale. The statue of Venus and the portrait of Mona Lisa attract a large number of visitors.

卢浮宫扩建工程玻璃金字塔外景
Exterior of the glass pyramid

下篇：外国建筑科学文化精品图说／法国

从玻璃金字塔入口进入室内
Interior when entering from the pyramid

展出的维纳斯雕像
Statue of Venus

《蒙娜丽莎》
Mona Lisa

实例124　巴黎杜伊勒花园

杜伊勒花园位于卢浮宫西面，其西端连接着协和广场。经过几个世纪的花园建设，到了17世纪路易十四时期，由勒·诺特进行改、扩建，将杜伊勒花园与卢浮宫结合为一体。杜伊勒花园有一突出的中轴线，在此轴线或两侧布置喷泉、水池、花坛、雕像等，轴线东端正对着卢浮宫建筑，以突出王宫，体现着君王的威严。这一中轴线后来向西延伸，并成为巴黎的城市中轴线而闻名于世。杜伊勒花园采用下沉式（Sunkun），既扩大了视野，又可减少周围环境对花园的干扰。1806—1808年，在杜伊勒花园东端建起一座卡鲁塞尔凯旋门，它模仿罗马君士坦丁凯旋门，其拱门上方还镌刻着纪念拿破仑战绩的浅浮雕。

Example 124 Tuileries Garden

The garden is integrated with the Louvre and has a prominent central axis, on which fountains, pools, flower beds and statues are arranged or on both sides. The east axis is facing the architectural center of Louvre, which highlights the palace and embodies the majesty of kings. This central axis later extended westward and became the central axis of Paris, which is famous all over the world.

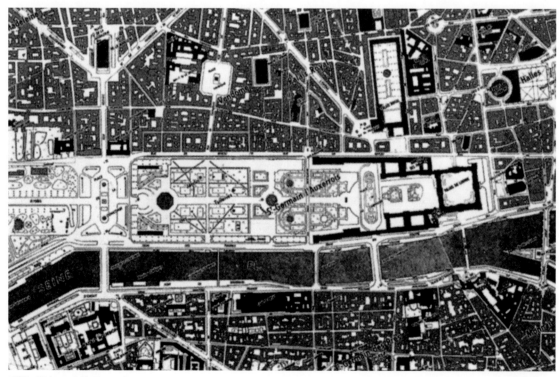

平面（Edmund N. Bacon）
Plan

现状鸟瞰（Yaun Arthus-Bertrand）
Aerial view

大水池
Big pool

实例125 巴黎协和广场

协和广场紧临杜伊勒花园，坐落在卢浮宫、杜伊勒花园向西延长的东西向中轴线上，19世纪30—40年代改建成了今天的模样。广场中央耸立着埃及卢克索神庙中的方尖碑，这是穆罕默德·阿里送给法国路易菲力普的礼物。方尖碑高23m，碑身刻着歌颂拉姆西斯二世法老光辉业绩的文字。在方尖碑的南北两侧还建有两座仿造罗马圣彼得教堂前广场的雕塑喷泉。

Example 125 Concorde Square

In the center of the square stands the obelisk of Luxor Temple in Egypt, with a height of 23 m. The obelisk is engraved with words praising the glorious achievements of Pharaoh Ramses II. Located on the north and south sides of the obelisk, there are two sculpture fountains imitating the square in front of St. Peter's Cathedral in Rome.

从东向西望，中为方尖碑，左为铁塔
From east to west, the obelisk is in the middle and the iron tower is on the left

从北向南望，左边建筑为塞纳河南岸的布尔绷宫（现为国民议会会址）
From north to south, Palais Bourbon is on the left

实例126　巴黎明星广场凯旋门

从协和广场沿中轴线西行，在香榭丽舍大街西端的沙右山丘上矗立着一座雄伟的建筑，它就是世界闻名的明星广场凯旋门。这座凯旋门是奉拿破仑命令用来纪念法国大军的，建于1836年。其规模宏大，高49.4m、宽44.8m、厚22.3m，墙上的浮雕尺度巨大，而且造型简洁。1920年在拱洞下还建了"无名战士墓"，每到傍晚这里便燃起不灭的火焰，以示悼念。

Example 126 Arch of Triumph

At the west end of the Champs Elysées stands the world-famous Arc of Triumph. It was built in 1836 by Napoleon's order to commemorate the French army. It has a large scale, with a height of 49.4 m, a width of 44.8 m and a thickness of 22.3 m. In 1920, "Tomb of Unknown Soldier" was built under the arch cave. Every evening, an immortal flame ignited here to show mourning.

凯旋门的正立面
The front of the Arc of Triumph

实例127　巴黎蓬皮杜文化中心和"中国建筑、生活、环境展览"

蓬皮杜文化中心是法国总统蓬皮杜决定在巴黎市中心兴建的一座重要文化建筑，该建筑设计面向全世界招标，最后著名建筑师皮阿诺和罗杰斯的设计方案被选中。工程于1977年建成，建筑外部像是一部机器，被漆成不同颜色，代表着不同的功能，其中蓝色是电气设施，红色为运输线，绿色是水处理系统。建筑内部为大柱网，其使用功能可灵活变动。由于建筑体量、建筑高度的控制，它与周围环境和谐统一。此建筑已选入《20世纪世界建筑精品集锦》环地中海地区卷中。

由中国建筑学会和法国巴黎蓬皮杜文化中心工业创作中心合办的"中国建筑、生活、环境展览"，于1982年5月18日在巴黎蓬皮杜文化中心隆重开幕。以中国建筑学会副理事长阎子祥为团长的中国建筑代表团出席了开幕式。我国驻法大使姚广等同志也出席了开幕式，并发表了讲话。据文化中心负责人拉霍什（Larroche）夫人讲，在蓬皮杜文化中心举行这类展览的开幕式中，本次展览是人数最多、气氛最为热烈的一次。这次展览使法国人民了解到广大中国人民的生活和城乡建设情况，它增进中法人民之间的友谊，同时也为中法文化交流开辟了一条民间交往之道。

蓬皮杜文化中心外景　Exterior

我国驻法大使姚广在中国展览开幕式上致辞
Yao Guang, Chinese Ambassador in French, is giving a speech at the opening ceremony

阎子祥先生（右一）在"中国建筑、生活、环境展览"开幕式酒会上
Mr. Yan Zixiang (first on the right) in the drinking party

Example 127 National Art and Cultural Center of Georges Pompidou and the Exhibition of Chinese Architecture, Life and Environment

The exterior of the building of the center is like a machine, painted in different colors to represent different functions, and the interior is a large column grid, which can flexibly change its use functions.

The Exhibition of Chinese Architecture, Life and Environment, was grandly opened on May 18, 1982, jointly organized by Architectural Society of China and National Art & Cultural Center of Georges Pompidou. This exhibition is to enable the French people to understand the living conditions and buildings of China, and enhance the friendship of the two countries.

"中国建筑、生活、环境展览"展品内容

城市生活图片展
Exhibition of city life

城市规划建设图片展
Exhibition of master plan

农村居室实物展
Exhibition of rural household goods

苏州网师园模型
Model of Wangshi Garden in Suzhou

实例128 巴黎新歌剧院

它坐落在象征法国革命的著名巴士底广场上，新歌剧院内有一个可容纳2700人的观众厅。该建筑设计方案通过国际设计竞赛选出的，从设计到建成是在20世纪80—90年代。其建筑形式注重体量分割与对比，以此来同原有街区、广场相适应，建立起一种整体体系的概念。新歌剧院外的巴士底纪念柱，建于1830年，它是为纪念1789年7月14日法国人民攻占巴士底狱而建的烈士碑。这是一座高52m的青铜柱，柱顶立一金翅自由神像，左手提着砸断的锁链，右手高举火炬。1880年6月，法国国民议会把巴黎人民攻占巴士底狱的日子——7月14日定为法国国庆日。

Example 128 Opéra de la Bastille

It is located in the Bastille Square, which has an audience hall that can accommodate 2,700 people. The Bastille Memorial Column outside the theater is a martyr monument built in 1830 to commemorate the French people's capture of the Bastille on July 14, 1789. It is a 52 m-high bronze column with a golden-winged statue of freedom standing on the top of the column, carrying a broken chain in the left hand and holding the torch high in the right hand.

外景
Exterior

门廊上部 Upper part of the porch

门厅 Lobby

观众厅一角 A part of the audience hall

巴士底纪念柱 Bastille Memorial Column

实例129　法国国家图书馆

　　法国国家图书馆位于巴黎东部13区，总建筑面积35万m²，1995年3月建成，这是密特朗最后一个也是最重要的工程。其建筑的特点是没有界限，它完全融于城市中，并向所有人开放空间。国家图书馆中心是座大花园，其建筑有如4本展开的书，它是图书馆的象征。设计人是法国著名建筑师多米尼克·佩罗。此建筑获法国国家建筑大奖，并已被选入《20世纪世界建筑精品集锦》环地中海地区卷中。

Example 129　National Library

　　It is located in the 13th District in the east of Paris, with a total construction area of 350,000 m². The characteristic of the building is that it has no boundaries and is completely integrated into the city. It is a space open to all, with a large forest-like garden in its center. The whole building is like four unfolding books, surrounding this natural space. This building won the French National Architecture Award.

外景
Exterior

2. 杜关市

实例130 弗雷斯诺国家当代艺术学校

它建在法国里尔城附近的杜关市代斯卡特大街，是法国著名建筑师屈米的一个新建筑创作。该地段原有20世纪20年代建造的综合娱乐中心，包括电影院、舞厅和转马亭建筑，设计人将这些原有建筑保留，采用一个新屋顶盖在新、旧建筑之上，新屋顶高出老屋顶，并设有云状的玻璃天窗，建筑自然采光、自然通风。设计师富有诗意地把新、老建筑组合为一个整体。

5.2 Tourcoing

Example 130 Le Fresnoy National Contemporary Art Center

The original comprehensive entertainment center built in the 1920s, including cinemas, dance halls, etc., was retained by the designer, and a new roof was used to cover the new and old buildings. The new roof was higher than the old one, with cloud-shaped glass skylights, natural lighting and natural ventilation, and played a poetic role in combining the new and old buildings into a whole.

带玻璃天窗的大屋顶，左侧为原有建筑
The roof with glass skylights and the original building on the left

3. 里尔市

实例131　里尔美术馆

这是一个扩建工程。原里尔老美术馆建于1895年，位于里尔市中心，是一座古典复兴式建筑，庄重严整。该建筑扩建时，首先清理和恢复原有建筑，并在其对面采用新技术建一附有薄片的建筑。其薄片立面上的透视景观由几层竖向材料组成，表面为透明玻璃，中间的小镜片布置成方格网，镜面映出老美术馆的形象。这一景象就成为新与旧的结合点，其创新手法得到大家的肯定。1997年该项目建成，并获法国银角奖，设计人是法国JM伊宝和M·魏达尔。

5.3 Lille

Example 131 Palais des Beaux-Arts

This is an expansion project. The former one was built in 1895, located in the center of Lille. It is a classical renaissance building, solemn and neat. Then a new building with a thin slice was built on the opposite side. The perspective landscape on the thin slice facade is composed of several layers of vertical materials, the surface is transparent glass, and the small lens in the middle is made into a square grid, reflecting the impression image of the old art museum on the mirror surface. This scene becomes the junction between the new and the old.

老美术馆外景
Exterior of the old art museum

新建筑镜面映出老美术馆印象形象
The reflection of the old art museum on the new building

六、西班牙

15世纪末,西班牙完成了国家的统一。16世纪初,查理一世(1516—1556年)当选为罗马帝国皇帝,并在意大利战争中打败法国。16世纪20至30年代,西班牙侵占美洲、北非一些地方,成为殖民大帝国。菲力普二世(1556—1598年)于1588年派舰队远征英国,并在英吉利海峡遭遇失败,从此,海上霸权让位给了英国,西班牙逐渐衰落。

1. 马德里

马德里是西班牙的首都,也是欧洲著名的历史文化名城。它地处梅塞塔高原,海拔670m,是欧洲地势最高的首都。1561年西班牙国王菲力普二世迁都到这里,19世纪时发展为大城市。马德里市内名胜古迹荟萃,旧城得到完整地保护,新区又紧邻旧城发展,两者相得益彰。

6. Spain

At the end of the 15th century, Spain completed the national unification. During the 1620s and 1630s, it became a colonial empire. In 1588, it failed in the expedition to UK. Since then, maritime hegemony has given way to UK, and the country has gradually declined.

6.1 Madrid

Madrid is the capital of Spain and a famous historical and cultural city in Europe. It is the capital with the highest topography in Europe. The city is full of scenic spots and historical sites, the old city is completely protected, and the new district is close to the old city.

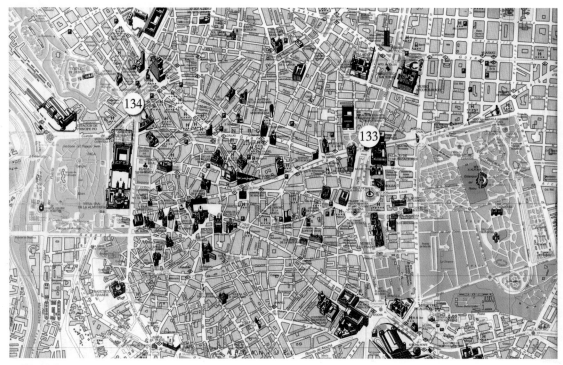

马德里城中心区平面　Plan of the central city

实例132 马德里埃斯库里阿尔宫

该宫建于1563—1584年，位于马德里城西北48km处。从西门正中进入，院南是修道院，院北是神学院和大学，往北正中是教堂。教堂地下室为陵墓，教堂北部突出部分是皇帝的居住之处。埃斯库里阿尔宫的平面布局，是从大的分区考虑的，由不同的大小院落组成建筑群，其中轴线凸出、规整有序，且利于采光通风。该建筑物全部采用大块花岗石、青瓦等材料，墙面简洁，酷似文艺复兴前期式建筑，其教堂的穹顶和四角的尖塔组成有气势的造型。该建筑的庄严雄伟曾轰动了欧洲宫廷，后路易十四在建造凡尔赛宫时明确了一定要超过埃斯库里阿尔宫。

Example 132 EI Escorial Palace

The palace is located 48 km northwest of Madrid. There is a monastery in the south, a seminary and a university in the north, a church in the center of the north, and a tomb in the basement of the church. The prominent part in the north of the church is the residence of emperors. The central axis is convex, regular and orderly, and conducive to lighting and ventilation. This building is made of large granite and blue tiles, with simple walls and looks like a pre-Renaissance building.

埃斯库里阿尔宫外景 Exterior

实例133 交通部门大楼

交通部大楼坐落在阿尔卡拉（Alcala）和Paseo街交叉口Cibeles广场的东南角上，其庄严壮观的外立面形成了纪念性风格，它也是19世纪政府各部门大楼的一个范例。在此建筑前面，即广场的中心有一座美丽的喷泉雕像，只见Goddess Cybele坐着车，由两头狮子牵拉，其形象生动。每当夜晚灯光照亮，雕像晶莹辉煌，这组喷泉雕像景观已成为马德里的象征。

Example 133 Transport Department Building

Located at the southeast corner of Cibeles Square at the intersection of Alcala and Paseo Street, the commemorative style formed by its spectacular and solemn facade is an example of various government buildings in the 19th century. In front of this building, there is a fountain statue. Goddess Cybele is pulled by two lions in a car. The image is vivid, illuminated by night lights and brilliant.

大楼夜景
Night view

实例134 马德里西班牙广场

马德里西班牙广场位于旧城的西北角处，广场中耸立着巍峨的塞万提斯纪念碑，碑前有堂吉诃德和桑科·潘扎的塑像。该塑像前为一方形水池，两侧有郁郁葱葱的树木，池中倒映着生动的塑像，构成了一幅优美的画面。

Example 134 Plaza de Espana

It is located in the northwest corner of the old city and the towering Cervantes Monument stands in the square with statues of Don Quixote and Sancho Panza in front of it. In front of the monument is a square pool with lush trees on both sides, and vivid statues are reflected in the pool, forming a beautiful picture.

塞万提斯纪念碑
Cervantes Monument

2. 巴塞罗那

实例135　巴塞罗那绿地系统和旧城

巴塞罗那是西班牙的历史文化古城和最大海港城市，位于东北部地中海岸。巴塞罗那市区依山傍海，海滩宽阔，已建成较完善的绿地系统。在巴塞罗那旧城区内，有古罗马城墙和公元6世纪的宫殿遗址。在东部中心矗立着哥特式天主教大教堂，巴塞罗那守护女神圣欧拉利亚的墓棺就放在教堂的地下室内。东部还保存着11—12世纪建起的堡垒式宫殿和雕刻精美的过街廊等。西部住区庭院保存着西班牙传统建筑的特色，最突出的是南北向的兰布拉斯（Rambles）大街，它连接着北面加泰罗尼亚广场，兰布拉斯大街现已成为闻名欧洲的步行街。在19世纪城市扩展时，这个巴塞罗那旧城被完整地保存下来。

6.2 Barcelona

Example 135　Green Space System and Old City

Barcelona is an ancient historical and cultural city and the largest seaport in Spain. It is located on the Mediterranean coast in the northeast, and a relatively complete green space system has been built. In the old city, there are ancient Roman walls and 6th century palace ruins. The Gothic Catholic Cathedral stands in the center of the east. Courtyards in western residential areas preserve the characteristics of traditional architecture. The most prominent street is Rambles Street, which has become a famous pedestrian street in Europe.

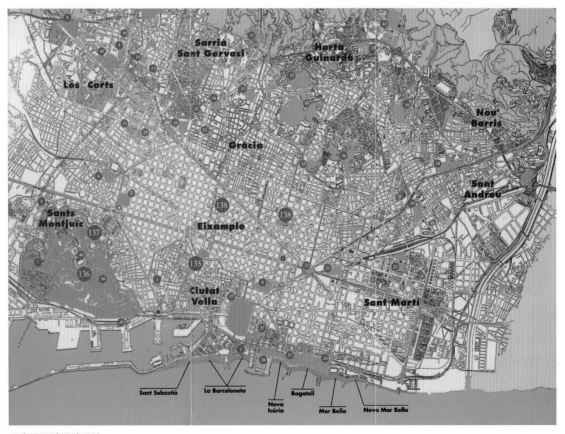

巴塞罗那城绿地系统
Green space system

旧城西部住区庭园
Courtyard in residential area in the old city

旧城东部雕刻精美的过街廊
Elaborately carved cross-street corridor in the old city

旧城兰布拉斯步行街
Rambles pedestrian street in the old city

实例136　巴塞罗那蒙特胡依克山

这里是个山丘，高213m，位于巴塞罗那旧城西面，最早称为朱匹忒（古罗马主神）山。17世纪在巴塞罗那蒙特胡依克山上建有城堡。1929年为举办国际博览会，在其山丘北边建起博览会艺术品陈列馆等。1992年为举办世界奥运会，又在其山丘中心部位修建了奥林匹克运动场等各项设施。此外，山丘上还建有植物园、游乐场以及高架缆车等，如今此处已是巴塞罗那最吸引人的一处观光游览地点。

Example 136 Montjuïc Hill

The hill is 213 m high and is located in the west of the old city of Barcelona. In the 17th century, there was a castle. In 1929, an art gallery was built on the north side of the hill for the International Expo. In 1992, an Olympic stadium was built in the center of the hill. In addition, there are botanical gardens, amusement parks and elevated cable cars on the hills.

下篇：外国建筑科学文化精品图说/西班牙

从博览会艺术品陈列馆向北望博览会会场及城市景观
Expo site and city landscape from the view of the gallery of artwork

博览会艺术品陈列馆
Art gallery

实例137　巴塞罗那国际展览会德国馆

这个馆在现代建筑史上有一定的重要作用，它是世界著名建筑师密斯·凡·德·罗的一个杰作。1929年修建在Montjuïc山脚，即国际博览会会场中轴线的西侧。展览馆规模为50m×25m，建筑采用钢结构，由8根钢柱支撑屋面。建筑与一大一小两个方形水池结合，小水池是露天的，它与室内联系紧密，其端部立着一尊人体雕像，整个室内空间是流动和变化的。此建筑室内外空间设计对现代建筑设计产生很大影响，以致1986年又在原址修建了一个原样复制品。该项目也已被选入《20世纪世界建筑精品集锦》环地中海地区卷中。

Example 137　German Pavilion of the International Expo

The exhibition hall is 50 m × 25 m in size. The building adopts steel structure, and the roof is supported by eight steel columns. The building is combined with two square pools, one large and one small. The small pool is open to the air, but it is enclosed in the building and closely connected with the interior. A guiding human body statue stands upright at its end, and the whole indoor space extends and changes with each other, resulting in the effect of "flowing space".

室内小水池与人体雕像
Small pool and the human body statue

流动的室内外空间设计
Inner space from the big and small pools

实例138　巴塞罗那高迪宅邸建筑、圣家族教堂

高迪是世界著名的建筑师，1852年在巴塞罗那出生，1878年毕业于巴塞罗那一所建筑学校，1926年因车祸逝世。高迪的40多年建筑创作给巴塞罗那留下了许多建筑佳作。下面介绍他的两座建筑。

米拉宅邸（Casa Milà）位于巴塞罗那市中心区Passeig de Gracia街的东面，1905年开始建造，1910年竣工。高迪的建筑创作理念是"自然"，即创造由曲线、曲面空间组成的"自然式建筑"。在建筑的入口、阳台、屋檐线条、烟道和室内装饰方面，高迪都重视同当地传统文化结合，并强调创新。米拉宅邸的出现让人焕然一新，这就是高迪的"自然风格"。

圣家族教堂（Sagrada Familia）具有8个高耸入云的圆形塔，它已成为巴塞罗那的标志性建筑，也是高迪创作规模最大、时间最长的一幢代表作品。此教堂矗立在巴塞罗那市中心，其周围建筑低平，而教堂的塔尖轮廓线却格外突出。教堂的曲线、曲面空间、高大塔尖雕刻装饰等体现着高迪建筑设计的自然风格。在色彩运用方面，圣家族教堂在其内部的右边采用白色和金色，以象征快乐；在其内部的左边则使用青莲色和黑色，以象征忧愁。高迪去世时，圣家族教堂尚未完工，1935年又因战争遭到破坏，但其后人还是继续修复、建造。高迪设计的米拉宅邸、巴特娄宅邸和圣家族教堂这三幢建筑皆被选入《20世纪世界建筑精品集锦》环地中海地区卷中。

Example 138 Residence Building of Antoni Gaudi and Sagrada Família

Antoni Gaudi is a world famous architect. During his 40 years of architectural creation, he left many outstanding architectural works for Barcelona.

Casa Milà is located in the downtown area. The architectural concept is "nature", which is composed of curve and surface space. Later, Casa Batlló was built. Sagrada Família, with eight towering circular towers, stands in the center of the city, and the surrounding buildings are low and flat, which makes the contour line of the spire stand out. In terms of color application, inside the church, white and gold are used on the right to symbolize happiness, while violet and black are used on the left to symbolize sorrow. When Gaudi died, the building had not been completed.

米拉宅邸外景
Exterior of Casa Milà

米拉宅邸局部
Part of Casa Milà

巴特娄宅邸外景
Exterior of Casa Batlló

圣家族教堂外景
Exterior of Sagrada Família

3. 格拉纳达

实例139　阿尔罕布拉宫苑

这座规模宏大的宫苑具有典型的西班牙伊斯兰式风格，其形式的形成要从伊斯兰教传入西班牙说起。公元711年阿拉伯人和摩尔人（摩尔人是阿拉伯和北非游牧部落柏柏尔人融合后形成的种族）通过地中海南岸侵入西班牙，占领了比利牛斯半岛的大部分。13世纪末，西班牙收复失地运动大体完成，而阿拉伯人只剩下位于半岛南部一隅的格拉纳达王国据点。1492年格拉纳达王国据点被西班牙收复，从此，西班牙结束了长达7个世纪被阿拉伯人占领的历史。

阿尔罕布拉宫苑建于公元1238—1358年，位于格拉纳达（Granada）城北面的高地上。此宫苑建筑与庭院把阿拉伯伊斯兰式和希腊、罗马式中庭（Atrium）有机结合在一起，并创造出西班牙式的伊斯兰宫苑，即帕提奥式。下面着重介绍此宫苑的特点。

（1）这组建筑是由4个"帕提奥"和一个大庭院组成的。"帕提奥"（Patio）的特征是：①建筑位于四周，围成一个方形的庭院，建筑形式多为阿拉伯式的拱廊，其装修雕饰十分精细；②位于中庭的中轴线上，有一方形水池或条形水渠或水池喷泉。在夏季炎热干燥地区，水极为宝贵，让人凉爽湿润；③在水池、水渠与周围建筑之间种以灌木、乔木，其搭配因地制宜；④周围建筑多为居住之所。此宫庭院的4个"帕提奥"具有上述的全部特征，十分典型。

（2）桃金娘庭院（Court of the Myrtle Trees）。这个庭院是45m×25m，正殿是皇帝朝见大使举行仪式之处，庭院南北向，柱廊是由白色大理石细柱托着精美阿拉伯纹样的石膏块贴面的拱券，轻快活泼，建筑倒影水池之中，形成恬静的庭院气氛。在水池两侧种植两条桃金娘绿篱，故此处名为"桃金娘"庭院。

（3）狮子院（Court of Lions）。此院为30m×18m，它是后妃的住所。此庭院属于十字形水渠的"帕提奥"类型，其水渠伸入四面建筑之内，在水渠端头还设有阿拉伯式圆盘水池喷泉，可使室内清凉。庭院四周由124根细长柱拱券廊围成，其中柱有3种类型，即单柱、双柱和三柱组合式，显得十分精美。最突出之处是在庭院的中央，立一近似圆形的十二边形水池喷泉，其下为12个精细石狮雕像，水从喷泉流下并连通十字形水渠。此石狮喷泉成为庭院的焦点，并形成高潮，故此院名为狮子院。原院中种有花木，后改为砂砾铺面，这更加突出石狮喷泉的形象。

（4）林达拉杰花园（Limdaraja garden）。从狮子院往北，即到此园，这处后宫属于中心放置伊斯兰圆盘水池喷泉的"帕提奥"类型。环绕中央喷泉布置规则式的各种花坛，花坛以黄杨绿篱镶边。在此花园西面还有一柏树院（Cypress court），它是17世纪时扩建的，同样属于"帕提奥"类型。

（5）帕托园（Partle garden）。这一花园不属于"帕提奥"型，它是一台地园，下面有一大水池。沿其轴线的台地上有一水渠，渠水沿台阶流下，同池水相连。在此眺望台下，视野开阔，景色尽收眼底。

6.3 Granada

Example 139 Alhambra Palace

Built in 1238 to 1358, it is located on the highland in the north of Granada, creating a Spanish-style Islamic palace, which is called "Patio".

(1) "Patio" is characterized by: ① The buildings are located around and form a square garden. Most of the buildings are Arabic-style arcades, and their decoration and carving are very fine. ② On the central axis of atrium, there is a square pool or strip channel or pool fountain. It feels cool and humid in summer. ③ Shrubs and trees are planted between pools, canals and surrounding buildings. ④ The surrounding buildings are mostly residential places, and in some places, several gardens are organized together.

(2) Court of the Myrtle Trees: The main hall is where the emperor meets the ambassador for a ceremony. Two myrtle hedges are planted on both sides of the pool.

(3) Court of Lions: This courtyard is the residence of the queen and maids, and is surrounded by arch corridors of 124 slender columns. In the center, there is a dodecagonal pool fountain with 12 stone lion statues under it.

(4) Limdaraja Garden: A variety of regular flower beds are arranged around the central fountain, and in the west, there is a Cypress Court.

(5) Partle Garden is not a garden surrounded by buildings. There is a large pool below it, and a canal on the terrace along its axis. The water is connected with the pool water along the cascade flow.

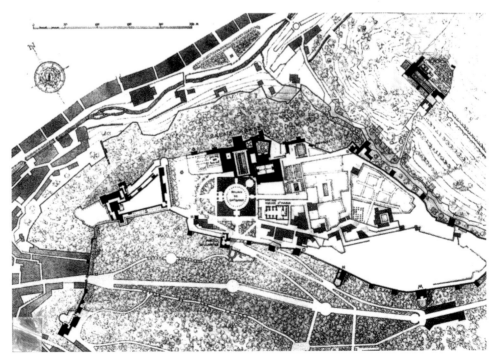

总平面 Site Plan

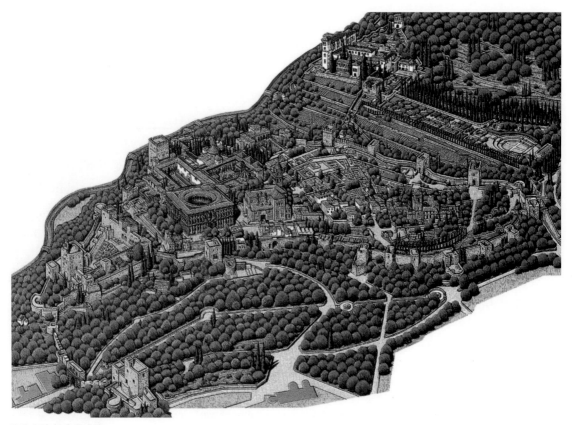

阿尔罕布拉宫苑鸟瞰
Aerial view

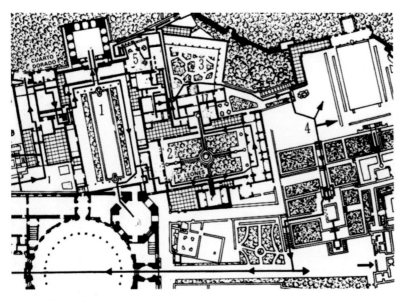

阿尔罕布拉宫苑平面
1-桃金娘庭院；2-狮子院；3-林达拉杰花园；4-帕托花园；5-柏树庭院
Plan
1-Court of the Myrtle; 2-Court of Lions; 3-Limdaraja Garden; 4-Partle Garden; 5-Cypress Court

桃金娘庭院
Court of the Myrtle Trees

帕托花园
Partle Garden

狮子院水渠伸入建筑中
Canal in the building

实例140 吉纳拉里弗宫苑

此园位于阿尔罕布拉宫东面,顺沿阿宫城墙左转即可到达。其特点是:

(1) 这处庭园比阿尔罕布拉宫高出50m,可纵览阿尔罕布拉宫周围景色,它们形成互为对景的关系。

(2) 在进入主要庭院之前,布置有一长条形花园和水池喷泉,其层次丰富、色彩鲜艳。此园具有明显的导向性。

(3) 从条形花园北端的庭院中再转一个方向,就进入这里的主庭园,它是一个中庭为条形水池的"帕提奥"式园林。其主体建筑为拱廊托起的二层坡顶楼,条形水池纵贯全院,池边喷出拱状水柱,两侧配以花木。在阳光照射下,这里五彩缤纷、灿烂夺目。

1996年7月,中国建筑师代表团在参加第19届国际建筑师协会大会之后,专程到格拉纳达城观看了这两个著名的历史宫苑。

Example 140 Generalife Garden

Features include:

(1) It is higher than the Alhambra Palace that it has an overview of the scenery around the Palace.

(2) Before entering the main courtyard, there is a long yard with a pool fountain of rich layers.

(3) From the courtyard at the northern end of the long yard, there is the main garden of typical "Patio" style. The main building is of a two-floor slope roof supported by the arch. The strip-shaped pool runs through the whole garden, with arched water column ejected from the pool and flowers and trees on both sides.

条形花园大水池喷泉 Pool fountain at the long yard

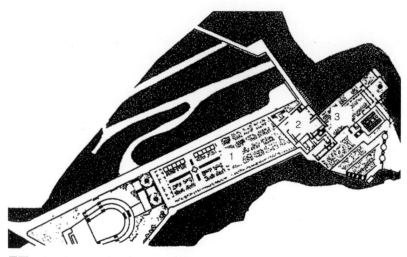

平面 1-条形花园；2-转折处庭院；3-主庭园
Plan 1-Long yard; 2-Courtyard at the turning point; 3-Main garden

进入主庭园 Entering the main garden

从主体建筑拱廊望主庭园 Main garden from the view of the main building

七、英 国

1. 伦 敦

伦敦是英国的首都，始建于公元43年，11世纪已发展成为商业和政治中心。伦敦市内历史文物古迹众多，但其整体保护与新旧建筑的结合还不如罗马和巴黎。伦敦市中心区完整地保存着5座原有的皇家贵族园林，这些皇家贵族园林后来都对公众开放，并成为城市公园。这5座公园位于泰晤士河北面，它们分别是摄政公园、肯辛顿花园、海德公园、格林公园和圣·詹姆斯公园，它们像绿宝石一般镶嵌在伦敦城市中心地区。

实例141　伦敦国会大厦

它是伦敦象征性的标志性建筑，矗立在泰晤士河畔。从11世纪到16世纪初，它是国王的宫室，后逐步供上议院、下议院使用。此处屡遭火灾，现存建筑是1840—1865年重建的。大厦立面长280m，是世界上少有的宏大哥特式建筑群。大厦西南角有高104m的维多利亚塔楼，东北角立有高98m的方塔钟楼。此钟楼即世界闻名的威斯敏斯特钟楼，始建于1859年，顶部装有重21t的特大时钟。时钟有4个面，钟面直径近7m，原为机械转动，1909年改为电动，1923年英国广播公司向全世界播送钟声。

沿泰晤士河全貌　View of Thames

7. United Kingdom

7.1 London

London, the capital of Britain, was founded in 43. There are many historical relics and historic sites in the city. Five original royal noble gardens are completely preserved in the downtown area, and then opened to the public as parks.

Example 141 Parliament Building

Standing on the bank of the Thames, it was a royal palace from the 11th century to the early 16th century, and then gradually changed to the upper and lower houses. There is a Victoria Tower with a height of 104 m in the southwest corner of the building, and a square tower bell tower with a height of 98 m in the northeast corner. This bell tower is the world-famous Westminster Bell, which weighs 21 tons and has a diameter of nearly 7 m. In 1923, the BBC broadcast the bell to the whole world.

2. 布赖顿

实例142　生活环境

布赖顿位于英国南部海滨，是一座海滨疗养城市。1987年7月我们来到布赖顿出席国际建筑师协会（UIA）第16次大会。这里原是渔村，后因有医生赞美此地海水、空气有益健康而出名。1783年，威尔士王子等前来考察建设项目，此后这里逐步成为世界闻名的海滨旅游疗养胜地。该处海滨建有东西两条伸入近海的长堤，音乐厅、剧场、娱乐宫等布置其中。城区街道整齐有序、安静宜居，并富人情味。1787年，威尔士王子修建了一座皇家宫殿，1815年又将其重修并增加东方建筑的特点。

7.2 Brighton

Example 142 Living Environment

Located on the south coast of England, It used to be a fishing village, and has gradually become a famous seaside resort. On the seashore, there are two long dikes extending into the offshore, where the concert hall, theater and entertainment palace are arranged. Urban streets are orderly, quiet and livable, and full of human touch. In 1787, the Prince of Wales built a royal palace, and in 1815, it was rebuilt to add the characteristics of oriental architecture, and the interior decoration also has Chinese style.

伸入近海的长堤
Long dike

城市住区一角　Part of the residential area

住区中公共活动生活广场　Square of public activity

1987年正在修缮的皇家宫殿　Royal palace under construction in 1987

八、爱尔兰

都柏林

实例143 "三一学院"

都柏林是爱尔兰的首都。"三一学院"位于都柏林的市中心，是爱尔兰最著名的高等学府。该学院创建于公元17世纪，由两个庭院组成，中间立一高耸钟楼，并将议会广场与图书馆广场分开。议会广场的主要建筑为4层，内有公共剧场和小礼拜堂。"三一学院"图书馆是爱尔兰最大的图书馆，存放有历史上著名的手稿和早期出版的书刊。这座学院的建筑特点是：整体布局严整、重视保留和尊重历史建筑、新旧建筑协调，并能融进摩尔的现代艺术雕塑。同时，它还重视建筑与自然的结合。

1987年7月国际建筑师协会（UIA）第十七次代表会就在这里召开。

8. Ireland

Dublin

Example 143 Trinity College

Dublin is the capital of Ireland. Trinity College, located in the center of the city, was founded in the 17th century. It consists of two courtyards, with a bell tower in the middle separating the Parliament Square from the Library Square. The main building of Parliament Square is four floors, with public theater and chapel. The library is the largest library in Ireland, storing famous manuscripts and early published books and periodicals in history.

入口（原摄政及议会用房，左为议会广场，现总称前广场）
Entrance, Parliament Square on the left

钟楼(后面为图书馆广场,右面为图书馆)
Bell tower, the library on the right

学友广场(左为老图书馆,中为新图书馆)
Fellows Square, old library on the left and new library on the right

实例144　都柏林圣·斯蒂芬公园

　　该公园位于都柏林的市中心区，在"三一学院"的南面。该园用地为长方形，但内部布局却十分自然。圣·斯蒂芬公园花草树木繁茂，湖面自由开敞，还有瀑布、建筑点缀其间，并创造出有层次的景观。成年人常到这里的草坪上晒太阳，少年儿童常到这里游戏或组织各项活动。在都柏林市中心地区还保留这样一块规模较大的绿地真是难得，这是值得称赞的。

Example 144 St. Stephen's Park

　　It is located in the downtown area. The land is rectangular while the internal layout is very natural. Flowers and trees are luxuriant, the lake is free and open, and there are waterfalls dotted with buildings to create a layered landscape. People often come for leisure and activities. It is rare to keep such a large green space in the downtown area for a long time.

少年儿童游玩在自然环境里
Kids are playing in the natural environment

大草坪与花卉景色
Grassland and flowers

观景建筑
Sight-view building

园中雕塑装饰
Statues

九、德 国

实例145 法兰克福

法兰克福位于莱茵河支流的美因河河畔，是德国的一个工商业大城市。它建于公元1世纪，19世纪中叶前已成为商业城市，后开始发展工业，并以化学工业为主。现除工商业发达外，法兰克福还是德国的一个金融中心和重要的航空港。1987年7月我们到达法兰克福，重点观看了其旧城的一个活动中心——罗马人广场。罗马人广场原有建筑保存完好，这里过去是罗马皇帝在此行加冕礼的地方，主要建筑有旧市政厅、大教堂等，罗马人广场中立有正义女神雕像。法兰克福旧城虽不大，历史建筑古迹也很少，但这里的罗马人广场已成为市民最喜爱和常到的地方，也是外地游人们必到之处。

9. Germany

Example 145 Frankfurt

Located on the Main River, it is a big industrial and commercial city in Germany. It is also a financial and economic center and an important airport in Germany. The main activity center of the old city is Roman Square. The original building of the square is well preserved. It used to be the coronation place of the Roman emperor. The main buildings include the old city hall, cathedral, etc. There is a statue of the Goddess of Justice in the square.

美因河畔夜景
Night view of Main River

罗马人广场入口
Entrance of Roman Square

美因河畔新建筑天际线
Skyline of the new buildings along the Main River

十、比利时

1.安特卫普

安特卫普位于斯凯尔特河下游,距北海岸86km,它是比利时最大的港口城市。公元13—16世纪,安特卫普为欧洲的一个重要贸易中心,后来衰落,直到19世纪中叶才逐渐复兴并成为现代化的港口城市。传说古罗马侵略者唆使一个巨人来此掠夺财富,后被一位名为布拉伯的勇士战败。该勇士断了巨人的左臂,并抛进河中,该城市据此称为"安特卫普",这是弗拉芒语"抛断手臂于河中"的意思。现市政厅前广场中心的雕像,就是一个勇士屹立船头,手举断臂并正在抛掷,一巨人伏其脚下,即是这个传说的描述。

实例146　安特卫普市旧城中心区

在城区中心矗立着16世纪初建起的哥特式安特卫普大教堂,教堂高123m,是比利时最大的教堂。从四面八方都能望到安特卫普大教堂,它是该市的一个重要标志。在安特卫普市中心广场及其周围还建有大量的行会大厦,它们以山墙作为正面,由于正面窄、进深大,因此加强了水平线的划分。山花墙以小尖塔作装饰。15世纪以来,这里以艺术中心闻名,陆续建起各类博物馆、工艺美术馆和艺术学院等。从17世纪开始,此地成了世界钻石加工和贸易中心。近代,在城边还修建了一个80hm^{2}的现代雕塑公园。

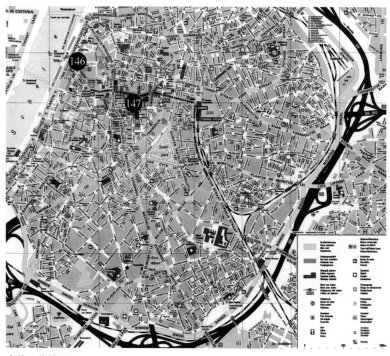

安特卫普城平面
Plan of the city

10. Belgium

10.1 Antwerp

Located in the lower reaches of the Scheldt River, it is the largest port city in Belgium. There is a statue in the center of the square in front of the city hall, which is demonstrating the legend describing the origin of the city name.

Example 146 Central Area of the Old City

Gothic Antwerp Cathedral stands in the center of the city with a height of 123 m, which is the largest church in Belgium. There are many guild buildings of gables with spires in and around the downtown square. In addition, there are various museums, arts and crafts galleries and art colleges. Since the 17th century, it has become the processing and trade center of diamond.

街巷联排山墙式大厦
Gable buildings in a row

市政厅前广场（左为"勇士抛掷断臂"雕像）
Square in front of the city hall (the statue of "a warrior throwing the broken arm" in the left)

安特卫普大教堂 Antwerp Cathedral

下篇：外国建筑科学文化精品图说／比利时

联排山墙式行会大厦
Gable industry buildings in a row

国家海事博物馆
National Maritime Museum

现代雕塑园一角
Part of the modern statue garden

实例147 鲁宾斯博物馆

安特卫普从公元15世纪创建画院开始,到16、17世纪画家辈出,这期间建有不少画家的博物馆,于是城市以艺术中心闻名。这里介绍的鲁宾斯博物馆就是其中之一。鲁宾斯博物馆位于安特卫普市中心区,原为画家鲁宾斯的住处和工作室,1640年他病逝于此。该建筑为意大利文艺复兴时期券柱式砖石木构造,其室内外装饰和家具与陈设等带有巴洛克样式,建筑室外有一花园。这里的一切建筑创作都出自鲁宾斯一人之手,这反映了他的人文主义思想与其绘画一样,都是那么自然、真实、和谐和统一。

Example 147 Rubens Museum

It is located in the downtown area and used to be the residence and studio of artist Rubens. This building is of Italian Renaissance arch type and masonry structure. The indoor and outdoor decoration, furniture and furnishings are influenced by Baroque style. There is a garden outside the building. All the architectural art forms here are created by Rubens alone, which reflects his humanism. Like his painting style, it is natural, real and integrated.

建筑与花园
Building and garden

拱形花架与券拱式门窗遥相呼应
Arch-shaped flower frame

鲁宾斯早期著名作品《亚当与夏娃》
Adam and Eve, a famous early work by Rubens

展室与庭院
Exhibition room and courtyard

2. 布鲁日

实例148 水乡

布鲁日位于比利时西北部，离北海岸14km，是一座历史文化古城。它始建于公元9世纪，到14世纪时已成为欧洲的一个商业港。后来衰落。到20世纪初，工商业又重新复兴，故布鲁日又有"北方威尼斯"之称。旧市区河道成网，河水清澈，河旁建筑修缮得整洁有序，它仍保留着欧洲中世纪水乡传统风貌。在布鲁日市中心广场矗立着一座公元13世纪末至14世纪初建起的哥特式市政厅塔楼，其高85m，这在世界建筑史上占有重要地位。它表明，早期一些市政建筑的重要性已超过了教堂建筑。这座高耸建筑起着标志性的作用，人们从很远处都能望到它，它丰富并统帅着城市的立体轮廓线。布鲁日水乡风貌的整体保护，使得世界各地到此旅游的人数日益增多，其整体保护水乡的观念值得我们学习。

10.2 Bruges

Example 148 Water Town

Located in the northwest of Belgium, it was founded in the 9th century, and is known as "Venice of the North". The rivers in the old urban areas are netted, the water is clear, and the buildings beside the rivers are repaired neatly and orderly, which still retains the traditional features of European medieval water towns. In the downtown square stands a Gothic city hall with a height of 85 m, which plays a symbolic role and enriches the three-dimensional outline of the city.

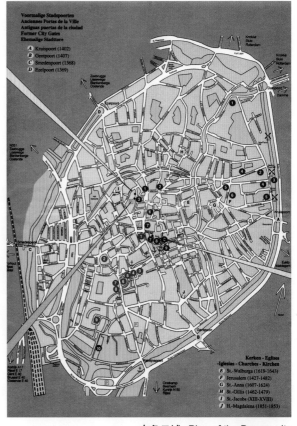

布鲁日城 Plan of the Bruges city

从河面上望市政厅塔楼 City hall from the view of the river

水乡风貌 Landscape of the water town

3. 布鲁塞尔

从1830年11月比利时宣告独立起至今，布鲁塞尔就一直是比利时的首都。这里原是一片沼泽地，公元10世纪末兴建要塞和码头，16世纪后被西班牙、奥地利、德国、荷兰占领。现布鲁塞尔为比利时的工业中心，而且还有700多处国际机构也设在这里，故素有欧洲首都之称。

实例149　布鲁塞尔市中心大广场

从城市的历史文化来看，布鲁塞尔市中心的"大广场"最为突出，其四周屹立着中世纪哥特式和尼德兰式古建筑，还有壮丽的皇家大厦、市政厅、历史博物馆以及1830年爆发革命的金融街剧场和举世闻名的"撒尿小孩"铜像等。

皇家大厦为典型的尼德兰哥特式建筑，建成于15世纪末，现为布鲁塞尔市博物馆。该博物馆展出的展品有布鲁塞尔市模型、宝石、绘画、艺术品等。此外，三楼有一展室专门展出各国送与"撒尿小孩"的服装。

"撒尿小孩"铜像，被称为"布鲁塞尔第一公民塑像"。传说古时候西班牙入侵撤离时拟用炸药毁城，幸好一个勇敢小孩夜出撒尿将导火线浇湿，布鲁塞尔城才得以保存，而小孩却中箭身亡，故立此铜像纪念。

10.3 Brussels

It is the capital and industrial center of Belgium, and is called the capital of Europe because there are more than 700 international institutions located here.

Example 149 City Center Square

Ancient medieval Gothic and Nederland buildings stand around the square.

The Royal Building was built at the end of the 15th century and is now a city museum. The bronze statue of the "Urinating Boy" in the square is called "the statue of the first citizen of Brussels". It is said that in ancient times, when Spain invaded and evacuated, it was planned to destroy the city with explosives. Fortunately, a brave boy urinated at night to wet the fuse, and the city of Brussels was preserved.

"撒尿小孩"铜像位置
Location of the statue of the "Urinating Boy"

"撒尿小孩铜像"
Statue of the "Urinating Boy"

皇家大厦正面
Front side of the Royal Building

实例150 布鲁塞尔原子球

这是一座比利时为1958年在布鲁塞尔举办万国博览会而设计建造的象征性建筑物，其位置在布鲁塞尔市西北面易明多公园内，设计人是比利时著名工程师昂德雷·瓦特凯恩。设计师的创作意图是以微小原子的概念表达人类和平利用原子能的美好愿景。原子球外观是9个巨大金属圆球以粗大钢管连接并构成正方体图案，这个图案正好是放大了的α铁的晶体结构。8个圆球位于正方体的8个角，1个圆球在正方体的中心。圆球直径18m，每根钢管长26m、直径3m，正方体总重量2200t、高102m。我们先从地面圆形大厅乘电梯直达100m高的圆球厅观赏市容，再通过钢管内的自动楼梯到其他圆球厅展室。

Example 150 Atomiun

The appearance is that nine huge metal balls are connected by thick steel pipes to form a square pattern, with the diameter of the balls being 18 m and each steel pipe being 26 m. You can take the elevator from the rotunda on the ground to the top round ball hall to enjoy the city, and then go to other round ball exhibition rooms through the escalator inside the steel pipe.

原子球及其环境外景　Exterior

十一、俄罗斯

莫斯科

实例151　莫斯科红场

它是俄罗斯首都莫斯科市中心广场，紧靠克里姆林宫东墙的东面，原用作商业广场、刑场使用。公元17世纪莫斯科从波兰贵族军队手中解放出来。17世纪中叶起莫斯科市中心广场称"红场"，俄语中红色有美丽之意，红场即美丽的广场。广场为长方形，南北长695m、东西宽130m，总面积9万m²，它是莫斯科重大历史事件发生的场所，也是10月革命后苏联人民和俄罗斯人民举行庆祝活动与集会的场所。广场主要面朝东，正中为列宁墓，列宁墓后是两端耸立着斯巴斯基钟楼和尼古拉塔楼的长条形克里姆林宫墙。红场南面是华西里·伯拉仁内教堂，建于16世纪中叶，是为庆祝战胜蒙古国的侵略而建，它是由9个墩式教堂组成，总高47m，中央帐篷顶上是一个小穹顶，周围8个小墩子排成方形，色彩强烈、装饰华丽，充分体现了俄罗斯民族独立的主题。东面是规模宏大的百货商场，北面是19世纪建起的具有俄罗斯风格的历史博物馆。

11. Russia

Moscow

Example 151 Red Square

It is the central square of Moscow, which is rectangular, with a length of 695 m from north to south, a width of 130 m from east to west and a total area of 90,000 m². In the center of the square is the Lenin's Mausoleum, and behind the tomb is the long Kremlin wall with the Spassky Bell Tower and Nikolai Tower standing at both ends. To the south of the Red Square is the Vasile Assumption Cathedral, which was built to celebrate the victory over Mongolia's aggression, to the east is a large-scale department store, and to the north is a Russian-style history museum built in the 19th century.

莫斯科红场模型
Model of the Red Square

夕照下的莫斯科红场 Red Square in the sunset

莫斯科红场夜景 Red Square at night

实例152　莫斯科马涅什广场购物中心

马涅什广场位于红场的北面，它与克里姆林宫墙相邻，并挨着亚历山大洛夫斯基公园。为了改善这里的环境和方便生活，20世纪90年代开始对此广场进行改造，并举办了设计竞赛，获胜方案是拟修建一个地下7层的多功能购物中心。后来方案将7层改为4层，并布置一条倾斜的拱形通道将购物中心与公园联系起来，创造了一个优雅、清新的休闲购物环境。4层的购物中心面积为7万m^2，其圆形彩色玻璃屋顶和其他装饰体现着浓郁的俄罗斯风格。此改造工程于1997年完成。1998年我到此参观，当时的感触是：作为首都的中心广场，最好根据发展需要增加设施，同时为了不影响原有广场面貌，建筑最好向地下发展。马涅什广场购物中心这个实例证明了这种做法是行之有效的。

Example 152　Manezhnaya Square Shopping Center

Located in the north of Red Square, it is adjacent to the Kremlin wall. In 1990s, this square was transformed. A multi-functional shopping center with four floors underground was built, and an inclined arched passage was connected with the park. The 4-storey shopping center covers an area of 70,000 m^2, and its round stained glass roof and other decorations embody a strong Russian style. This renovation project was completed in 1997.

购物中心内景
Interior of the shopping center

十二、瑞 士

1. 伯尔尼

实例153　旧城保护

伯尔尼是瑞士的首都，位于瑞士高原中央高地上，莱茵河支流阿勒 (Aare) 河从中穿过。伯尔尼从公元12世纪末开始沿河湾处逐步发展，至17世纪形成了河湾内3条街组成的旧城。原此地是"熊"出没之地，"熊"在德语中就是伯尔尼，故城市取名伯尔尼，如今"熊"已成为这里的标志。瑞士的现代建筑在旧城外面发展，而旧城却被完整地保存下来。重要的历史文化建筑，如100m高的明斯克大教堂就是原状保护、原状陈列的。沿阿勒河的古老宅邸也是原状保留，但会加以使用，如其中一座上层作为市长官邸，下面辟为城市历史陈列室。其余大量的沿街建筑，仅保留建筑的外貌，其内部则按照需要改作商店、旅馆，并全部更新装修与设备。1979年6月，我们应邀来瑞士进行建筑学术交流，他们的整体保护旧城的做法以及重视环境建设的理念给予我们代表团很大的感触。

12. Switzerland

12.1 Berne

Example 153　Protection of the Old City

It is the capital of Switzerland and located on the central highland of the Swiss Plateau. It formed an old city consisting of three streets in the river bay in the 17th century. Modern architecture developed outside the old city, and the old city was completely preserved. Important historical and cultural buildings, such as Minsk Cathedral, have been preserved in their original state, and the ancient residence along the Ale River has been preserved in its original state as well. The upper floor serves as the mayor's official residence, while the lower floor serves as the city history showroom. The rest buildings along the street were only kept the appearance and the interior was updated with decoration and equipment.

伯尔尼旧城　Old city

2. 日内瓦

实例154　湖滨保护园林化

日内瓦是日内瓦州首府，傍日内瓦湖而建，素有国际大都市之称。其旧城区依然被完整地保护和使用着，新区街道宽阔，绿地成荫。自19世纪末起至20世纪，这里修建了大量的国际机构，但其高层建筑都建在城市里面，并未沿湖建设，从而保护了湖岸的自然景观。这种做法值得学习与推荐，这是我想介绍的第一点。第二点是环境建设。瑞士政府提倡大量植树，而每砍一棵树都要经过有关部门批准。日内瓦市除在建筑之间栽有各种花木外，沿湖还修建了激流公园、玫瑰公园、珍珠公园、植物园、英式花园等，其城市与环境融为一体。

12.2 Geneva

Example 154　Landscaping of Lakeside Protection

It is built beside Lake Geneva and is known as an international city. The old city is completely protected and used, and the new city has wide streets and shaded green spaces. Since the end of the 19th century, there have been a large number of international institutions here, but their high-rise buildings have not been spread along the lake, thus protecting the natural landscape of the lake shore. In addition to various flowers and trees planted between buildings, the city of Geneva has built Rapids Park, Rose Park, Pearl Park, Botanical Garden and English Garden along the lake.

日内瓦湖畔建筑
Building along the river

下篇：外国建筑科学文化精品图说／瑞士

日内瓦湖滨玫瑰公园
Rose Park

3. 阿尔卑斯山

实例155 雪景

访问瑞士，要欣赏阿尔卑斯山的雪景。在主人的精心安排下，我们乘车到南部的博瑞格（Brig）山城观光。山城整洁幽静、空气清新，我们在这里参观了先进的卫星站。之后，再从小城乘车至半山的则尔玛特（Zermatt）旅游点，从这个旅游点可乘电气火车或缆车登山，也有马车供游人乘坐。这里的体育、娱乐、旅馆、餐饮等设施齐全，是人们休闲、登山的重要枢纽中心。最后，我们乘爬山火车登上阿尔卑斯山，观看到欧洲最辽阔的山脉雪景。在瑞士境内有许多山地湖泊，如苏黎世湖、图恩湖等，这些地方湖水晶莹、远山白雪皑皑。瑞士最著名的特殊景观当属白雪皑皑的阿尔卑斯山。这次瑞士登山之行给我最大的感受是：开发旅游风景区一定要重视基础设施的完备，特别是交通设施，不仅要有汽车，还要辅以其他交通工具，以便游人在最短时间内到达各大景点。

12.3 Alpine Mountain

Example 155 Snowscape

When visiting Switzerland, you should enjoy the snow in the Alps. You can take a bus to the mountain city of Brig in the south, where it is clean and quiet and the air is fresh. After that, take a bus to Zermatt in the middle of the mountain. From here, you can climb mountains by electric train or cable car, and there is also a carriage for tourists to watch around. Finally, take the mountain climbing train to climb the Alps, and you can see the snow scene of the largest mountain range in Europe. In Switzerland, there are many mountain lakes, such as Zurich Lake, all of which have beautiful pictures of crystal clear water and snowy mountains.

Brig 卫星站
Brig satellite station

阿尔卑斯山雪景
Snow scene

Zermatt车站的马车,乘车者为前辈建筑师王华彬先生
Carriage at the Zermatt station, the passenger is Architect Mr. Wang Huabin

Zermatt旅馆与娱乐服务设施
Hotel and entertainment facilities at Zermatt

4. 北部和南方

实例156 住宅

我们曾参加在瑞士举办的中瑞建筑学术交流会，当时的主题是住宅。瑞士建筑师、学者布拉则先生很欣赏中国合院式住宅方式，他曾多次访问中国北京、苏州等地，拍了这方面内容的照片，并出版了专著。同时，在瑞士还举办了这方面专题展览会，受到瑞士建筑师们的欢迎。我们从瑞士的北部到南方，看到瑞士城乡住宅的特点是：特别重视环境保护。其住宅层数不高，多为1~2层和4~5层的住宅，很少有高层住宅，而且重视节能和利用太阳能。对于乡镇的原有木构住宅仍然予以保留，并注意修缮和增添内部的现代化设施。

12.4 North Part and South Part

Example 156 Residence

Switzerland's urban and rural residential buildings are characterized by paying special attention to the environment. The buildings are always with low floors, mostly 1-2 floors and 4-5 floors, and paying attention to energy saving and solar energy utilization. For the original wooden houses in villages and towns, they still keep them, and pay attention to repairing and adding modern internal facilities.

北部乡村住宅
Residence in north village

中部木构住宅
Wooden residence in middle area

住宅前庭院
Courtyard at the front

南部多层住宅
Multi-floor residence at south area

十三、美　国

1. 华盛顿

华盛顿是美国的首都，全市面积174km²，有一条波托马克河（Potomac River）像碧蓝色的晶莹宽带从市中心飘过。其人口不到百万，这里原是印第安人住地，公元17世纪初欧洲移民在此建起烟草种植园，18世纪末由首任总统华盛顿委任参加过独立战争的法国工程师皮埃尔·夏尔·朗方负责华盛顿新都的规划设计工作。华盛顿市区为一正四边形，主要轴线为东西向，中心有国会大厦，南北向轴线对着国会大厦的侧面。国会大厦为华盛顿市制高点，其他建筑都不能超过它的高度，因而全城没有塔式高楼。城市街道从国会大厦向四周放射，形成方格加放射斜路的网状样式。东西向道路以阿拉伯数字命名，南北向的道路以英文字母命名。华盛顿市中心区的布局十分有序，建筑同大片自然绿地相结合。国会大厦坐东朝西，其前有对称的四条路，两条斜行路中向西北方向延伸的宾夕法尼亚大街与白宫连接，而两条垂直路中北面一条为宪法大道，南面的一条为独立大道。在这两条垂直大道的内侧布置国家艺术博物馆、自然历史博物馆、美国历史博物馆、国家航天博物馆、弗利尔艺术馆、维多利亚艺术与工艺品陈列馆、里香博物馆与雕塑花园等，所以华盛顿以博物馆闻名世界。在这两条垂直大道的外侧安排了联邦政府各部、白宫直属单位以及其他行政机构。在其中轴线上布置一条宽阔的林荫绿带，供人们休息散步和集会使用。此绿带中心矗立着高耸的华盛顿纪念碑，碑的南、北、西三面为开敞的大片绿地，碑北面绿地端部是白宫，碑南面绿地端部是杰弗逊纪念堂，碑西面绿地的末端是林肯纪念堂。华盛顿市没有大型工业企业，居民以公务人员为主，其他的就是文化、教育、娱乐、商业、服务和旅游业人员。下面将专项介绍中心区的几个重要建筑。

实例157　美国国会大厦

美国国会大厦，始建于1793年。1814年英、美战争时期，英军占领华盛顿，烧毁了美国国会大厦。1819年国会大厦重新修建，由渥尔特（Thomas U.Walter）设计，主要是扩展了两翼大楼和中央主楼的穹顶，建筑为古典复兴式。国会大厦高大的穹顶及其鼓座系仿照巴黎先贤祠的模样，但其更加雄伟，所用材料和结构为铁构架，既轻且高大。国会大厦整体为白色，在广阔的绿地衬托下显得格外庄重、典雅、壮丽、突出。国会大厦亦为国会参众两院所在地，四年一度的总统就职典礼就在国会大厦主楼的平台上举行。国会大厦后面是美国最高法院和国会图书馆。

13. United States

13.1 Washington

It is the capital of the United States, with an area of 174 km^2 and a population of less than one million. It was originally an Indian residence. The urban area is a regular quadrangle, with the main axis being east-west and the center being the Capitol. The Capitol is the highest point in the whole city, and no other building can exceed its height, so there are no tower-type hiyh-vise buildings in the city. The city streets radiate around the Capitol building, forming a net pattern of squares and radial ramps. The east-west roads are named after Arabic numerals, and the north-south roads are named after English letters. The layout of the central area is very orderly. and The building is combined with a large area of natural green space. The Capitol faces east to West, and There are four symmetrical roads in front of it. A wide tree-lined green belt is arranged on the central axis for people to rest, walk and gather. The towering Washington Monument stands in the center of the green belt, with the White House at the green end in the north, Jefferson Memorial at the green end in the south and Lincoln Memorial at the green end in the west. There are no large-scale industrial enterprises in the city, and the residents are mainly public servants, while others are cultural, eduertional, entertainment, commercial, and tourism workers.

Example 157 Capitol

It was built in 1793, burned down during the Second War between Britain and America in 1814 and rebuilt in 1819. It mainly expands the domes of the two-wing building and the central main building, which is a classical revival style. The materials and structures used are iron frames, which are light and tall. The whole building is white, which is particularly solemn, elegant, magnificent and prominent against the vast green space. This building is the seat of the Senate and House of Representatives. Behind the building are the Supreme Court and the Library of Congress.

华盛顿城中心区鸟瞰
Aerial view of the downtown area

国会大厦正面外景
Front view of Capitol

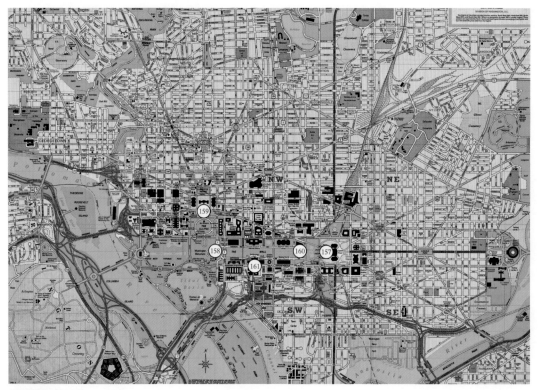

华盛顿城中心区
Plan of the downtown area

下篇：外国建筑科学文化精品图说／美国

鸟瞰国会大厦全景
Top view of Capitol

实例158　华盛顿纪念碑

华盛顿纪念碑坐落在华盛顿东西向主轴线和中心绿带的核心位置上，是为纪念美国首任总统华盛顿而建的。1885年该工程落成，1888年10月华盛顿纪念碑对外开放。其整体造型是一座白色大理石方尖形石碑，顶部酷似埃及金字塔，顶部高约17m，整座碑高169m，全部用花岗石建造。华盛顿纪念碑内设898级铁梯和一部电梯，顶部为观光厅，游人在观光厅可饱览华盛顿全景风光。华盛顿纪念碑的四周是一片碧草如茵的开阔草坪，这里已成为人们集会游行的场所。

Example 158 Washington Monument

It is located at the core of Washington's east-west main axis and central green belt, and was built in memory of George Washington, the first president. The overall shape is a white marble obelisk, the top of which resembles the Egyptian pyramid. The whole monument is 169 m high, all of which are made of granite. There are 898 iron steps and an elevator inside, and the top is the sightseeing hall, where visitors can enjoy the panoramic view of Washington. Around the monument is an open lawn covered with green grass, which has become a place for people to gather.

从林肯纪念堂望华盛顿纪念碑
Washington Monument from the view of Lincoln Memorial

实例159 华盛顿白宫

华盛顿白宫为美国总统府,它坐落在宾夕法尼亚大街华盛顿纪念碑的北面,占地7.29hm^2。修建于18世纪90年代,1800年美国第二届总统约翰·亚当斯住进使用,1814年同国会大厦一起被英军烧毁,1815年重新修复,后不断扩建。白宫建筑为古典希腊式,北面入口采用爱奥尼克柱廊、三角形山花墙,因外墙为白色砂岩石,故称"白宫"。白宫主楼底层大厅为总统进行外事活动的地方,在此可以接见外国元首和使节等,而大厅前方的南草坪就是举行欢迎仪式的地方。白宫楼内还有图书室、地图室和金银瓷器陈列室。其一层北面是正门,进门后有门厅、内厅,其中东厅作举行酒会、文艺演出、记者招待会使用,西厅为举行国宴的宴会厅。二层是总统家人居住之所。主楼的西翼是办公区,其内侧为总统椭圆形办公室;东翼则供游客参观,每星期二至星期六对外开放。

Example 159 White House

Located in Pennsylvania Avenue, it is called "White House" because its outer wall is white sand and rock. The hall on the ground floor of the main building is where the president conducts foreign affairs activities. There are also library, map room and showroom in the building. The east hall on the first floor is used for holding cocktail parties, the atrical performances and press conferences, and the west side is the banquet hall for holding state banquets. The second floor is where the president's family lives. The main building has two wings, the west wing is the office area, the inner side is the oval office of the president, and the east wing is for tourists to visit.

白宫北立面
North facade

实例160 华盛顿国家艺术馆东馆

华盛顿国家艺术馆东馆位于国会大厦前西北向、紧靠国家艺术馆老馆的东面，1978年建成，设计人是著名美籍华裔建筑师贝聿铭。总统卡特曾出席华盛顿国家艺术馆东馆开幕式剪彩，他称赞这座建筑与城市相协调，是一座公共生活同艺术结合的象征性建筑。设计人采用一条对角线将建筑分为两个三角形，其中大三角形是以多层大厅为中心的新艺术馆，而小三角形则是艺术研究中心。新馆大厅采用新技术有序地布置高低桥廊与自动扶梯，新馆顶部为开有天窗的三棱锥体钢网架空间，并当空悬吊着可转动的挂雕展品，显得生动活泼。大厅周围是4层陈列室。艺术研究中心的四周是办公室、研究室和图书馆、阅览室，共8层，高21m。新馆与旧馆的连接是通过地面广场喷泉、水斗、天窗和地下通道水景（是地面广场的底层）组合在一起的。贝聿铭先生的大胆创新精神和新旧馆相结合的手法很有参考价值。华盛顿国家艺术馆东馆这个项目已被选入《20世纪世界建筑精品集锦》北美卷中。

Example 160 East Building of National Gallery of Art

There is a diagonal line used to divide the building into two triangles. The big triangle is the new art museum and the smaller triangle is the art research center. In the hall of the new museum, high and low bridge corridors and escalators are arranged in an orderly manner by adopting new technology, and the top is a triangular pyramid steel grid space with skylights, in which rotatable hanging carving exhibits are suspended. The art research center is surrounded by offices, research laboratories, libraries and reading rooms. The connection between the old and new pavilions is through the ground square fountain, water bucket, skylight and underground passage waterscape.

展览大厅有序布置着桥廊与扶梯
Bridge gallery and escalator in the hall

西入口与新老馆间广场
West entrance and the square between the old and new building

连接新老馆的地下通道，左侧为地上广场的水斗
Underground passage and water bucket on the left

面向中心绿地的国家艺术馆东馆外景
Exterior of the green space

实例161　华盛顿史密松非洲、近东及亚洲文化中心

该文化中心工程是全美国瞩目的重点工程，位于史密松学会总部的南部。该设计1981年竞标，最后是美国SBRA公司获胜并取得设计权。此文化中心工程于1986年建成，其建筑面积为6.2万m^2，它建在史密松总部及其属下两个博物馆间的一块4.2英亩的地段上。史密松总部是建于1849年的维多利亚城堡式建筑，它的西面是建于1923年的意大利文艺复兴式的弗利尔艺术馆，东面是建于1881年的古典尖顶式的维多利亚艺术与工艺品陈列馆，这3组建筑都是美国重点保护文物。在如此特定条件下，设计者大胆采用地下空间方案，只将3个入口亭建在地上，将96%的面积放入地下。该文化中心入口建筑总高12m，由于体量小而不影响原来的整体空间环境。此文化中心工程获得1988年托克优秀设计奖（Tucker Award of excellence）。

Example 161 Cultural Center of Africa, Near East and Asian of Smithsonian

The cultural center is built in a 4.2-acre area between Smithsonian headquarters and its two museums. The headquarters is a Victoria castle-style building, with Freer Gallery on one side and Victorian Art and Crafts Exhibition Hall on the other side. These three groups of buildings are key cultural relics protected in the United States. Therefore, the designer boldly adopted the underground scheme, only three entrance pavilions were built on the ground, and 96% of the area was placed underground. The entrance building was 12 m high and small in volume, which did not affect the original overall space environment.

文化中心地面入口设计
Entrance pavilion of the cultural center

下篇：外国建筑科学文化精品图说／美国

地面鸟瞰。该文化中心建成后，仍维持原有3组建筑的造型
Exterior of the ground, the same as before of the three buildings

地下展室一角
Part of the underground exhibition

2. 纽　约

纽约是美国第一个大城市，1686年建成，共有5个区，其市中心在曼哈顿区。1825年伊利运河通航和后来铁路的兴建，使纽约城市迅速发展，并成为世界大港口和经济、金融中心。纽约中心区闻名的华尔街是美国最大的银行集中地，也是美国金融中心的象征。其百老汇街聚集了许多剧场、戏院、歌舞厅和夜总会，它是一条闻名的娱乐场所大街。在自由岛上还矗立着高46m的世界著名自由女神像。纽约中心区内虽高楼林立，但其中心地带于1857年还兴建了340hm²的美国第一个城市大公园——中央公园。

实例162　洛克菲勒中心下沉式广场等

洛克菲勒中心位于纽约曼哈顿中心区，建造时间是1929—1939年，由3个事务所合作设计的。洛克菲勒中心是由9座不同规模的建筑组成，共占有3个街区，是一个集写字楼、商店、娱乐、交通枢纽于一体的综合公共活动场所。这个综合体项目已成为城市公共服务空间发展的一个较好范例。

洛克菲勒中心有一下沉式广场十分出名，该广场中心侧面立有镀金雕塑喷泉，冬季这里可以作冰场，夏季则改作室外餐饮休闲之地，深受广大市民的喜爱。

13.2 New York

New York is the first big city in the United States. Wall Street is the largest bank concentration in the United States, and Broadway is a famous entertainment street. The Central Park covering an area of 340 hectares has been built in the downtown area.

Example 162　Sunken Plaza in Rockefeller Center

Rockefeller Center, located in the downtown area, is a comprehensive public activity place, which consists of nine buildings of different sizes. There is a gilded sculpture fountain on the side of the center of Sunken Plaza, which becomes an ice rink in winter and an outdoor dining and leisure place in summer, and is deeply loved by the general public.

洛克菲勒中心下沉式广场
Sunken Plaza in Rockefeller Center

3. 芝加哥

芝加哥是仅次于纽约的美国第二大城市，人口300多万。1830年时芝加哥人口不及1万，此后运河、铁路兴建起来，芝加哥城市迅速发展，并成为世界重要谷物交易市场，也是美国钢铁冶炼工业的重要基地以及河、湖、海运的港口。1871年芝加哥曾遭大火，城市建筑被烧毁，后重建。1886年5月1日，芝加哥几十万工人举行罢工，它是"五一国际劳动节"的发源地。芝加哥市还是世界高层建筑的发祥地，1974年建起的西尔斯塔楼已成为当时世界最高的建筑。芝加哥市区位于密执安湖西南岸。密执安湖汪洋千顷，酷似大海，其湖滨布置着大片绿地公园。芝加哥河蜿蜒流经芝加哥市区，芝加哥河两岸则建有许多知名建筑师设计的著名建筑。1993年6月，在芝加哥市举办了国际建筑师协会（UIA）第18次大会和第19次代表会，在这次代表会上，我国北京成功地获得1999年国际建筑师协会大会的举办权，其意义重大非凡。

13.3 Chicago

It is the second largest city with a population of over 3 million. It is an important grain trading market in the world and the birthplace of high-rise buildings. The urban area is located on the southwest bank of Lake Michigan, with large green parks on the lakeside.

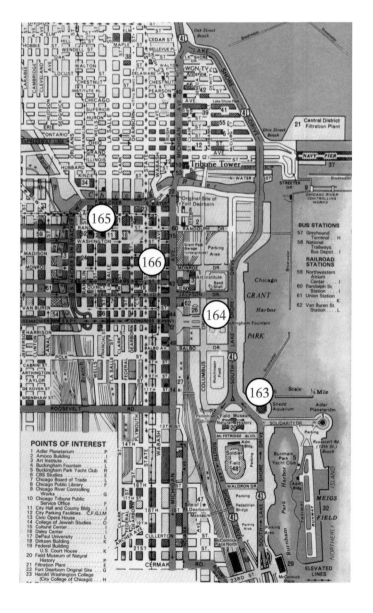

芝加哥城中心
Plan of downtown area

实例163 芝加哥滨湖建筑立体轮廓

滨湖建筑是芝加哥城的标志。为了建设好这一市中心区的城市环境,政府部门在密执安湖滨修建了一个较大规模的长条形公园绿地,建筑群都退到绿地后面。其高层建筑排列有序,左边是西尔斯塔楼,往右是第二高的石油塔楼,再右边是第三高的汉考克大厦,芝加哥湖滨展现出一幅有韵律节奏的建筑立体轮廓画面。芝加哥的高层建筑也仅集中在芝加哥市中心区的中部地带,而在其南部、西部,特别是郊区,建筑楼层低、园林绿地多,环境十分清幽。

Example 163 Three-dimensional Outline of Lakeside Building

Lakeside architecture is the symbol of Chicago. A large-scale strip park has been built on the lakeside of Michigan, with buildings behind it. High-rise buildings are arranged orderly. Sears Tower is on the left, the second tallest oil building is on the right, and Hancock Tower is the third tallest, showing a three-dimensional outline with rhythm. Chicago's high-rise buildings are only concentrated in the central part of the downtown area, and in other areas especially in the suburbs, there are many green spaces and the environment is quiet.

下篇：外国建筑科学文化精品图说/美国

阳光耀眼的芝加哥滨湖建筑雪初霁景
A sunny view of lakeside buildings as snow begins to fall

灯光晶莹的芝加哥滨湖建筑仲夏夜景
A midsummer night view of lakeside buildings

实例164　芝加哥密执安湖畔格兰特公园

　　这是一座沿密执安湖从北部伦道夫街至南部罗斯福路之间的条形绿地公园。它正对国会街，设置有世界最大的照明喷泉，其中央泉池占地55.74m²，中央水柱高达30m，水柱交汇形成壮观水帘，夜晚还有万束灯光照射，瑰丽多彩。沿湖滨有浴场和游艇区，这是市民嬉水、划船的场所。公园内鸟语花香、林木葱郁，还有精美雕像点缀其中。湖中点点帆船与公园幽美的自然景色相映成趣。这条长长的绿带公园改善了芝加哥市中心的环境质量，也为芝加哥市民提供了一个很好的憩息环境。

Example 164 Grant Park

　　It is a strip-shaped green park along Lake Michigan, with the largest lighting fountain in the world. The fountain covers an area of 600 square feet with central water columns as high as 30 m. The water columns meet to form a water curtain, which is illuminated by lights at night. There are bathing and yachting areas along the lakeside, which are places for citizens to paddle and row. There are flowers and birds in the park, lush trees and beautiful statues. The little sailboats in the lake are in harmony with the beautiful and natural scenery in the park.

从南向北望公园全景
Panorama

花卉、喷泉与雕塑景观
Fountain sculpture

大喷泉景观
Big fountain

滨湖景色
Waterside scene

实例165　芝加哥伊利诺伊州政府大楼

芝加哥伊利诺伊州政府大楼位于芝加哥北部鲁道夫街与克拉克街交叉口的西北面，由建筑师Helmut Jahn设计，1985年建成。设计师在高技派风格基础上借鉴古典穹顶模样，以表达伊利诺伊州政府大厦的含义。首先体现政府建筑的公众性要平易近人。其底层是商店和大厅，供大众使用；办公楼层在上面且均敞向中庭，为开放性办公室（Open Office）。建筑外立面采用沿街斜向后引曲面，以及选用底层柱面材料等，同周围建筑协调并以取得整体和谐的效果，这是其平易近人的内涵所在；建筑外墙采用反射玻璃和一段透明玻璃幕墙，使得中庭拥有自然采光。中庭大厅结构如同蛛网般一样，电梯箱、楼梯、结构构件与管道等制作精良，反映了新结构、新构件的现代美，也突出了其高技派的风格。

Example 165 Illinois Government Building

It was built and used in 1985. First of all, it reflects the publicity of government buildings. The ground floor is shops and halls for public use, and the open offices are above them, all toward the atrium. The exterior wall adopts reflective glass and a transparent glass curtain wall, which makes the atrium get natural lighting. The atrium hall is wrapped by a cobweb-like structure, with elevator boxes, stairs, structural components and pipes, etc. The manufacturing process is fine, reflecting the modern beauty of new structures and components.

入口外观
Entrance

从上俯视中庭圆形地面
Top view of the atrium

实例166 芝加哥卡森·皮里·斯科特百货公司

这是美国著名建筑师沙利文表现现代建筑初期的优秀作品,建于1899—1904年,地点在芝加哥老城政府街和麦迪逊街转角处。其建筑外部形式为3段,其中基座段为两层高浮雕铸铁部分,中段为芝加哥式横向窗和白瓷砖贴面,顶部是出檐和退进的敞廊,它体现了现代框架结构的特点。这座建筑也可以看作是沙利文的代表作品。沙利文提出"形式随从功能"的著名建筑理论。其高层公共建筑的功能与底层和二层功能相近,并形成整体;二层以上房间是相同的,均按柱网排列;顶部有设备层,可作不同处理。芝加哥学派在促进现代建筑发展方面起到了推动作用。这个项目已被选入《20世纪世界建筑精品集锦》北美卷中。

Example 166 Carson Pirie Scott Department Store

Built in 1899 to 1904, the exterior form is of three sections, the base section is a 2-storey high-embossment iron part at the bottom, the middle section is Chicago-style transverse window and white ceramic tile veneer, and the top section is a veranda with eaves and retreats, which fully reflects the characteristics of modern frame structure. The functions of the ground floor and the second floor are similar, which should form a whole. Rooms above the second floor are the same, which should be arranged according to the column grid, and there are equipment floors at the top, which can be treated differently.

百货公司底座段与中段外景
Exterior of the ground and middle parts of the building

节日期间政府街百货公司前景
Exterior during festivals

百货公司内景（冷色调大厅与保持原貌的柱头花饰）
Interior, cool-tone hall

百货公司暖色调大厅
Interior, warm-tone hall

实例167　芝加哥弗兰克·劳埃德·赖特住宅和工作室

　　弗兰克·劳埃德·赖特住宅和工作室于19世纪末建于芝加哥郊区，它完全体现了弗兰克·劳埃德·赖特的"有机建筑"理论。弗兰克·劳埃德·赖特认为，有机建筑是一种由内而外的建筑，与外部自然环境是一个整体。比如住宅，它是那个环境的一个组成部分，为环境添彩，也与环境融为一体，可以说有机建筑就是"自然的建筑"。该住宅建筑的"自然"反映在环境优美、宁静，所用材料是传统的砖、木和石头，其平面布局自然、灵活。它以壁炉为中心，精心安排门厅、起居室、餐厅、工作室等，室内外空间通透。建筑外部以水平横线条为主，长屋檐、连排窗，给人以舒展、安定、自然的感觉，而且这种连续成排的窗与门增加了室内和室外空间的联系。该住宅作为文物建筑已被保护起来，由弗兰克·劳埃德·赖特基金会负责管理。

Example 167　Residence and Studio of Frank Lloyd Wright

　　Built in the suburb of Chicago at the end of 19th century, it fully embodies Wright's thought of "organic architecture", that is, "natural architecture". "Nature" is reflected in the beautiful and quiet environment; the materials used are traditional bricks, wood and stone. The layout is flexible, with the fireplace as the center, and the foyer, living room, dining room and studio are arranged. The exterior of the building is dominated by horizontal lines, with long eaves and rows of windows, giving people a feeling of relaxation, stability and nature. Moreover, the continuous rows of windows and doors increase the connection between indoor and outdoor natural spaces.

室内与室外融为一体
Connection between indoor and outdoor natural spaces

下篇：外国建筑科学文化精品图说/美国

弗兰克·劳埃德·赖特住宅和工作室街景
Street scene

中部外景
Exterior of the middle part

453

4. 波士顿

实例168　波士顿中心绿地公园区

波士顿是美国东北部沿海大城市和海港，为马萨诸塞州首府，始建于1630年，其历史与美国的独立、发展密切相关，现为美国历史文化城市。波士顿市中心绿地公园区，为美国最早的公园区之一，其北面山丘上有烽火台，并作为17、18世纪时航海船只通信的信号，后来在此修建马萨诸塞州议会大楼。议会大楼中心为一镀金穹顶，数里之外都能望到这闪闪发光的金圆顶。几百年来这块中心绿地一直被保留着，并成为市中心的一块宝地。

13.4 Boston

Example 168　Central Green Park

Boston is a coastal city in the northeastern United States and the capital of Massachusetts. Its history is closely related to the independence and development of the country. Central Green Park, one of the earliest parks in America, has beacon towers in its northern hills serving as a signal for communication between sailing vessels. Later, the Massachusetts Parliament Building was built here. The center of the building is a gold-plated dome, which can be seen from miles away.

市中心公园旁的州议会大楼
Parliament Building

市中心公园鸟瞰
Aerial view

实例169　波士顿查理河两岸

从汉考克大厦顶部可以观赏到查理河两岸优美、开阔的城市景观。查理河南面成片整齐的住宅区有绿地穿插其间，而查理河北面有著名的麻省理工学院等。宽敞的查理河及其绿化带将两岸的建筑连为一个整体。总之，城市规划与设计，一定要结合自然，尊重历史建筑，保护自然环境。

Example 169 Banks of Charles River

On the south side of Charlie River, a neat residential area is interspersed with green spaces. On the north side of Charlie River stands the famous Massachusetts Institute of Technology. The shapes of the buildings are simple and varied, but they are not towering but echo the riverbanks orderly. The wider Charles River and its green belts connect the buildings on both sides as a whole.

查理河旁的住宅区，绿带穿插其间
Residential area along the river

下篇：外国建筑科学文化精品图说／美国

查理河北岸的
麻省理工学院主楼
Main building of MIT

查理河北面沿河景观
Riverside view at the
north part

查理河河岸绿地
Green space

实例170 波士顿市府办公楼

在改建波士顿市府广场和设计市府办公楼时，充分考虑了与原有建筑的协调。波士顿市府办公楼外观简洁朴素，市府办公楼上凸出部分为市长办公室，平时市民可以随便进出，十分开放。市府办公楼前右侧有一活动场地，系著名建筑师贝聿铭先生设计，市民可在这里举办各项活动。市府办公楼后原有一个市场，现仍保留并修缮一新，已成为市民购物、游逛的地方。波士顿市府办公楼体现了社会公正，它既开放，又接近市民、方便市民。

Example 170 Boston Government Building

Its appearance was simple and plain, and the prominent part of the upstairs was the mayor's office, where citizens could enter and exit at any time. There is an activity venue in front of the building on the right, where citizens can hold various activities. There was a market behind the building, which is still preserved and renovated, and has become a favorite place for shopping and wandering.

市府办公楼及其广场活动场地
Government building and the activity venue

实例171　波士顿公共图书馆

　　该图书馆老馆是1888—1895年Mckim等人设计的，为文艺复兴时期样式，它有一精美的券柱围廊式古典庭园。20世纪60年代，图书馆老馆根据发展需要进行扩建，由著名建筑师菲力普·约翰逊设计，于1973年建成新馆。新馆采用现代结构，其跨度大，中央大厅高20m，带有玻璃天窗，而且自然采光，其外观十分重视与旧馆的协调一致。弦月窗与拱券呼应，取得了新旧馆和谐的效果。波士顿公共图书馆重视新旧建筑结合的设计理念和手法，值得大家学习与借鉴。

Example 171 Boston Public Library

　　It is a Renaissance style and has a beautiful classical garden with an arch column and a corridor. In 1960s, according to the development needs, the new building was built in 1973. The new one adopts modern structure, with a large span. The center is a 20-meter-high hall with glass skylight, which has natural lighting. However, its appearance attaches great importance to the coordination with the old one. The height is uniform, the external wall materials are consistent, and the string moon window echoes the arch, thus achieving the harmonious effect between the old and new.

新馆与老馆外景
Exterior

5. 费　城

实例172　费城中心区下沉式广场

　　费城始建于1682年，1790—1800年作为美国的首都，现是宾夕法尼亚州的首府。20世纪上半叶，由著名的规划建筑师Edmund N. Bacon提出的费城发展规划很有特色，其纵横向中轴线十分突出，建筑与自然紧密结合，绿化融入建筑中，建筑亦在绿化中。半个多世纪前费城就规划并建设了市中心区下沉式广场，此广场一边联系服务大楼，另一边连接市场街上带有玻璃顶的3层建筑，并发展人行交通系统，使之与地下铁等公共交通系统连为一个整体。

13.5 Philadelphia

Example 172 Downtown Sunken Plaza

　　Founded in 1682, it is now the capital of Pennsylvania. The vertical and horizontal central axis of the city is very prominent, and architecture and nature are closely combined. Sunken Plaza in the downtown area is connected with the service building on one side and the 3-storey building with glass roof on the market street on the other side. It develops a pedestrian transportation system and connects with public transportation such as subway as a whole. In this way, it is very convenient for people to go to and from the city center, and they are free from car interference and feel safe when moving in the center area.

费城市市府大楼前的下沉式广场
Sunken Plaza

6. 纽黑文

实例173　纽黑文耶鲁大学

纽黑文是美国东部的一个重要海港和文化古城，始建于1640年，1701—1895年为康涅狄格州的双首府之一。18世纪后期与19世纪初期，纽黑文为兴盛的海港，它在历史上一直以生产枪炮、弹药、橡胶而出名。1701年在纽黑文创办的耶鲁大学，现已世界闻名，它也是美国第三所最古老的高等学府。耶鲁大学校园内有希腊、罗马古典复兴式的庭院建筑群，亦有20世纪下半叶添建的新建筑，新建筑与老建筑融为一体。耶鲁大学校园有25座建筑由美国有名建筑师设计创作的，其中1917年建起的由詹姆斯·干伯·罗杰斯（James Gamble Rogers）设计的哈克尼斯钟楼（Harkness Memorial Tower）最为有名。该楼高67m，为哥特式，它也是耶鲁大学建筑的制高点。又如，1977年由路易斯·康（Louis I. Kahn）设计的耶鲁英国艺术中心（Yale Center for British Art and British Studies），完全采用新材料、新结构，其顶部为玻璃天窗，可以自然采光，外部墙面由不锈钢和薄玻璃组成。同时，设计人重视建筑的高度与线条的分割，并与其周围建筑相协调。还有扩建的图书馆新馆，采用白色大理石墙面，并控制建筑高度，它与周围砖石建筑和谐统一，且又有新意，一看便知道它是20世纪下半叶扩建的新图书馆（1961年）。

13.6 New Haven

Example 173 Yale University

New Haven is an important seaport in the eastern United States. It has always been famous for producing guns and rubber in history. Yale University, founded in 1701, is the third oldest institution of higher learning in America. There are old classical revival buildings and new buildings on campus. The Harkness Memorial Tower is the commanding height of the three-dimensional outline of campus buildings. Yale Center for British Art and British Studies is completely made of new materials and new structures, with a glass skylight at the top and natural lighting. The outer wall is made of stainless steel and thin glass. There is also an expanded new library with white marble walls, which is innovative but also in harmony with other buildings.

哈克尼斯钟楼
Harkness Memorial Tower

古典复兴式老学院的建筑与庭院
Old buildings and yards of classical revival style

右侧建筑为耶鲁英国艺术中心
Yale Center for British Art and British Studies on the right

图书馆新馆
New library

7. 洛杉矶

实例174　洛杉矶市区

洛杉矶位于美国西部海岸、加利福尼亚州南边，其城市的名称来自西班牙语，意思是"天使之城"。原印第安人在这里居住，20世纪初因铁路通到这里，后又发现了石油，并创建了好莱坞影业，促使城市迅速发展起来。洛杉矶城市布局的特点是分散式的，由宽阔的道路将分散的城镇连接成网，一些高层建筑集中于中心区，我们从夜景照片上可以看出这一分散式布局的特点。洛杉矶现为工商业海港城市，其工业有石油、电子、宇航、飞机等。1984年在洛杉矶举办了第23届奥运会，我国也是第一次参加奥运会。在本届奥运会上，国际奥委会主席萨马兰奇还为中国射击运动员许海峰颁发了奥运会金牌，极富有历史意义。好莱坞影城在洛杉矶市西北郊，它素有"世界影都"之名，因而成为著名的游览胜地。

13.7 Los Angeles

Example 174 Downtown Area

Located on the western coast of the United States, the city name comes from Spanish, meaning "city of angels". The urban layout is characterized by dispersion, where scattered towns are connected into a network by wide roads, and some high-rise buildings are concentrated in the central area. Los Angeles is now a port city for industry and commerce, with industries such as oil exploitation, electronics, aerospace and aircraft manufacturing. Hollywood Studios is located in the northwest suburb. It has the name of "the world's movie capital", so it has become a famous tourist destination.

洛杉矶夜景鸟瞰，远处为市中心城区
Top view of the night scene

1984年举办的第23届洛杉矶奥运会体育场外景
Sport center for the 23rd Olympic Games in 1984

好莱坞影城入口处
Entrance of Hollywood Studios

好莱坞影城一角
Part of Hollywood Studios

实例175　洛杉矶亨廷顿文化园

洛杉矶亨廷顿文化园是洛杉矶市的一个国家文化和教育中心。它位于洛杉矶市的东北部，建于1919年，占地207英亩，这是亨廷顿先生根据自己的喜爱和收藏而创建的，属于私人所有、非盈利的一个单位。亨廷顿先生不仅拥有一个金融财团，包括铁路公司和在南加利福尼亚的不动产，而且他还非常喜欢书籍、艺术和花园，任凭自己的财力建造了这个地方。亨廷顿文化园图书馆有珍贵的书籍以及英国和美国历史的手写原稿，共400多万部。亨廷顿文化园艺术品收藏馆位于图书馆的西南面，原是亨廷顿先生的住所，里面收藏着法国与英国18—19世纪的艺术品，还有美国1730—1930年期间的绘画作品，以及欧洲文艺复兴时期的绘画和法国18世纪的雕刻、瓷器与家具等。亨廷顿文化园植物园占地130英亩，共建有12个花园，包括亚热带花园、草本植物园、丛林园、棕榈园、玫瑰园、日本花园等。

Example 175 Huntington Cultural Park

Now, it is a national cultural and educational center in Los Angeles. Covering an area of 207 acres, it was founded by Henry E. Huntington in 1919, according to his favorite and collection, and belonged to private ownership at that time. The library has precious books and handwritten manuscripts, and the art collection house has paintings, porcelain, furniture, etc. The Botanical Garden covers an area of 130 acres and has 12 gardens, including subtropical garden, herb garden, jungle garden, palm garden, rose garden and Japanese garden.

亨廷顿艺术品收藏馆
Art collection house

图书馆
Library

专题陈列馆
Exhibition hall

雕塑园
Sculpture garden

实例176　洛杉矶迪士尼游乐园

　　洛杉矶迪士尼游乐园位于洛杉矶郊区，于1955年建成开放，占地30hm^2，是世界闻名的卡通式大游乐园。它是由美国动画片米老鼠形象创作者沃尔特·迪士尼创立。整座游乐园由中小古城、美国西部市镇、古堡、卡通馆、立体电影厅、游乐场和溪流、瀑布、雪山、原始森林等自然景观组成，并以道路、广场、水路、铁路等相连。迪士尼游乐园内有马车、双层公共汽车、火车、高空缆车和船只等交通工具将游人送往各分区景点。迪士尼游乐园创建时有18个游乐点，1980年增加到57个，这些游乐点深受广大儿童的喜爱，也是儿童心目中的天堂。迪士尼游乐园内外餐厅、商店、停车场、旅馆等服务设施齐全，每年接待游客超过千万。迪士尼游乐园的现代卡通式奇特形象早已名传全球，许多国家都纷纷仿建。

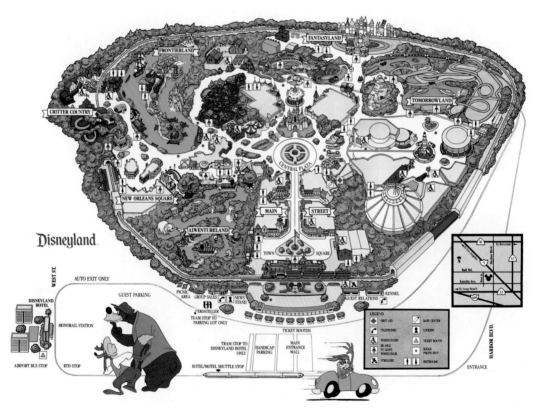

洛杉矶迪士尼游乐园平面
Plan of Disneyland

Example 176 Disneyland

Located in the suburb of Los Angeles, it was opened in 1955 and covers an area of 30 hectares. It is a world-famous cartoon amusement park. There are small and medium-sized ancient cities, western American towns, castles, cartoon halls, three-dimensional movie halls, amusement parks, streams, waterfalls, snow-capped mountains, virgin forests and other natural gardens, which are connected by roads, squares, waterways and railways. There are carriages, double-decker buses, trains, high-altitude cable cars and boats in the park to send tourists to scenic spots in various districts. The restaurants, shops, parking lots, hotels and other service facilities inside and outside the park are complete, receiving more than 10 million tourists every year after 1980.

游乐场
Amusement park

卡通园
Cartoon amusement park

雪山湖泊
Snow-capped mountain and lake

卡通人物游行表演
Cartoon figures' parade

圣诞节游行
Parade at Christmas

实例177 洛杉矶加登格罗夫社区教堂（水晶教堂）

该教堂位于洛杉矶市南郊，靠近迪士尼游乐园，是1980年建立在高速公路旁的一座高大水晶体，故又称其为水晶教堂，设计人是美国著名建筑师菲力普·约翰逊。该设计创新了教堂建筑的平面布局和空间环境。为了使每个座位都接近圣坛，设计师将古典希腊十字形教堂平面改为四角星状，圣坛在长边的中心，其前为长方形座池，共有1778个座位；东西耳厅挑台各有403座，南面挑台有306座；大厅高43m、长138m、宽69m，屋顶由白色网架和反射玻璃组成，仅有8%的光线透入，并形成晶莹透明、与天呼应的宁静氛围。3个挑台下部设3个2.5m高的出入口，使低矮入口同高大圣堂形成强烈对比，让人感到自己的渺小。

Example 177 Garden Grove Community Church

Located in the southern suburb of Los Angeles, it is a tall crystal body built beside the expressway in 1980, so it is also called Crystal Church. In order to make each seat face and approach the altar, the plane of the classical Greek cross-shaped church is changed into a star shape with four corners. The altar is in the center of the long side, with a rectangular pool in front of it. The hall is 43 m high, 138 m long and 69 m wide. The roof is white grid and reflective glass, which only allows 8% of light to penetrate, forming a quiet feeling corresponding to the sky. Three entrances and exits with a height of 2.5 m are set at the lower part of the three platforms, which makes a strong contrast between the low entrance and the tall temple.

外景
Exterior

内景
Interior

长边挑台及其下入口
Platform at the long side and the entrance at below

8. 西雅图

西雅图是美国西北部的一个海港城市，背山面水。西雅图原是一位印第安酋长之名，他帮助过第一批来此移民的白人，因此，1869年西雅图建市时取其名"西雅图"。1889年西雅图发生大火，城市被烧毁，后来又重建。1916年在西雅图建立了波音飞机公司，1962年在西雅图举办了"21世纪博览会"。西雅图全市有公园和游乐园140多处，其绿地系统完善。

实例178　西雅图市中心区沿海滨建筑群

西雅图市中心区沿海滨从北到南分为3段，北部是著名的西雅图中心，这里有西雅图最高的标志性建筑——"太空针塔"（Space Needle）。太空针塔高185m，其所在地区为低层建筑群。中部退后的东面为新建高层建筑群。南部是开拓者广场和历史保留区，这北部、中部、南部三部分共同构成了西雅图优美的海滨天际线。

13.8 Seattle

It is a seaport city in the northwestern United States, named after an Indian chief who helped immigrants. Now there are more than 140 parks and amusement parks in the city.

Example 178 Seaside Buildings in Downtown Area

The downtown area is divided into three sections along the coast from north to south. The north is the famous Seattle center. There is the highest landmark building of 185 m in Seattle, the "Space Needle". To the east of the middle is a new high-rise building complex. In the south are Pioneer Square and Historic Reserve. All the sections rhythmically form a three-dimensional outline along the coast.

优美的海滨天际线　Seaside skyline

下篇：外国建筑科学文化精品图说／美国

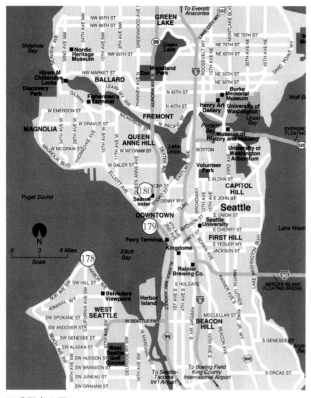

西雅图中心区
Plan of downtown area

太空针塔，其顶部像悬于空中的飞碟
Space Needle

北部为低层区，后面中部为高层区
Lower part at north and higher part in middle

从太空针塔南望中部高层区，其后面是南部低层区
Higher part in middle from the view of Space Needle

实例179　西雅图公共市场中心

　　公共市场中心位于市中心区中部临普吉特（Puget）海峡水面上，这是个老市场，现被完整地保存下来。市场里面供应市民需要的海鲜以及其他食品，其食品新鲜且有地方特色，深受广大市民欢迎。西雅图这一繁荣的传统市场居然还被保留在现代化城市的中心地带，这值得我们大家学习。

Example 179 Public Market

　　It is located in the middle of the downtown area near Puget Strait. It is an old market, which has been completely preserved. It serves seafood and other foods that citizens need. It is fresh and has local characteristics and is well received by the general public.

从Stewart街眺望市场中心
Public market from the view of Steward Street

实例180　西雅图北部展览馆建筑

这组建筑是为1962年在西雅图举办21世纪博览会而设计建造的，设计人是美籍日裔著名建筑师山崎实。展览馆建筑为浅色尖拱券柱式设计，采用自然采光通风，并设有水池喷泉。展览馆入口处还立有高大轻巧的拱架标志物，十分新颖醒目。这种轻快、明亮、向上的建筑风格以及善于利用阳光、水和尖拱的设计手法是雅马萨奇建筑创作的特色。其"走向自然，建筑与自然相结合"的设计理念值得我们大家关注与学习。

Example 180 Exhibition Hall at North Seattle

The designer is Minoru Yamasaki, a famous Japanese-American architect. The building of the exhibition hall is of light-colored arch pillar type, with natural lighting and ventilation, and is also equipped with a pool fountain. There are tall and light arch markers at the entrance, which are novel and eye-catching. This light, bright and upward architectural style, good use of sunshine, water and arch are the characteristics of Minoru Yamasaki's architectural creation.

展览馆建筑鸟瞰
Aerial view

展览馆前绿地
Green space at front

展馆外景
Exterior

十四、加拿大

1.蒙特利尔

蒙特利尔位于加拿大魁北克省南部，人口近300万，居民中60%以上为法国移民后裔。1760年蒙特利尔被英国占有，1825—1849年曾作为加拿大政府所在地，现为加拿大重要的工业、商贸和文化中心城市之一。

实例181 蒙特利尔市中心区

蒙特利尔市中心区已建成为建筑、交通连为一体的"室内城市"（Indoor City）模式。其中心区有3片连接的综合建筑群，这3片综合建筑群同地下铁环状网连通，并与中心区主干道垂直交叉。

14.Canada

14.1 Montreal

Montreal, located in the south of Quebec Province and with a population of nearly 3 million, is now one of the important industrial, commercial and cultural centers in Canada.

Example 181 Downtown Area

Its central area has realized the integration of architecture and transportation, which is called "Indoor City". There are three connected complex buildings in the central area, which are connected with the subway ring network and cross vertically with the main road in the central area. People's activities in the central area are not affected by car interference and rain and snow.

蒙特利尔市中心区模型
Model of downtown area

东部片区Place-des-Arts建筑外景
Ground look of the Place-des-Arts in the east

西部片区北端Cours Mont-Royal商场
地上出入口
Entrance of the Cours Mont-Royal
shopping mall

西部片区北端Cours Mont-Royal商场地下层、地下铁
与建筑连通
Underground part of the Cours Mont-Royal shopping
mall is connecting with subway and other buildings

实例182　蒙特利尔Desjardins建筑

该建筑位于市中心区东部片区环形地铁线北站Place-des-Arts和南站Place d'Armes之间，其地下层将东部片区南北两站间的建筑群连接起来，并形成地下空间网。该建筑地上地下商业服务设施齐全，十分方便市民购物、出行。在此建筑的地下层、地上各层均布置有喷泉、花木，它创造了自然舒适的环境。1990年国际建筑师协会第十七次大会的各国建筑图片展就安排在此一楼大厅内，其地上、地下开敞明亮，自然采光。通过地下层南行就可以到达蒙特利尔会议中心。

Example 182　Desjardins

It is located in the east of the downtown area. The underground layer connects the buildings between them to form an underground space network. The above-ground and underground commercial service facilities are complete, making it very convenient for citizens to shop and travel. Fountains, flowers and trees are arranged in the basement and above, creating a natural and comfortable environment. The aboveground and underground halls are open and bright, and make full use of natural light.

Desjardins建筑一楼大厅
Ground hall

Desjardins建筑的地下层
Underground floor with fountain and tress

国际建筑师协会建筑图片在Desjardins大厅展出
The picture of UIA on exhibition

实例183　蒙特利尔加拿大建筑中心

该中心是北美用作建筑展览和研究的首要机构，它于1986—1988年建在蒙特利尔，设计人是P·罗斯。P·罗斯的设计以往都是采用钢和玻璃设计现代建筑，但这次对加拿大建筑中心的创作，因为要保留一座公元19世纪建造的肖内西住宅以及该中心要反映蒙特利尔建筑遗产保护的要求，设计师一改原有的风格，新建筑采用U字形，从开长条窗、砌石灰石外墙并使得建筑体量、高度与原有建筑融为一体，仅在女儿墙处做成一圈铝制漏空檐口，使整体统一，并体现现代建筑的细部。其内部布局合理安排、装饰简洁，充满现代气息。此项目已被选入《20世纪世界建筑精品集锦》北美卷中。

Example 183 Canadian Centre for Architecture

The center is a modern building designed with steel and glass, which is used as the leading institution for architectural exhibition and research in North America. However, in order to reflect the spirit of protecting architectural heritage, the new building changed its original style. U-shape is adopted and used long window as well as limestone exterior wall. It only makes a circle of aluminum leaky cornice at parapet, which reflects the details of modernity. The interior layout is reasonably arranged according to the needs of architectural exhibitions, and the decoration is simple and modern.

肖内西住宅，其后为新建建筑中心
New building of the architecture center

建筑中心报告厅
Report hall of architecture centre

肖内西住宅圆形厅
Round hall inside

2. 多伦多

多伦多是加拿大第一大城市，是安大略省省会，位于安大略湖西北岸边，其意是"富饶的土地"。1787年，英国人廉价从印第安人手中买了这块土地。1900年、1953年先后两次合并附近市镇，并扩建了城市，人口达300万，现已成为加拿大的工业、金融、商业和文化中心之一。

实例184　加拿大多伦多电视塔和滨湖建筑群

位于滨湖西弗伦特街的加拿大国家电视塔，造型优美，像一支利剑直插云天，在其塔旁还建有可开启屋顶的大型体育场馆。以这组建筑为统领的滨湖建筑群形成了城市天际线，它也成了多伦多城市的标志。1976年建成的这座高553m的电视塔，其旋转餐厅就高达350.5m，曾居全球之冠。

14.2 Toronto

Toronto, the largest city in Canada and the capital of Ontario, has become the industrial, financial, commercial and cultural center of Canada.

Example 184　TV Tower and Lakeside Buildings

There is Canada's National TV Tower, which is the second tallest in the world. Its revolving restaurant is as high as 350.5 m, ranking first in the world. There is also a large stadium with an openable roof beside the tower. The three-dimensional outline formed by the lakeside buildings is led by this group of buildings.

多伦多市中心区湖滨建筑群构成城市天际线
Skyline of the lakeside buildings

多伦多市中心
Plan of downtown area

电视塔与体育馆外景
TV tower and stadium

实例185　多伦多市中心区靠近湖滨休闲岛区

该岛区位于多伦多市中心区湖滨的对面，它与多伦多电视塔隔湖相望。这里有奥林匹克岛、岛上飞机场、中心岛、南岛以及蛇岛等，是一处自然休闲游览区。此处体育设施齐全、场地宽阔，还有许多游艇等，真是游憩健身的幽静之地。各个岛之间联系方便，住宿、餐饮、娱乐设施配套齐全，游人在此可享受清新、静谧的自然美景。

Example 185　Leisure Island near the Lakeside

Located opposite the lakeside in the city center and across the lake from Toronto TV Tower, it is a natural scenic and leisure tourist area. Here, sports facilities are complete, the venue is wide, and there are many yachts. The rest accommodation, dining and entertainment facilities are well equipped, so visitors can enjoy the quiet natural beauty here.

休闲岛区鸟瞰　Aerial view

岛上球场　Soccer field

实例186 多伦多伊顿中心

伊顿中心位于多伦多市中心商业区，1967—1981年共三期工程全部完成，总建筑面积56万m^2。这个商场以一条长274m的商业街为中心，周围布置3层楼的300家店铺。商业街通过天桥和地道同公共交通、周边建筑相连。其顶部为拱形玻璃天窗，悬挂着栩栩如生的群鸟。下面不同层面均布置有喷泉和花木，当阳光照进大厅，室内生机盎然、满园春色。

Example 186 Eaton Center

It is located in the downtown business district with a commercial street as the center, and 300 three-floor shops are arranged around it. Commercial streets are connected with public transportation and surrounding buildings through overpasses and tunnels. Its top is an arched glass skylight. Fountains and flowers and trees are arranged at different levels below, and sunlight can be sprinkled into the hall.

伊顿中心内景（其建筑底层与地下铁相连）
Interior (the ground floor is connected with the subway)

实例187 尼亚加拉大瀑布

该景区位于多伦多东面的尼亚加拉瀑布城,是世界著名的大瀑布景观。"尼亚加拉"来源于印第安语"雷格隆",意思是雷声隆隆。此瀑布呈半圆形,宽约800m,平均落差51m。该景区建设得比较完美,中间有一条宽敞的道路作为分隔,路的一边是近观大瀑布区,路的另一边是服务区。服务区安排有旅馆、商店、娱乐场等各种服务设施以及丛林、草地等,可供游客使用与休息。尼亚加拉大瀑布水电站可发电181万kW。

Example 187 Niagara Falls

It is a world-famous waterfall landscape with a width of about 800 m and an average drop of 51 m. There is a spacious road in the middle of the scenic spot. One side of the road is a close-up waterfall area, and the other side is a service area, various service facilities and parks such as jungles and grasslands for tourists to use and rest. The hydropower station in this waterfall can generate 1.81 million kilowatts.

大瀑布全景
Panorama

大瀑布位置
Location

大瀑布近景
Close shot

十五、巴基斯坦

1. 拉合尔

拉合尔是巴基斯坦第二大城市，是一座历史文化名城，始建于公元1世纪末，公元1021—1186年迦兹纳维王朝在此建都，1525—1707年为莫卧儿王朝都城。老城在现有市区的西北部，并已作为老城区保留下来。

实例188 真珠大陆旅馆

该旅馆有300间客房，餐厅、宴会厅、会议室、电信、商店、咖啡店等设施齐全，室外还有游泳池、网球场等体育设施。其场地宽阔自然，超大的后花园里可举办盛大的集体活动。拉合尔真珠大陆旅馆自然环境十分突出。

15. Pakistan

15.1 Lahore

It is the second largest city in Pakistan and was founded at the end of the first century. The old city has been preserved in the northwest of the existing urban area.

Example 188 Pearl Continental Hotel

The hotel has 300 guest rooms, with complete facilities such as restaurants, banquet halls, conference halls, shops and so on. Outdoor swimming pools, tennis courts and other sports facilities have wide space, and the oversized back garden can hold grand group activities.

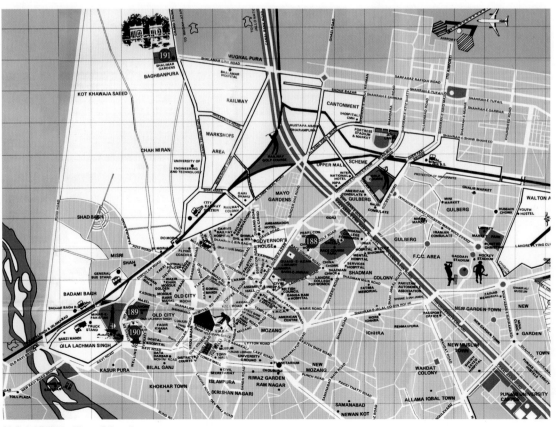

拉合尔城平面 Plan of the city

下篇：外国建筑科学文化精品图说／巴基斯坦

主楼及后面的游泳池
Main building and the swimming pool

可举办集体活动的大花园
The oversized garden

493

实例189 拉合尔城堡

拉合尔城堡位于拉合尔老城的西北角，是公元16世纪中叶至17世纪莫卧儿王朝修建的王宫。它始建于1566年，阿克巴（Akbar）王在拆除旧有土城后在这里修建了王宫，主要建筑在现城堡的北部，即北部高台上褐红色拱廊围合的建筑，当时阿克巴王在这里处理朝政和接见外宾。到了贾汉吉王时期，则沿此建筑轴线在北边修建一组建筑和花园，并作为贾汉吉王办公休息和沐浴之处。再往后到了沙加汗王时期，在平行最北边的这组建筑处建造了自己的宫室，其规模较大。此外，还建有一处供皇后使用的豪华镜宫，此宫墙壁用白色玉砌成，而且在拱顶上还镶嵌宝石、玻璃球等，五彩缤纷。这些宫室花园，基本上都是莫卧儿阿拉伯式园林，有自己的特色。

Example 189　Lahore Fort

First Akbar Palace is built on the northern terrace, surrounded by a maroon arcade, and used as a place to handle the government affairs and meet foreign guests. After that, along the axis, a group of buildings and gardens were built in the north as a place to rest and bathe. Later, the palace was built, and there was a luxurious mirror palace, with white jade walls and jewels and glass balls inlaid on the vault, sparkling brightly.

阿克巴王宫（阿克巴王在此处理朝政）
Akbar Palace

贾汉吉王宫
Jahangir Palace

沙加汗王修建的豪华宫室
Luxurious palace built by Shah Jahan

实例190　拉合尔巴德夏希清真寺

巴德夏希清真寺位于拉合尔老城西北角，其东紧临拉合尔城堡，是巴基斯坦最大的清真寺，其建造年代为公元1673—1674年（即莫卧儿王朝时期）。巴德夏希清真寺寺院由高大红砂岩石墙和四角耸立的宣礼高塔围成，在东西向主轴线的东端建有壮丽的入口大门。从大门登上22级台阶，进入院内是一160m见方的大广场院落，广场中心又有一大理石砌筑的16m见方的水池，水池可供穆斯林礼拜前戒斋沐浴之用。沿此中轴线往西，尽端即是礼拜寺殿，殿顶立有3个银球穹顶，穹顶在阳光下闪闪发光。进入殿内，穹顶、墙面上的图案花纹以及涂上金粉的古兰经文等室内外装饰工艺都极为精湛，它体现了莫卧儿王朝时期伊斯兰式的建筑风格。

Example 190 Badshashi Mosque

Located in the northwest corner of the old city, it is the largest mosque in Pakistan. The mosque is surrounded by tall red sandstone walls and minaret towers standing at four corners. It is the entrance gate at the east end of the main axis. When climbs 22 steps and enters the courtyard, there is a marble pool in the center for Muslims to fast and bathe before worship. The Temple of Worship, which is topped with three silver domes, is shining in the sun. Entering the temple, the dome and wall are engraved with patterns and Koran coated with gold powder.

礼拜寺殿外观
Exterior of the Temple of Worship

入口外观
Entrance

礼拜寺殿中心入口处
Central entrance of the Temple of Worship

礼拜寺殿内景
Interior of Temple of Worship

实例191 拉合尔夏利玛园

夏利玛园是国王沙加汗的庭园，1643年开始建造，它修建在巴基斯坦拉合尔市东北郊，并以其父贾汉吉在克什米尔的别墅园夏利玛取名，并仿其别墅园布局样式。这时期的拉合尔城市规模比当时的伦敦、巴黎还大，十分繁荣。该园的特点是：

(1) 突出纵向轴线。该园是长方形，南北向长、东西向短，地势北低南高，顺南北向做成3层台地，由南北纵向长轴线将3块台地贯穿在一起，并形成对称规则式的整体格局。

(2) 中心为全园的景观高潮。中心在第二层台地的中间，布置一巨大的水池，水池中立一平台，有路同池旁东西两亭相通。池北面设凉亭，水穿凉亭流下至第三层台地的水渠中。池南面设一大凉亭，其底部设一御座平台，人们在此可观赏到大水池中144个喷泉及其周边的美丽景色，并形成景观高潮。

(3) 通过十字型划分作四分园。其第一层高台地和第三层低台地都采用了十字型，它可划分作四分园。在每块分园中，又可以十字型再分成4片。

二层平台大水池
Big pool at the second platform

Example 191 Shalamar Bagh

Located in the northeast suburb of the city, it is characterized by:

(1) Prominent longitudinal axis. The whole garden is rectangular and has three terraces along the north-south direction, forming a symmetrical overall pattern.

(2) The center is the climax landscape. In the middle of the platform on the second terrace, there is a huge pool in the middle. Water passes through the pavilions on each side and flows down to the canal of the third terrace. At the bottom of the pavilion in the south of the pool, you can enjoy the 144 fountains in the pool and the beautiful scenery around them.

(3) The garden is divided into four sub-gardens, and each one is also divided into four pieces.

这是中轴线上第二层平台大水池
Central axis

从大水池北凉亭眺望北面最低层平台
Lowest platform at north

2. 伊斯兰堡

实例192　伊斯兰堡市区

伊斯兰堡是巴基斯坦的首都，1961年始建，1970年基本建成，人口有20多万。该市北依玛格拉山，东临拉瓦尔湖，城市路网为东西向长条状棋盘式布局，其东部是总统府、议会大厦和政府各部门的行政办公区。行政区南面为大使馆区，西面为银行、电视广播、通讯等公共事业区。西部为住宅社区，住宅社区内有商店、学校和清真寺等基础服务设施。西南部则为轻工业区和大专院校区。伊斯兰堡市区内无高层建筑，主要是以绿带、公园穿插在各区，并配合路网将各区连为一体。伊斯兰堡是一个花园城市，其绿地覆盖率高、自然环境好且又是一座新建城市。伊斯兰堡重视自然、重视社会生态功能的规划和做法，值得我们大家学习。在市区南面夏克帕利山上建有玫瑰、茉莉公园，这里种植了250多种玫瑰和几十种茉莉，山顶上还专门开辟了一块为来访外国首脑植树留念的园地。1964年初周恩来总理在这里亲手栽植一棵乌桕树，现已枝叶繁茂，这是中巴友谊的象征，让后人缅怀。

15.2 Islamabad

Example 192 Downtown Area

It is the capital of Pakistan with a population of over 200,000. The urban road network has a checkerboard layout in east-west direction, with administrative office area to the east, embassy area to the south and public utility area to the west. The west part of the city is residential community, while the southwest is of a light industrial zone and a college zone. There are no high-rise buildings in the urban area, and green belts and parks are interspersed in all districts. There are Rose and Jasmine Parks on Shakar Parian Mountain in the south of the city. On the top of the mountain, a garden dedicated to planting trees for visiting foreign leaders has also been opened.

行政区街景
Street scene of the administrative area

从玛格拉山南望绿海中的市区
Downtown area from the view of Margalla Hill

1964年周恩来总理在夏克帕利山顶上栽植的乌桕树
Chinese tallow tree planted by Premier Zhou Enlai in 1964

实例193　伊斯兰堡费萨尔清真寺

该寺位于伊斯兰堡市北部玛格拉山山脚，占地20hm²。建造时间是1970—1986年，设计人是土耳其建筑师V·达洛开侬，他是参加国际设计竞赛而被选中的。这组建筑群包括礼拜堂、沐浴区院落和一所国际伊斯兰大学。主体建筑礼拜堂平面为正方形，堂内可容纳1万人做礼拜，其屋顶是用空间构架形成巨大帐篷，高达40m，大堂四周还立有4个高80m的拜克楼。礼拜堂前是个大院落，"麦加朝圣墙"和"圣龛"位于大堂和院落的主轴线上。建筑采用钢筋混凝土结构，屋顶饰面为大理石，地面为花岗岩。设计人突破了传统常规的拱券、穹顶、伊斯兰图案的形式，而是应用新材料、新结构体现传统的意味，但这种创新而不失传统的设计精神，这是该建筑的一大特点。此项目已被选入《20世纪世界建筑精品集锦》南亚卷。

Example 193 Faisal Mosque

The temple is located at the foot of Margalla Hill, covering an area of 20 hectares. This group of buildings includes the chapel, the courtyard of the bathing area and an international Islamic university. The main building is a square chapel, which can accommodate 10,000 people for worship. The roof is a huge tent. There are four Minarets with a height of 80 m around the lobby. There is a big courtyard in front of the chapel and the "Mecca pilgrimage wall" and "holy shrine" are located on the main axis. The building adopts reinforced concrete structure, with marble roof and granite floor.

礼拜堂内景
Interior

下篇：外国建筑科学文化精品图说／巴基斯坦

礼拜堂全景
Panorama

礼拜堂正门入口处
Main entrance

十六、日 本

1. 东 京

东京是日本的首都，人口约1200万，面积2155km²。历史上此地称江户，1868年明治天皇从京都迁都到江户，遂改名东京。1943年通过颁布法令改作东京都，并扩大其管辖范围。东京是日本第一大城市、第四大港口，其工业发达，大专院校有190多所，银座、新宿、浅草、涩谷等区商业繁华，因而东京都又是日本的政治、经济、文化、交通中心。1979年，东京和北京市结为友好城市。

实例194 东京都新都政厅

20世纪80年代，东京为减轻旧区的压力和发展的需要将都政厅迁往新宿区，并新建新都政厅。新都政厅建筑总面积19.56万m²，1986年由日本著名建筑师丹下健三先生中标设计。这组建筑包括办公塔楼、低层议会议事堂和一个市民广场，其设计合理、使用方便。新都政厅内部设施具有高科技和现代化，其水、电、暖、空调、办公完全由智能电子系统导入；其外观亦充满现代化气息，似集成电路图案装饰，两边直立双塔的造型，也似哥特式教堂。它将传统与现代有机结合在一起，特别是墙面门窗、线条等制作精细，这充分体现了日本高科技与现代化的水平。

16. Japan

16.1 Tokyo

It is the capital of Japan with a population of about 12 million and an area of 2,155 km². It is also the largest city in Japan, with developed industry, more than 190 colleges and universities and prosperous commerce.

Example 194 New City Hall

The new city hall covers an area of 195,600 m², including office tower, low-rise parliament hall and a citizen square. The internal facilities are rich in high technology and imported the intelligent electronic systems. The appearance is full of modern flavor. The walls, doors, windows and lines are carefully made, which fully reflects the level of high-tech industrialization.

塔楼两侧围廊
Corridor on both sides of the tower

下篇：外国建筑科学文化精品图说／日本

办公塔楼外景
Exterior of the tower

505

实例195　银座

银座位于东京都中央区，北起京桥、南至新桥的一条长1.5km的大街及其东西两面的区域，也是东京都最繁华的商业区。德川幕府时代，银座为铸造银币的官署，即银座官署，于是在明治三年（公元1870年）将这里定名为"银座"。银座大街两旁是各类商店、百货公司，它们专门销售和陈列高档名贵商品。银座大街以东为东银座，以西为西银座，并建有许多酒店以及娱乐场所。从1970年8月起，每逢星期日下午和节日午后，这里禁止一切车辆通行，并改作步行街，以方便游客。夜晚，这里的霓虹灯五光十色。

节假日银座步行街
Pedestrian street in holiday

Example 195 Ginza

Located in the central area, it is the most prosperous business district in Tokyo. There are various shops and department stores on both sides of Ginza Street of 1.5 km long, which specialize in selling and displaying high-grade precious commodities. There are many restaurants, hotels and entertainment places. From August, 1970, every Sunday afternoon and holiday afternoon, vehicles were forbidden to pass, and pedestrian streets were used instead, making it more convenient and safe for visitors. At night, the neon lights here are colorful.

银座夜景
Night view

日本的"Nissan""Maxell""日本酒中心"等大公司形象展示
Exterior of buildings of Nissan, Maxell and Wine Center

实例196　上野公园

上野公园位于东京都西部，占地52.5hm^2，是东京都最大的公园。上野公园的北面是历史文化区，有东京国家博物馆、国家科学博物馆、国家西洋美术馆、东京美术馆、东京文化会馆等，这些文化馆建筑是19世纪和20世纪时期建造的，并配以规整的自然绿化设施，共同构成了一个具有历史文化氛围的游览区，可以称其为日本的"文化之林"。上野公园的西北面是动物园，它建于1885年，也是日本最早的动物园，中国赠予的大熊猫欢欢和飞飞就饲养在这里。上野公园的西南部是不忍池，池内栖息着野生雁、黑天鹅、鸳鸯、野鸭等水禽数千只。1964年在不忍池旁还建有水族馆，动物园与不忍池之间通过桥与空中缆车相连接。

Example 196 Ueno Park

Covering an area of 52.5 hectares, it is the largest park in Tokyo. To the north is the historical and cultural area, including Tokyo National Museum, National Science Museum, National Western Art Museum, etc. Northwest is the zoo, which was built in 1885 and is the earliest zoo in Japan. In the southwest of the park is Shinobazu Pond, where 5,000 to 6,000 waterfowl such as wild geese and black swans inhabit. In 1964, an aquarium was built beside the pool. The zoo and Shinobazu Pond are connected by a bridge to the aerial cable car.

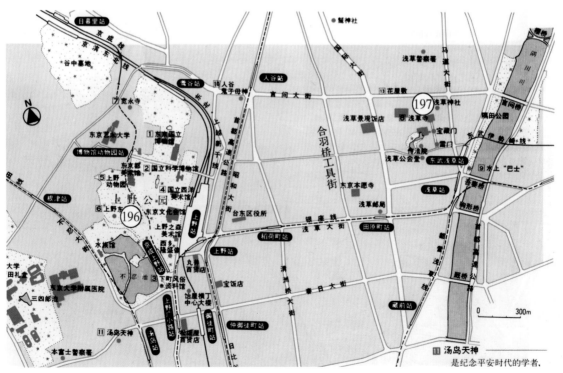

东京都上野公园与浅草区
Plan

国立西洋美术馆（馆前立有罗丹的雕塑）
National Western Art Museum with the sculpture of Rodin in front

博物馆前绿地景观
Green space in front of the museum

生机勃勃的动物园入口
Entrance of the zoo

实例197　浅草寺

浅草寺位于上野公园的东面，这里保留着江户时代形成的浓厚文化风格，它也是"老东京"们回忆往事的地方。浅草寺寺院庭园形成于公元17世纪，其一年当中有许多节日活动，如1月1日的"初拜"、2月3日的"节分"、7月10日的"卖酸果集市"、12月17日的"羽毛毽木板（正月玩的玩具）集市"等。浅草寺周围还有花市、商场、游戏机设施等，并已成为一个老东京商业繁华区，它也是游客到东京必去观光的地方。

Example 197　Asakusa Temple

There is a strong cultural style and history formed during the Edo period, and it is the place where "old Tokyo" people recall the past. The garden of the temple was built in the 17th century, during which many festivals were held here. There are flower markets, shopping malls, game rooms and other facilities around the temple, which has become a bustling commercial area in old Tokyo and a must-see place for foreign tourists visiting Tokyo.

浅草寺旁商场
Shopping mall

浅草寺入口景观
Stone sculptures of animals at the entrance

浅草寺主殿
Main hall of the temple

浅草寺主殿与塔
Tower and main hall

浅草寺旁花市
Flower market

2. 名古屋

实例198　名古屋市中心大街

名古屋是日本第四大城市，为日本中部行政、经济、工业、文化中心，也是日本第三大港口城市。名古屋市中心南北向轴线大街宽达100m，其街的两旁有松下电器、住友商事、爱知文化讲堂、县美术馆、三越百货、大和银行等。街心也是100m宽的绿地公园，偏北方向矗立着180m高的名古屋电视塔。此外，还有喷泉、不锈钢雕塑等，其环境清新、视野开阔，这里已成为名古屋的重要标志。大街下方还开辟有地下街道，地下街道商店齐全，又有地下铁与其相连，市民和游客来去都非常方便。

16.2 Nagoya

Example 198　Central Street

It is the fourth largest city in Japan. The north-south axis of the city center is 100 m wide. To the north stands the Nagoya TV Tower with a height of 180 m, as well as fountains and stainless steel sculptures. The environment is fresh and open. Under the main street, there are underground streets with complete shops and subway connections, making it very convenient for citizens and tourists to come and go.

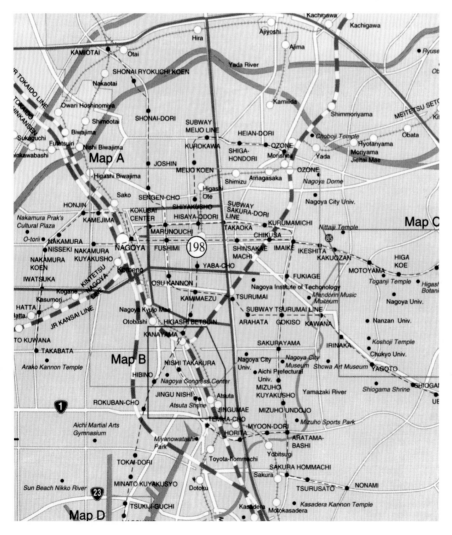

名古屋城图
Plan of the city

北望全景
Panorama of the north

电视塔南侧地下街出入口
Entrance and exit of the underground street to the south of the TV tower

3. 奈 良

实例199　奈良东大寺

奈良是日本古都，于公元710—789年在此建都，名平城京。现平城京遗址位于奈良市西郊，为方形，它仿中国唐长安城建造。奈良建筑古迹甚多，其中东大寺是现存世界上最大的木结构建筑。奈良东大寺始建于公元8世纪前后，它按中国寺院建筑结构建造，1980年完成了第二次整修。寺内大佛殿高46m、宽57m、进深50m，殿内供奉着金铜奈良大佛。奈良大佛佛体与石座高21.45m，重达452t，为世界第二大佛，它仅次于中国西藏扎什伦布寺的"未来佛"。东大寺寺旁古杉、古松参天，这里还有千头养鹿可供游人观赏，人们在此可领略到生机勃勃的大自然美景。

16.3 Nara

Example 199　Todaiji Temple

Nara, an ancient capital of Japan, was built in imitation of Chang' an in Tang Dynasty, with many architectural monuments, among which Todaji Temple is the largest existing wooden temple building in the world. It was built according to the architecture of Chinese temples. The Great Buddha Hall is dedicated to Nara Buddha of gold and bronze. Its body and stone pedestal are 21.45 m high and weigh 452 tons, making it the second largest Buddha in the world. There are towering trees beside the temple and thousands of deer are raised for visitors to watch.

东大寺旁的二月堂　Nigatsu-do

主殿正面景观　Exterior of the main hall

东大寺前方鹿池　Deer garden

4. 京 都

京都位于日本本州中西部，公元794—1869年曾为日本首都，也是日本古都和宗教、文化的中心，面积610km², 人口150万。京都市内建筑、绘画、雕刻、园林被定为国家级文物的有21处，占日本国家级文物总数的15%。由于京都历史文物古迹多，1950年京都被定为国际文化观光城市。京都城市东、西、北三面环山，市内街道为棋盘式，市中心为皇城、宫城，这是仿中国唐长安城、洛阳城的格局建造的。早在1000多年前中日就进行了文化交流。1974年，日本京都市同中国西安市结为友好城市。

实例200　京都平安神宫

此宫在京都市左京区，于公元1895年（明治28年）为纪念桓武天皇建都平安京（京都）1100周年而建造，总面积3.3万m²，它是日本最宏伟的一座神宫。其建筑雄伟壮丽、碧瓦红柱，正殿高17m，进深30m，屋脊两端用镀金鱼尾形构件来装饰，显得格外醒目。殿两边以走廊连通东面的苍龙楼和西面的白虎楼。神苑分东、西、中三部分布置在神宫周围，林木繁茂。每年10月22日都会在此举行作为京都三大祭之一的时代祭活动。

正殿全景
Panorama of the main hall

16.4 Kyoto

It is the ancient capital and religious and cultural center of Japan. The city is surrounded by mountains on three sides from east to west. The streets in the city are chessboard-shaped, and the city center is modeled after Chang' an City in Tang Dynasty.

Example 200 Heian Shrine

With a total area of 33,000 m^2, it is the most magnificent shrine in Japan. The building is magnificent, with green tiles and red columns and gold-plated fishtail parts at both ends of the roof. Shrine garden is divided into three parts: east, west and middle, which are around the shrine with lush forests.

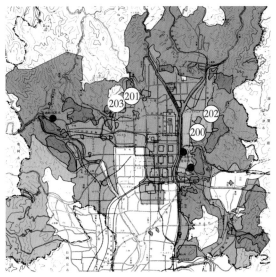

京都城
Plan of the city

正殿与东面苍龙楼
Main hall and the pavilion on eastside

东苑景观
Landscape of the east garden

实例201　金阁寺庭园

该园位于日本京都市北部，始建于1397年，当时为幕府将军足利义满的别墅，后改为寺院，占地9hm^2。此园的特点是：舟游回游混合型；寺阁展立湖岸旁；建筑镀金金光闪。金阁寺共3层，在建筑外部镀金箔，故名金阁。其第一层为法水院，第二层供奉观音，第三层供三尊弥陀佛。该阁1950年被大火焚毁，1955年又复建，其造型轻巧舒展，做工十分精细，已被列为日本的名胜古迹。

Example 201 Courtyard of Kinkakuji Temple

Located in the north of Kyoto, the courtyard covers an area of 9 hectares. The temple pavilion is located on the lakeside, with three floors, and is plated with gold foil outside. It was destroyed by fire in 1950 and restored in 1955. Its shape is light and stretched, glittering in the sunshine.

金阁寺庭园一隅
Temple at the lakeside

实例202　银阁寺庭园

该园位于京都东部，是幕府将军足利义政（1436—1490年）仿金阁寺的造型在东山修建的山庄，占地1公顷多，设计人为著名造园艺术家宗阿弥。银阁寺庭园的特点是：舟游式和回游式的混合型，原计划建筑外部涂银箔，后因主人去世改涂漆料，故俗称银阁。此银阁共2层，一层为心空殿，二层供观音。庭园虽小，但却小中见大。园内有月台，为赏月之地。

Example 202 Courtyard of Ginkakuji Temple

It is located in the east of Kyoto, covering an area of more than 1 hectare. It was originally planned to paint the outside of the temple with silver foil, and then it was painted with paint. Although the area of the courtyard is small, it is exquisitely arranged and imitates famous places to create landscapes. There is a scene similar to the West Lake scenery, which is used for enjoying the moon.

银阁寺外景
Exterior of the temple

实例203 龙安寺石庭

该庭位于日本京都西北部,邻近金阁寺,建于15世纪。此石庭面积为330m^2,是一座矩形封闭庭院。其特点是:自然朴素,并具抽象美。日本庭园受中国禅学思想影响,于是创造出抽象的自然景观。用白砂象征大海,用山石象征岛群。龙安寺石庭就是在白砂地面上,布置了15块精选之石,并依次按5块、2块、3块、2块、3块分成五组均衡摆放,象征5个岛群。这就是禅学精神在日本庭园中的体现。

Example 203 Stone Yard of Ryoanji Temple

Located in the northwest of Kyoto, it was built in the 15th century and covers an area of 330 m^2. There are 15 selected stones on the white sand ground, which are arranged in five groups according to 5, 2, 3, 2 and 3 blocks in turn, symbolizing five island groups, achieving balanced and perfect effect. The white sand at the bottom is raked into water stripes to symbolize the sea. This group of abstract sculpture stone garden has reached the natural landscape symbolizing the sea islands.

石庭
Stone Yard